The
Tyranny
of the
Straight
Line

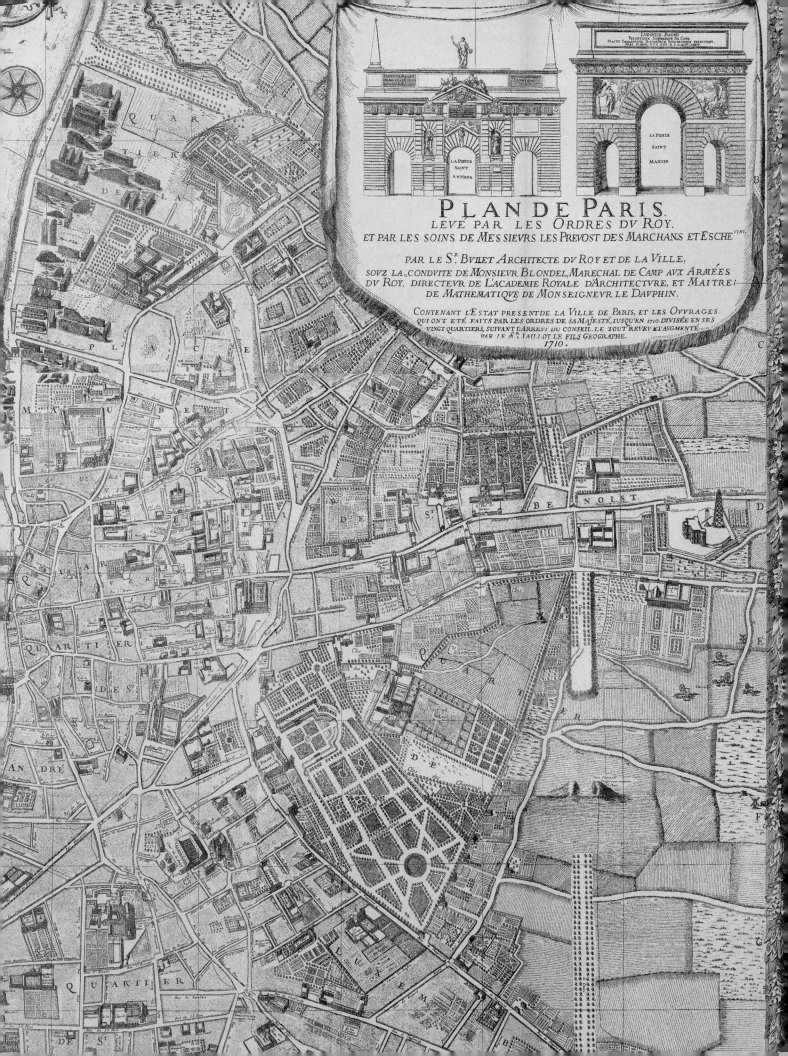

Min Kyung Lee

The Tyranny of the Straight Line

Mapping Modern Paris

Yale University Press

New Haven and London

yalebooks.com/art

Designed by Leslie Fitch and Jeff Wincapaw

Cover designed by Jeff Wincapaw

Set in Crimson and Cronos Pro

Printed in Malaysia for Imago Group

Library of Congress Control Number: 2023947221

ISBN 978-0-300-26764-8

A catalogue record for this book is available from the British Library.

This paper meets the requirements of ANSI/NISO Z39.48–1992 (Permanence of Paper).

10 9 8 7 6 5 4 3 2 1

Jacket illustration: (front) fig. 3.16 (detail); (back) fig. 1.4.

Pages ii–iii: fig. 2.8 (detail); page iv: fig. 1.9 (detail); page viii: fig. 2.40 (detail); page x: fig. 5.8 (detail); pages xiv and 9: fig. 1.8 (details); page 152: fig. 5.14 (detail); pages 156 and 168: fig. 5.15 (details); page 186: fig. 2.4 (detail).

For Andrea, Lelio, and Folco

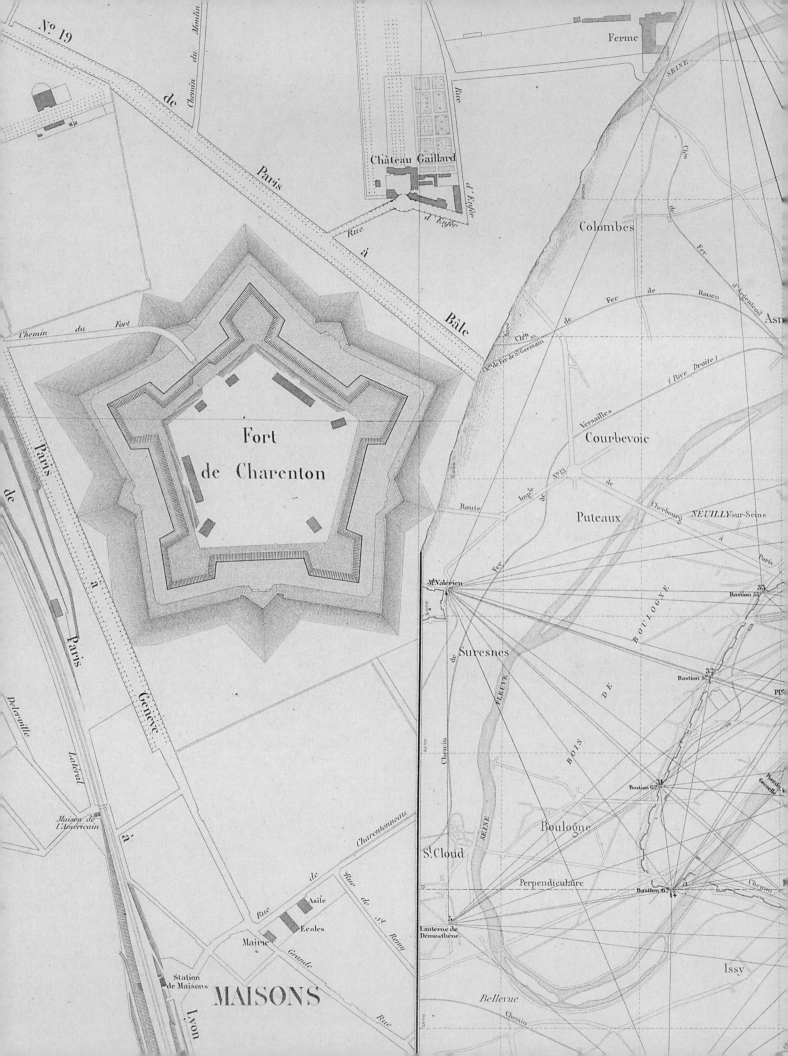

Contents

XI   Preface

XII   Acknowledgments

1   Introduction

11   Chapter 1: Master Plans

29   Chapter 2: Triangulating the City

63   Chapter 3: Drawing Grids

105   Chapter 4: The Bureaucracy of Plans

127   Chapter 5: Cartographic Presentations

153   Conclusion: The Total View of the City

157   Notes

169   Bibliography

187   Index

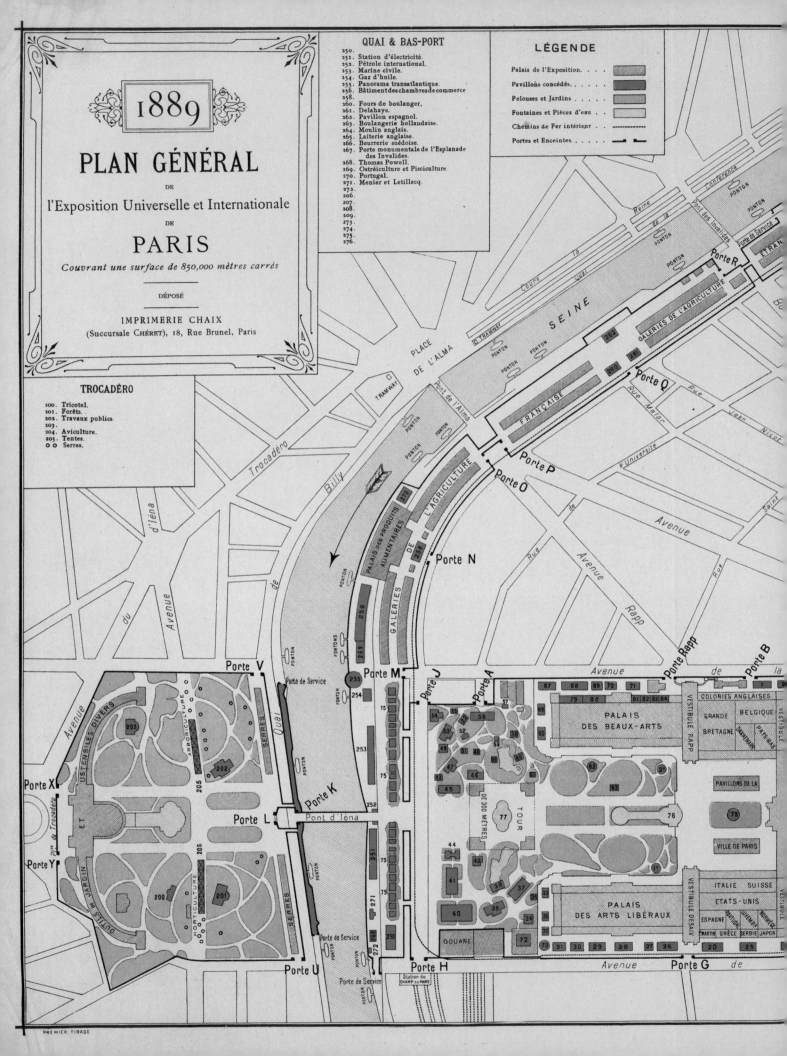

1889

PLAN GÉNÉRAL

DE

l'Exposition Universelle et Internationale

DE

PARIS

Couvrant une surface de 850,000 mètres carrés

DÉPOSÉ

IMPRIMERIE CHAIX
(Succursale CHÉRET), 18, Rue Brunel, Paris

QUAI & BAS-PORT

250.
251. Station d'électricité.
252. Pétrole international.
253. Marine civile.
254. Gaz d'huile.
255. Panorama transatlantique.
256. Bâtiment des chambres de commerce
258.
260. Fours de boulanger,
261. Delahaye.
262. Pavillon espagnol.
263. Boulangerie hollandaise.
264. Moulin anglais.
265. Laiterie anglaise.
266. Beurrerie suédoise.
267. Porte monumentale de l'Esplanade des Invalides.
268. Thomas Powell.
269. Ostréiculture et Pisciculture.
270. Portugal.
271. Menier et Letillacq.
272.
206.
207.
208.
209.
273.
274.
275.
276.

LÉGENDE

Palais de l'Exposition. . . .
Pavillons concédés. . . . . .
Pelouses et Jardins . . . . .
Fontaines et Pièces d'eau . .
Chemins de Fer intérieur . . -------
Portes et Enceintes. . . . .

TROCADÉRO

200. Tricotel.
201. Forêts.
202. Travaux publics.
203.
204. Aviculture.
205. Tentes.
o o Serres.

PREMIER TIRAGE

Preface

My desire to understand the power of maps in nineteenth-century Paris derives from another moment in history in another part of the world. In 1945 one American and one Soviet general placed a ruler along a grid to draw a straight line on a map. The line followed a latitude that is thirty-eight degrees north of Earth's equatorial line, roughly at the midpoint of the Korean peninsula. This line would divide the country until it was breached by North Korean troops five years later, initiating the Korean War. The border that divides the Korean peninsula is not the subject of this book. Yet, having grown up in the aftermath of that drawn division and having crossed another set of mapped borders to immigrate to the United States, mapped lines form the background to my identity and country of birth. Lives are most often shaped by unknown people who make decisions in mundane offices with maps. Whether it is the Demilitarized Zone in Korea or redlining in the United States, drawing lines is a practice of power.

The history of Paris's modernization centers on Napoléon III's map, which has largely been left unexamined. This oversight speaks to the ways maps and their forms of visualization have been thoroughly normalized: the ways in which they frame our experiences and spaces and shape institutions are taken for granted, and as such, they are unfortunately often assumed to be outside of history. If there is one argument that readers should take away from this study, then, it is to see how maps and plans as visual representations are contingent artifacts that have their own histories, both conceptual and material.

The relationship between these visualizations and urban space takes on a particular valence in the case of nineteenth-century Paris, insofar as the experience of modernization centered on cultural products as privileged sites of political, social, and psychological contestation. What varied writers and scholars of modernity—from Walter Benjamin to T. J. Clark to Hollis Clayson—all share is an argument that images were not mere descriptors, and that forms and practices of describing modern experiences were simultaneously intangible and central. Images represented power and authority as well as critique and resistance—and the fraught discourse between them. Maps were no different from other visual representations in this respect, but in the hands of the government, their presentation as universal and constant defined them as objective and apart from the vicissitudes of politics and society. This effort was largely successful. My intention then is to delineate the historical context that gave these maps and plans the power to define the modern city of Paris.

Acknowledgments

I would like to begin by thanking Katherine Boller, Rachel Faulise, Alison Hagge, and Elizabeth Searcy at Yale University Press, as well as Laura Hensley, who copyedited the text, Scott Howard, who corrected earlier versions, Noreen O'Connor, who proofread the layouts, Chelsea Connelly, who reviewed the layouts, David Luljak, who indexed the book, Jeff Wincapaw, who designed the interior and jacket, Jillian Lunoe, who helped with images, and the external readers who made this a better book.

My research began with an undergraduate thesis at the University of Pennsylvania that became a doctoral dissertation, and then evolved into a book manuscript. The undergraduate thesis compared Adolphe Alphand's Buttes Chaumont to Bernard Tschumi's Parc de la Villette. The doctoral dissertation focused on Haussmann and his administration's use of maps and plans; this research was expanded for the current study before you now. This long trajectory thus has indebted me to many generous people. One could not have asked for better teachers than David Brownlee, Elaine Simon, and Anne Whiston Spirn, who introduced me to questions and methods that I continue to carry with me. For my doctoral studies, Northwestern University offered an exceptional nexus of scholars. Courses and advising from Peter Carroll, Hannah Feldman, Stephen Eisenman, Sarah Fraser, and Ken Alder at Northwestern, and from Katherine Taylor at University of Chicago, were deeply influential. My appreciation also extends to a larger Northwestern family: Hector Reyes, James Glisson, Shaoqian Yang, Zirwat Chowdhury, Leslie Ureña, Justine DeYoung, Ana Croegaert, Gergely Baics, Christopher Mead, Kevin Murphy, Jésus Escobar, Paul Jaskot, and Sheila Crane.

I would especially like to thank Holly Clayson and David Van Zanten. Holly's acumen, inspiring insights, and discerning critiques offered during my studies in Chicago and beyond wholly frame this manuscript. And David Van Zanten, as my mentor and supervisor, has inflected every part of this book with his method, scholarship, teachings, and guidance. His commitment to his students is unparalleled. He generously read drafts of this manuscript and, as always, provided immensely helpful perspectives. I was fortunate to have participated in the scholarly community that they both created and nourished.

I spent some years at the École des hautes études en sciences sociales, during which time I was advised by Michaël Darin, Christian Topalov, and Karen Bowie. I owe each of them immense gratitude for sharing their time and energies to mold a chaotic tangle of thoughts into a researchable question that could be studied. There, in Paris, I was fortunate to meet and work with generous colleagues who continue to be important interlocutors and friends: Sean Weiss, Sony Devabhaktuni, Matthew Wendeln, Peter Sramek, Deborah Goldgaber, Miriam Jerade, Mario LaMothe, Jean Beaman, and Ellen McBreen.

My many stays in Paris have been supported by several fellowships and institutions while I was writing my dissertation and book. These include the Bourse Chateaubriand, the Bourse Jeanne Marandon, the Camargo Foundation, the Paris Program in Critical Theory, Deutsches Forum für Kunstgeschichte, and, most recently, the Banister Fletcher Global Fellowship with the University of London Institute in Paris, the Queen Mary University of London, and the Bartlett School of Architecture. I am grateful to Dario Gamboni, Jérémie Koering, Andreas Beyer, Tobias

Vogt, Gitta Ho, Dominic Brabant, Samuel Weber, Michael Loriaux, Anna-Louis Milne, Hannah Williams, and Peg Rawes. I am also thankful to the librarians and staff at the Bibliothèque historique de la Ville de Paris, Bibliothèque administrative de la Ville de Paris (now the Bibliothèque de l'Hôtel de Ville), Bibliothèque national de France, Archives nationales, Archives de Paris, and Archives de la Police. A special mention of gratitude must be extended to the late Pierre Casselle, who came out to meet me one day when I requested an obscure carton. He offered his help early in my archival research, and I am eternally appreciative of his assistance over the many months spent in his library.

I continued to work on these issues through fellowships and positions at U.S. institutions: the Getty Research Institute, Swarthmore College, College of the Holy Cross, and Bryn Mawr College. These institutions supported me with leaves and grants for research and travel. Individuals I would like to thank include Steven Nelson, Pushkar Sohoni, Leah Clark, Kristine Neilson, Kivanç Kilinç, Swati Chattopadhyay, Alice Friedman, Zeynep Celik, Marta Gutman, Logan Grider, Michael Cothren, Randall Exon, Tariq al-Jamil, Amanda Luyster, Nancy Burns, Cristi Rinklin, Michael Beatty, Tom Landy, David Karmon, Jeff Cohen, Homay King, Alicia Walker, and Sharon Ullman. The insights of these scholars, artists, and teachers found their way into much of this work.

Portions of chapters 2, 3, and 4 have appeared in modified forms in the following publications: "Constructing Nineteenth-Century Paris through Cartography and Photography," in Peter Sramek, *Piercing Time: Paris after Marville and Atget, 1865–2012* (2013); "An Objective Point of View: The Orthogonal Grid in Eighteenth-Century Plans of Paris," in the *Journal of Architecture, RIBA* 17, no. 1 (2012); and "The Bureaucracy of Plans: Urban Governance and Maps in Nineteenth-Century Paris," in the *Journal of Urban History* 46, no. 2 (March 2020).

For too long, I've discussed this book with friends and family, who are likely relieved to hear that the project is finally finished. Your encouragement kept me going until the end. Thank you to Mollie Rubin; Lori Szudarek; Marion Enjolras; Pietro Saitta; Catalina Meija; Tilo Arnhoff; Olga Sendra-Ferrer; Anna Vallye; Alex Arteaga; Farah Chami; Etienne Gillabert; my Italian family, Ave and Guido Borghini; my parents, Cha Nam and Jong Seung Lee; and my aunts, Heasoon and Heasook Rhee. My most important and dearest supporter has been Andrea Borghini, who has never known me without this appendage. I could not have completed this project without his love. Words cannot fully express the profound happiness and purpose that he and our two children have brought to my life and that have sustained me while writing this book.

Porte S. Antoine

La Bastille

La place Royalle

Marets du Temple

Porte du Temple

Le Temple

Fauxbourg S. Martin

R. S. Martin

S. Martin

Les Halles

Le Ponceau

R. S. Denis

R. Mont-orgueil

R. Capucilliere

Mont-Marthe

Faubourg mont-Marthe

Le Palmail

S. Honoré

Les marets

La grange Bateliere

La Bibliothèque Municipale

Ceste ville est un autre monde
Dedans un monde florissant,
En peuples et en biens puissant
Qui de toutes choses abonde

Matheus Merian. Basiliensis. Fecit

Introduction

As regards the images themselves, one cannot say that they reproduce architecture. They produce it in the first place, a production which less often benefits the reality of architectural planning than it does dreams.

—Walter Benjamin, "The Rigorous Study of Art" (1933)

As recounted in 1875 by Charles Merruau, the loyal secretary to Napoléon III, the transformation of the French capital was realized through "a precious plan of Paris, on which lines were successively traced, rectified, coordinated that would determine the city's ensemble."[1] By his account, one that was shared by many of his contemporaries, the modernization of Paris during the nineteenth century was driven by this map, which was ultimately lost in the fire of the Hôtel de Ville in 1871.

Merruau described this plan of Paris as straightforward and precise, delineating the urban interventions that would realize the city's greatness.[2] And while he credited Napoléon as its author, Merruau nonetheless lent agency to the plan itself: its drawn lines run, traverse, straighten, divide, connect, and stop.[3] The subtle shifts in Merruau's discourse between Napoléon III and the plan as protagonist reveal a changing political culture in the public administration that emerged over the course of the nineteenth century. This culture increasingly assigned value to specific images and documents to describe accurately new spatial relations, while the individuals whose hands composed those lines and used those documents became simultaneously invisible and thereby less accountable.

As claims about the objectivity of these administrative documents, especially maps and plans, became louder, the documents were bestowed more authority. If there was a problem or an injustice in the urban interventions—and there were many—it was not the fault of a particular person, let alone the emperor, but an error in the representation itself.

In this historical moment of the nineteenth century, maps were found at a particular nexus in which ideas of expertise, urbanization, and modernity itself were materializing. But as they became central to how Paris was transforming into a modern metropolis, there was a way in which maps as a tool within these processes and relations became less obvious. Merruau's deferral of responsibility to an object, and the associated sublimation of intentionality, were central mechanisms of the orthographic map as it was used in the government administration of nineteenth-century Paris. These maps and plans composed of straight lines and right angles obscured the process of their creation as well as their address to specific audiences. Administrators presented them as autonomous representations that offered a view from nowhere. And by denying the contexts of their production and use, this concealed the limits and insufficiencies of the map's own representational and symbolic systems, the institutions built around them, and the epistemological values of objectivity invested in them.

This question of agency and authorship has long dominated the historiography of Paris's nineteenth-century modernization. Who was responsible or deserved credit for both the vast infrastructural improvements and the violent social reorganization of the city? Who wielded the power to enact and enforce these physical and political changes? The strong centralization of France's authority in Paris seems to provide an answer, but the labyrinthine administrative apparatus of governance that grew around that authority complicates it, as many of the memoirs of the city's administrators and the archival record attest.[4] Moreover, historiographic focus on Napoléon III and Georges-Eugène Haussmann, the prefect of the Département du Seine during the Second Empire, has deflected understandings of how representations—images and texts—used by the expanding urban bureaucracy were loaded with responsibility.[5] Maps described the territory and plans controlled building; all protest and recourse to urban development was regulated through these media. Orthographic maps and plans became trusted not only by authorities but also by the public at large, and they played a multifaceted role in representing the city or an urban process as well as in negotiating social relations to and within the city.

The title of this book, *The Tyranny of the Straight Line*, asks how a particular modality of representation and its most significant element—the orthographic line—gained such authority and became so thoroughly pervasive in the modernization of Paris. These straight drawn lines became physical straight lines, used to cut open the sick body of the city to cure it of its social and physical ills.[6] As Haussmann recounted: "It meant the disemboweling of the Old Paris, the *quartier* uprisings and barricades, by a wide central street piercing through this almost impossible maze, and provided with communicating side streets, whose continuation would be bound to complete the work thus begun. The subsequent completion of the Rue de Turbigo made the Rue Transonian disappear from the map of Paris!"[7] The easy slippage between city and map commonly found in these bureaucratic discourses reveals how embedded the plan was in the material processes of urban modernization. The city's "rationalization" could not even have been imagined without a map.[8]

Thus, by contextualizing the orthographic plan, this study seeks to understand how its surveyed and drawn lines acquired the capacity to mediate urban spatial and social relations. My hermeneutic perspective is not meant to recover a structuralist or post-structuralist analysis, even if my theoretical orientation is influenced by it. This inquiry does not take the position that meaning is produced solely through social conventions and symbolic systems, but it also does not align with the argument that meaning is created by autonomous individuals. Instead, the question of the agency of the map is situated in a materialist argument that its conceptualization as objective—that is, universal, normative, and given—obscured not only the context of its production and the social and political structures that allowed for its creation and use, but also the means by which authorities wielded these representations to gain and hold spatial power. It attempts to contextualize the plan's ideologies and cultural and political valences that ultimately shaped the meaning of modern space.

The book's title recalls a specific moment in French history when Napoléon III, having crowned himself emperor after the revolution of 1848, met with the minister of the interior, Jean-Gilbert-Victor Fialin, duc de Persigny, and ordered him to form a planning committee. Headed by Henri Siméon, the Commission des embellissements de Paris (Commission for the improvements of Paris) was charged with making recommendations for solving the city's infrastructural problems. In his directives, Napoléon III commanded that Paris not be enslaved by the straight line: he wanted to retain the picturesque beauty of the city, its past monuments and buildings, and encouraged the committee's architects to preserve angles and corners in the urban fabric. Ultimately, straight lines did come to define modern Paris. However, they did so not only in the obvious forms of the city's famous boulevards, but also in the installation of a geometric logic that determined the ways in which Paris was mapped and built.

The plans drawn by the commission served as the template for the map that the emperor would famously hand over to Haussmann. A map with straight lines drawn in various colors would be credited as the basis for Paris's modernization program. This lost map was claimed to have prefigured the modern city, and the succeeding years were merely concerned with building out the lines that were drawn. This narrative, birthed by Haussmann's self-serving memoirs and rehearsed in numerous contemporaneous accounts, is complicated by the publication of the commission's papers by Pierre Casselle.[9] Yet the scene's historiographic longevity reveals broader and more fundamental misconceptions of how modern space is shaped. It is a formulation that assumes the stability and legibility of a map and that does little to account for the actual iterative practices of shaping a city.

During the Enlightenment, maps became contested sites of scientific and political negotiation.[10] Cartographic representations had long been used by states to represent cities, the seas, and the world. Thus, the totalizing attempts of nineteenth-century maps were not what distinguished them from previous eras. What was radical was the invention of a totalizing *system* that equated the map to the rest of the world based on the metric measure. Scientific communities had raced to triangulate Earth's surface and determine a universal standard measure, representing efforts to quantify space.[11] The attempts to quantify space meant translating qualities that had formerly been described in units specific to particular bodies, temporalities, and sites into new, universally accepted numeric quantities.[12] This was an ambitious attempt to rationalize space into a geometric grid, which in large part succeeded.[13] A standardized system presented a radically different logic: one that was not tied to social hierarchies and the idiosyncrasies of physical labor, the environment, or materials, but one that allowed economic transactions to move unhindered across a frictionless space. The system meant that trust in an exchange was not dependent on a personal relationship, closed communities, or specific locale. Instead, it offered an absolute and fixed external reference that could be shared by strangers, regardless of rank or place.

Paris became the center where intellectuals gathered and argued for the necessity of quantification, where the institutions that maintained the authority of these numbers were instated, and where the effects of this extensive work were most readily implemented and assimilated in the following centuries.[14] More than simply a new measure, the use of the metric system in urban maps and plans dovetailed with their status as objective and scientific representations, generating and representing a conception of space as regular and neutral. These representations, practices, and discourses became the basis for Paris's claim to modernity, in which space became simultaneously inert (in the sense that it was understood as a backdrop to new social functions) and active (insofar as its divisions and definition into property lines and urban blocks became the basis on which new social practices arose).

These quantitative efforts had a direct and profound effect on how cities were mapped. Previously, maps had mostly rendered the city pictorially or perspectively. By the end of the eighteenth century, however, urban maps were drawn orthographically, and blocks and roads were drawn planimetrically. The meaning of accuracy changed and was evaluated not by the pictorial verisimilitude between image and terrain but by their geometric correspondence.[15] With the shift from the perspective to the orthographic, the map sacrificed pictorial legibility and became aligned with a growing and broader diagrammatic culture, most extensively manifest in Denis Diderot and Jean le Rond d'Alembert's *Encyclopédie ou dictionnaire raisonné des sciences, des arts, et des métiers* (Encyclopedia, or a systematic dictionary of the sciences, arts, and crafts) (1751–80). The "scientific" iconography of diagrams emptied the page of human figures and reduced all relations to lines on a paper surface. These lines also offered an image that could be consumed in a single glance. Immediate reception was part of the plan's condition as objective.[16] It was not a representation that offered depth and unfolded across pictorial space, revealing different elements over time. The orthographic plan, diagrammatic in its composition, equalized all lines onto the same surface of the page, assuming an atemporal space.

The graphic line's claim to universality, however, was based on more than the determination of units of measure; the thorough institutionalization, systematization, and internationalization of a standard set of references was what marked this quantitative endeavor apart from all others. Agencies, associations, educational pedagogies, and histories were created and organized to establish a modern bureaucracy that could bolster the new functions that the orthographic map afforded.[17] Cartographic documents and their associated material practices, such as marking and drawing, pervaded the government bureaucracy. Due to Paris's special status as both a French capital and municipality, architectural offices multiplied through functional replications such as the Conseil des bâtiments civils (Council for civic buildings) on the state side and the Direction du service de l'architecture (Direction of architectural services) on the municipal side, or the École des Beaux-Arts, the Académie des Beaux-Arts, and the Société centrale des architects. They contributed to the burgeoning of the administrative infrastructure and gained power in wielding this new graphic language.

These administrative bodies sought to retain the architect's status as artist against the ascendant engineering culture favored by Haussmann, who was partial to the arguments of science over the arts.[18] Ultimately, the technical culture of urban administration prevailed due to the growing dominance of the engineering profession and the engineering schools of the École des ponts et chaussées (School of bridges and roadways) and the École polytechnique, where mapping practices were consolidated through new curricula.[19] One of the surveying techniques taught in the schools' curricula was geometric triangulation that was combined with the drawing techniques of descriptive geometry. The resulting representations would essentially entrench their epistemological and representational modalities within the French administration, where many of the engineering students would establish their careers. That the orthographic plan became prevalent in engineering, architecture, and administrative bodies spoke to how these institutions and their cultures were mutually supportive, linked through shared media.

These agencies argued against the classicizing aesthetics preferred by the École des Beaux-Arts and instead espoused the functional needs of a modernizing city. Haussmann was a fervent critic of the École des Beaux-Arts, and the men he placed at the head of the major urban divisions, including the Service des promenades et plantations (Service for promenades and green spaces) and Service des eaux et égouts (Water and sewage works), all had engineering pedigrees, with the sole exception of the Service du plan.[20] His administrative culture leaned toward assertions of objectivity and quantification, and the number of auditors burgeoned in order to make the required calculations and keep track of the progress of their work and the developments of the city. These scientific methods were believed to be more objective, but they went hand in hand with administrative and institutional practices that were formed through political negotiations.

As many historians of science have argued, science has no value outside of its social use and is not divorced from the social context in which it is used.[21] Accordingly, science's employment in administrative and institutional practices was how quantitative representations, such as numbers, gained authority in the public sphere. These geometric maps featured tables and coordinates instead of royal cartouches and crests, constituting a different epistemological and social discourse for seeing and representing the city. Rather than being divorced from the political

realities of the nineteenth century, scientific methods and representations were made legible to the public precisely through political intentions. Moreover, as the architects in France sought to gain credibility and to establish norms against the increasing power of engineers during the long nineteenth century, they also often turned to the rhetoric of science, functionality, and objectivity to justify their work and value to the public.[22] Representations were integral to their public address, and while the number of images diversified, all claimed scientific status in the use of an orthographic modality.

As orthographic images acquired a symbolic meaning as scientific and objective, it was not necessarily in opposition to perspectival ones. In fact, the well-known *Plan de Turgot* (1739) was a scrupulously measured map of Paris that was considered the most accurate of its time. Moreover, the increasing dominance of orthographic cartographic modes was not marked as a linear progression. Perspectival or pictorial maps persisted and even gained more popularity than orthographic ones, especially among the public and notably during the Third Republic, when cartography and geography were fully incorporated into the public education curricula. The particular advantage of orthography and its adoption among urban administrators was that it allowed lines to be drawn on maps and plans. An orthographic plan could describe and project simultaneously.

In studying the drawn lines on plans, then, this book focuses on how geometry was not an iconographic symbol but an instrument that conditioned actual spatial practices, thereby shaping architectural work—and possessing agency within it.[23] As Robin Evans has theorized, the plan is at once an abstract and depthless space, as well as a very material two-dimensional surface.[24] Like Evans, this study does not assume the transparency of the medium. It attempts to provide an analysis of architectural drawing as an iterative process and analyzes drawing's representational limits. However, this book also expands on the notion of drawing as more than a confrontation of a singular architect with a blank sheet of paper, accounting for the collectivities and multiple actors and institutions involved in and answerable to the plan.[25]

Yet the same basic problems of translation endemic to drawing and building extend to the urban plan: something is lost in the space between. There always remains a chasm between representation and object to be filled in with "blank spaces" that bring to the foreground new graphic objects while occluding others. Moreover, what is not represented are the negotiations in marking one point as opposed to another; the labor of men who carry out the arduous tasks of surveying and hand-coloring prints; the production of paper and the expensive and meticulous printing from engraved copper plates; the viewers' responses to these plans; the institutions that supported the authority of the measure, that financed the map and its distribution, and that managed the vast paperwork supporting these plans. To ask questions about how these urban representations mediated spatial production is thus to ascertain who held power over and through these images, in the drawn lines as well as their blank spaces.

How were maps and the terrain made to correspond in the first place? Scholars of cartography have been dedicated to dispelling the positivist view of the map as an accurate representation of the world at least since the 1980s. Influenced by French theorists Michel Foucault and Jacques Derrida, J. B. Harley's canonical "Deconstructing the Map" asserted that a map is a value-laden image, and as such, is a product of its particular historical moment.[26] The assertion was meant to jettison the limiting assumption that maps were merely "communication devices."[27] In the proceeding decades, scholars of geography have interpreted maps and plans as socially constructed texts that carry contradictions and fragmentations, and represent power relations.[28] This critical approach was itself historicized by geographer Anne Marie Claire Godlewska, who placed the study and practice of geography within a trajectory that stretched to the Enlightenment, especially during Napoléonic France and through its imperialist ambitions.[29] Outlining how ideological motivations were concealed by the rhetoric of science, her historical work shows how maps lie, and how those lies justified violent forms of domination in the modern world through surveying, mapping, and drawing.

Coupled with histories of paperwork, the proliferation of maps and plans beginning in the late eighteenth century and through the nineteenth century was not merely a consequence of printing technologies and advancements in scientific instrumentation.[30] These urban maps and plans as well as the specificity of their graphic language were part of a growing bureaucratic infrastructure that marked the modern state. The claim of many media historians and theorists is that the medium of paper, its organization into stacks and files, and the document's circulation and organization are not an outgrowth of social relations

but, rather, conditioned them. The documents and its paper material compose a vital infrastructural element that undergirds systems of governance and the formation of our modern subjectivities. Yet, in analyzing this paperwork, the specific role of images has been ignored. Documents are assumed as text, when in fact many of the period's discourses sought to distance and distinguish maps and plans from words.

In addition to the proliferation of documents and their new uses, cartographic and diagrammatic images created new users. Often formatted into atlases for the public, orthographic maps exemplified a scientific genre that Lorraine Daston and Peter Galison argue was concerned with standardizing "the observing subjects *and* the observed objects."[31] These atlas compositions removed the observer's subjective idiosyncrasies by training and habituating the eye, offering phenomena as standardized—and not necessarily accurate—representations. Their critical insight is that the mechanical reproduction of scientific images was linked to the devaluation of subjectivity and the subsequent desire to eliminate all variability in scientific practices. Scientific images—such as botanical atlases—constructed the object in an ideal and thus fictive state while also arguing for its mimetic correspondence. Maps were no different, and, as Denis Wood writes, a map is a "tissue of fictions."[32] The border between water and land, the structure of a block, the outline of shadows, the conditions of foliage, and so on, were all features where no one fixed and constant state exists.[33] Yet the form of the atlas, in its repetitive configuration, worked to naturalize a complex and contingent idea into a fact, giving it an ideal shape that remained constant across the paper sheets.[34] Thus, not only the practices of measuring and surveying the terrain but also the drawing, printing, and viewing of new spaces consolidated an image of Paris, constructing it as a stable object that was able to be observed, controlled, and shaped.

More broadly, the production and use of these maps and plans were aligned with the ways in which vision and visuality radically changed in the first half of the nineteenth century. As asserted in Jonathan Crary's classic *Techniques of an Observer* (1992), vision is a historically contingent practice, constituted as much by the things seen as the people who see them. The nineteenth century offered a heterogenous landscape of new visual objects and instruments, including orthographic maps, with their diagrammatic compositions. They were integral to what

Crary has described as a scopic regime of modernity, in which avant-garde painting and positivist scientific culture were not opposed to, but conditioned by, each other.[35] As a product of both scientific labor and artistic skill, maps and plans represented an object located precisely at these categorical borders, and thus were exemplary sites on which these ideas were negotiated.

The maps and plans under consideration in this book are official governmental documents. Thus this study only addresses a small part of a growing mapping culture that pervaded France beginning in the late eighteenth century, and how it was aligned to a "governmentalization of the state."[36] Other phenomena, from the numbering of houses and assignment of street addresses to the proliferation of national border mapping, reflected not only a "cartophilia" of French institutions and the public but also a belief that particular and local spaces could relate to universal and global space.[37] The map's adoption of a universal measure created the possibility of comparison among any and all objects subjected to the meter. Orthogonality offered a means to maintain a fixed measure at various scales, and when applied to the urban form, the city became understood as a combination of elements, both large and small, that could be composed in various ways, making it into an ensemble. Architects became responsible for more than simply buildings, but also urban furniture such as benches and lampposts that were then linked in conception and representation to the street, the urban block, and the city as a coherent and linked whole, thus establishing a system. A drawn line on a map related to both the detail and the grand and was scaled in ways that assumed frictionless movement on the plan's blank sheet of a paper.

The significance of this level of standardization was beyond simple functional necessity or efficiency. It was part of a broader belief that sought mastery, or at least the possibility of mastery, that manifested not only in regularizing instruments but also methods of measure, inquiry, and examination.[38] This impulse was manifest from the surveys of the Parisian streets to the atlases of new sewers, and even the Expositions universelles, such that the city became "a space of calculability."[39] In these instances, administrative bodies found in maps a method to organize their world and submit that world to their gridded logic, whose tactics of governmentality could be applied at all scales.[40] This concept of governmentality is central to Foucault's course at the Collège de France, "Sécurité, territoire, population," in which he traced the shifting role of the state from securing the territory to ensuring the population. His objective was to demonstrate how "the sovereign of territory became an architect of disciplinary space" and, as Stuart Elden explains, "how the architect also became the regulator of a milieu where he or she did not so much fix the limits and frontiers, or the sites, but allowed circulation."[41] If securing the territory of the sovereign concerned spatial enclosure, then safeguarding the population entailed movement and passage. While Foucault juxtaposes discipline and security and their opposing characteristics of circumscription versus circulation, or isolation versus incorporation, there is sparse explanation of how these values of enclosure and movement were facilitated through specific tools.[42]

Foucault wrote about certain architectural typologies that emerged in the modern era, including psychiatric hospitals, asylums, and prisons, yet their architectonic quality was not his focus. Instead, these types informed an analysis of how a disciplinarian space was produced and defined modern life. They were part of a broader project to examine the political control of the body, as well as how urban space was conditioned by biopolitical practices.[43] The techniques of surveillance, classification, quantification, and even exclusion were central to the profession of urban planning that institutionalized a certain relation between "forms and norms" in which space was neither inert nor symbolic, but a *dispositif* that shaped socio-spatial relations.[44]

The translation of "dispositif" in English has often been "apparatus," and in the planning literature, the term has mostly been understood along functionalist imperatives that have often denoted "tools and devices."[45] The term as defined by Foucault, however, is much more fluid and heterogeneous, encompassing a matrix of "discourses, institutions, architectural forms, regulatory decisions, laws, administrative measures, scientific statements, philosophical, moral and philanthropic propositions."[46] Significantly, a dispositif describes a historically contingent relationship between discourse and material, the said and unsaid, the public and private, from which emerges a specific social body, and from which that social body also affects knowledge and conceptualizations of that very relationship. Thus, in this book, a map is not defined as a tool for biopolitical functions, nor is it a mere representation of government intentions. Instead, the map is located within a matrix from which a modern conception of space was negotiated on overlapping symbolic, epistemological, political, *and* functional grounds.

My aim in this study is to understand the modes of representation that inform architectural and planning practices by tracing their epistemological and social values, while also accounting for the materiality from which actions were drawn. In this respect, my methodological orientation moves beyond Harley's idea of a map as social construction and pursues mapping as a way to understand the "intermediality" of architecture and urbanism within a complex social, political, and cultural infrastructure.[47] Thus, more than a "discourse analysis" of maps, what follows is a study of discursive techniques, drawing from the disciplines of art history, media theory, the history of science and technology, and critical geography. References to German media theory are also clear in this study, especially in the ways that the built form is not its telos.[48] By centering the story of Paris's modernization through the production of its maps and plans, the city becomes a contingent artifact itself negotiated through urban representations, concentrating on the collectivities and processes involved in its conception and building.

The normative role of the plan in architectural and urban planning practice often obscures the historical specificity of its conception. Building does not require a plan, and yet plans became required. It is taken for granted that the design of buildings and cities begins with a drawing, on which lines describe the terrain, the situation, and the organization of the projected structure and its surroundings. Thus, in historically outlining the ways in which urban maps and plans became diagrammatic, this book also tries to outline how the plan's capacity to describe a given space was established through shifting descriptive and temporal values in architecture and in society as a whole.

The ways in which spatial and social experiences were mediated through an orthographic map that claimed to be removed from the very things that it sought to represent and control tracked with the double-faced nature of modern life described by Walter Benjamin.[49] For Benjamin, the ephemeral and transitory denoted *la modernité*, analyzed through Charles Baudelaire's "Le peintre de la vie moderne" (The painter of modern life) (1863).[50] Yet, unlike Baudelaire, who opposed the modern with the ancient, Benjamin saw in the emergence of the capitalist metropolis the concomitant development of a consumer culture that linked the fervent and insatiable desire for novelty

to a fetishization of its commodity form. Spatial experiences were then suspended between these temporalities—the transitory and the ever-repeating—such that the new, modern urban spaces were valued by the appearance and circulation of objects found in exhibition halls, department stores, arcades, and streets.[51] Orthographic maps in their schematic and drawn forms rendered this spatiotemporal tension in a graphic medium. The map made claims on the universal while sustaining projective and descriptive modalities in which the straight line created a kind of temporal suspension. Thus, in its form and function, these plans of Paris both produced and were products of its modern spaces and experiences, characterized by the promise and violence of an all-encompassing cartographic grid.

The book is organized into two parts. Chapters 1 and 2 are focused on the production of urban plans. The first chapter examines the myth of a master plan—a myth perpetuated in historical accounts of Paris's modernization—by analyzing the scenographic power of Napoléon III's cartographic gesture. Contextualizing urban maps within the tradition of portrait painting, the central focus of this chapter is the way that the myth of Haussmann's map gained authority and staying power. The second chapter outlines the survey of Paris by Edme Verniquet and his methods of triangulation, and also discusses the second survey by Eugène Deschamps under Haussmann, the only survey during the city's modern transformation. The history of surveying the city of Paris has important connections to the "scientific mission" in Egypt under Napoléon I, to the shifting social value of surveying labor (and expertise more generally), and to the growing publication and reception of surveying manuals outside of the Académie royale des sciences.

The second part of the book centers on the use of maps by urban administrators. Chapter 3 studies the cartographic grid. This device became an essential reference point for drawing new plans for the city, functioning to determine many angles of proposed streets, districts, and monuments. The fourth chapter proceeds from the simple fact that plans are not necessary for construction, and contends that their value lies in aggregating and linking various actors into the creation of a centralized bureaucracy. Many areas and buildings in Paris were erected without the aid of drawings. Yet in 1837 a law was passed requiring a plan to be drawn for all building projects.

The argument is that, more than just useful for conceiving of new spaces, maps and plans were important tools to consolidate administrative activities related to the built environment and its governance. The final chapter considers how government maps that were initially only for privileged viewers were eventually presented to the public at the Expositions universelles of 1878 and 1889. In two atlases, the government set out to narrate the history of Paris's modernization and to sell its legacy to new audiences.

Ultimately, the book aims to delineate the modes of cartographic representation that were deployed for building a city that would come to exemplify modern urban space. This book does not cover all of the maps that proliferated during the nineteenth century. Due to technological innovations and intellectual and cultural shifts, a broad cartographic culture flourished in France at the time; however, this study concerns only a small subset of official governmental maps—many of them technical. Moreover, this book also does not venture beyond Paris in its scope. There are excellent studies of the use of maps in the colonial and imperial conquests by the French (and British), but there remains a question of how Paris's own modernization—the modern metropole—was shaped by certain modalities of drawing and mapping.[52]

The starting point is the specific emblematic urban plan held by Napoléon III that defined the modern French capital. What was the significance of drawing the city in such an abstract and geometric manner? Extending from this image, the basic assertion is that the plan as it is used today in contemporary architecture emerged from specific historical conditions related to the architectural, scientific, and intellectual cultures of the long nineteenth century. These cultures were tied to the Enlightenment and to the political contexts of French imperialism and urban modernization. Beyond being a description, the plan was an image laden with political, social, and cultural values that aggregated diverse practices into a standardized and highly codified schema supported by new institutions and laws. Furthermore, the plan's eventual standardization went hand in hand with the formation of new practices and viewers that would define modern Paris and, ultimately, modern space itself.

The elusive quality of maps and plans is that they reveal and determine the forms of the world, while simultaneously concealing how those forms are shaped. The power of the map lies in how it presents its graphic forms as given and complete. In this respect, maps make invisible the ways in which they do not account for, and do not concern themselves with, social relations, paradoxically occluding that which they set out to determine. This book's inquiry into urban mapping practices thus attempts to trace historically how the map acquired such power in France's capital, by revealing the context that ensured its legibility. While French imperial and colonial power extended well beyond Paris, this study concentrates on the French capital and those in power, to examine how the composition and elements of nineteenth-century urban plans defined the city as modern.

1 Master Plans

A map is not the territory it represents.
—Alfred Korzybski, *Science and Sanity* (1933)

Georges-Eugène Haussmann's role in the transformation of Paris is often understood as synonymous with its modernization. During his tenure as prefect of the Seine, he oversaw the construction of new boulevards framing new public monuments, new parks for new forms of leisure and entertainment, new public spaces for consumption and performance, new infrastructure for transportation and circulation, and a restructuring of the municipal government. Coined in 1892, after his death, the term "Haussmannization" credited the prefect with developing the specific method of urban development favoring large-scale demolition.[1] It soon came to be shorthand for urban modernization in many new contexts, from Rio de Janeiro to London, Dakar, Cairo, and Brussels, and its original critical connotations were replaced with positive associations of efficiency and efficacy.[2] Paris became the model for modernization programs well into the twentieth century, and by extension, many cities around the world claimed Haussmann's legacy.[3] This link between modernization and Haussmann, however, begs examination, especially regarding the extent and nature of Haussmann's role in Paris's transformation.

HAUSSMANN'S COLORED MAP

As recounted in Haussmann's memoirs, Paris's modernization program began in June 1853, when Emperor Napoléon III handed him "a map of Paris on which were traced by the Emperor himself, in blue, red and green depending on the degree of urgency, the different new roads that he proposed to be executed."[4] The description of this cartographic gesture suggested that Haussmann's administration simply went about fulfilling those plans: Napoléon III had a project in mind, and Haussmann was the ideal bureaucrat to carry it out. The story's seeming clarity is seductive: the narrative assumes that first, there was an idea; then, the representation of that idea—in this case, through a map; and finally, a building process based on that representation.

Apart from Haussmann's own memoirs, the myth is perpetuated in Arsène Houssaye's supposed eyewitness recollection of Napoléon III in Saint-Cloud. He recalls the emperor, having returned from exile, unrolling "a great scroll of paper on which he had designed a city for twenty thousand inhabitants, with churches, fountains, squares, monuments, and, of course, a stock exchange. . . . A real city of the future, such as we shall have here in France . . . no longer will one house be built at a time, he said, but the whole will be begun on the same day and all shall be finished at the same time."[5] This scene was cited in several other accounts from the Third Republic, including Haussmann's. He was not present in Saint-Cloud, but also recounted a similar scene in his memoirs, which he penned after the collapse of the Second Empire, in an attempt to recuperate his reputation at a moment when the excesses of that period were under critical revision.

The mythic status of the scroll is perpetuated by the fact that this colored map was destroyed in the 1871 fire of the Hôtel de Ville during the Paris Commune, along with many of the city's archival records.[6] Some versions were recuperated later, however. One well-known version was retrospectively drawn by Charles Merruau, the secretary-general of the prefect, and was supposedly checked by the emperor and published in 1875 in Merruau's own memoirs (fig. 1.1).[7] Another plan, from 1867 and given to Friedrich Wilhelm of Prussia by Napoléon III, was discovered by André Morizet in 1932 in the state archives of Berlin.[8] This plan was later credited with being instrumental in France's swift defeat by the Prussians in the Franco-Prussian War that spectacularly ended the Second Empire. The map that was used to modernize and open the city to greater movement supposedly helped enemy forces surveil and conquer the city.

Descriptions of Napoléon's map continued to be repeated in the historiography of the twentieth century, starting from Gérard-Noël Lameyre's account in *Haussmann, "préfet de Paris"* (1958).[9] While offering different perspectives on the city's modernization, each of these studies gave the map a central place in the interpretation of Paris's transformation. David Pinkney squarely placed responsibility for the origins of modern planning on Napoléon III, tracing the emperor's encounters with maps and models that may have influenced his conceptions. For Pinkney, whose writings on Paris from the 1950s were important for introducing the subject to English-speaking scholars, urban planning was an inherently paternalistic practice, justified by the hygienic advantages gained by modernization. The map then served as proof of authorship and intention. Alternatively, Leonardo Benevolo's 1967 history of modern urban planning hinged on the revolution of 1848; he argued that the revolution's failure resulted in the hollowing out of town planning's explicitly political goals. After 1848, planning became understood as a technical practice, not without ideological motivations but obscured by the rhetoric of scientific neutrality. In Benevolo's Marxist interpretation, the map fit into a general argument about planners framing their work as technical in order to evade giving political justifications, done in service to the ruling class. Accordingly, David Harvey's 2003 study of Paris, offering a materialist approach to the city's forms, sought to demonstrate the vacillations of authoritarian imperialism and capitalist liberalism in creating the conditions to produce modern space, not only in economic and physical terms but also socially and psychologically. Harvey cited the map of the city from which Haussmann directed this spectacle of European modernity, using the representation to make claims about the vast new scale of finance capital.[10]

These historiographic repetitions parallel the logic of the story itself. The city is portrayed as a completed and legible image—a colored map—that includes the elements of modernity already prefigured. The modern city is conceived in its entirety. Its image suggests a synoptic coherence in which the immediacy provided by the cartographic projection links to its instantaneous execution. The assumption is that there is no mediation between idea and representation and building; the process appears frictionless—and therein lies its visual power. The modern city is ever and only present.

If repetition characterizes the historiography of Paris's modernization, repetition also forms the underlying conception of modern urban planning. On the one hand, a map is understood to be the origin of the modern city; it is projective, proposing new spaces and buildings. On the other hand, detailing the city as it is (and will be) built is a descriptive act. This duality of the plan's function is canonized in the *Histoire de l'urbanisme à Paris* (1975), in which Pierre Lavedan characterizes urban planning as a therapeutic practice.[11] The city becomes afflicted with chaos and squalor, necessitating a new plan from which the city is then reorganized—only for a succeeding plan to be drawn and the cycle to begin again. The plan simultaneously describes the problem and generates a solution that demands a new plan be drawn, eliding the solution of an urban problem with its representation.

This inextricable link between the formation of the modern city and the manufacture and consumption of images was part of the larger visual culture of the French Empire during the nineteenth century.[12] As art historians have astutely argued, it is in visual representations of

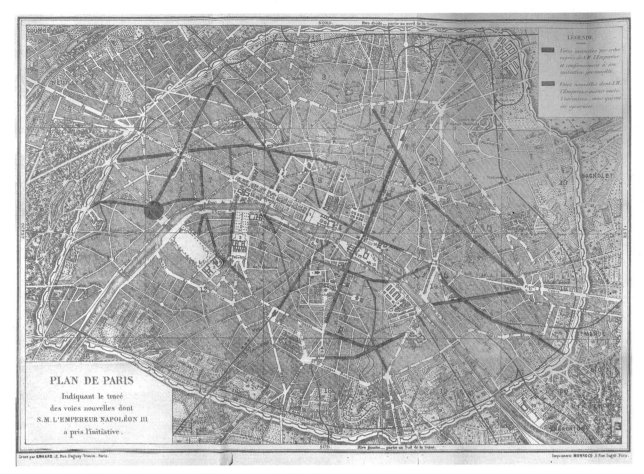

FIGURE 1.1 Plan of Paris indicating traces of the new streets initiated by Emperor Napoléon III, 1875. From Charles Merruau, *Souvenirs de l'Hôtel de Ville de Paris, 1848–1852* (Paris: Plon, 1875). Bibliothèque historique de la Ville de Paris.

modern life that the very terms of life were negotiated. Painting, whether avant-garde or rearguard, became a site in which self-fashioning and social identity were both situated and disputed. Yet while painting was a central medium for political and cultural discourse, maps and plans during the same period diverged from these discursive spheres. If earlier maps were understood to carry symbolic meaning and to acknowledge the limits of the known world, nineteenth-century maps disavowed any such mediation or philosophical baggage, even as they came to determine more and more of modern life. These images became defined in positivist terms, emptied of their social valence, and read as politically neutral. By the time Haussmann came to power, the orthographic method of representing the city had become normative of governance practices and integrated into all aspects of urban building. Over time, these images required less and less commentary about how they determined modern life through new cartographic forms and the new institutions that supported these norms.

These geometric forms diverged from the symbolism of earlier, more painterly mapping traditions and, most significantly, were composed using a graphic language that obscured a sense of temporality and context. Their synoptic quality was deployed to provide historical explanations, blurring together causes and effects. At the same time, urban administrators took advantage of the atemporality of these forms to justify the terms of a problem through its solution.[13] On the most basic level, the repeated scenes of Napoléon's map proposed that a city was prefigured before it was built and, moreover, that this city was based on a master plan composed by a single individual. They suggested that the city and its image were equivalent.

The idea of a master plan and the belief in it were enforced then historiographically, beginning from unquestioned personal accounts, whether it was Houssaye's description of the party in Saint-Cloud, Haussmann's memoirs, or Merruau's reconstruction based on Napoléon's memories. These subjective testaments are an incongruous start to a history of modernity, given the scientific and intellectual values of empiricism and objectivity that made these maps and plans trustworthy in the first place. This historiography has collapsed a critical distinction between the discourse on the modern project and the practices and institutions that came to define modern society. The map occluded this difference by standing in for both: as an expression of, and also as an instrument for, modernization. Yet this is only part of the story. Its grip lies in a broader cartographic culture as it relates to the map's privileged visuality of surveillance. Maps were instruments to define, see, and control the sites and terrain that they represented, which were acquired and maintained through imperial rule and military force; and the appeal of this origin story was situated in a historical claim of authority.

PORTRAITS OF MAPS

The scenographic power of the map is supported by a long iconographic tradition in French portraiture in which the map (or globe) is sign and symbol of a ruler's geographic reach, serving to delineate both actual descriptions and desired projections. It links the body and office of the ruler to a spatial form and territorial identity. In a 1739 portrait of the young dauphin, the artist Louis Tocqué surrounded the future king with cartographic props: a globe, rolled maps, and a geography book (fig. 1.2). The painting implies the study of the territories under his royal hand as he prepares for his future rule.[14] The dauphin points casually to the globe, which is turned to reveal Europe and the Mediterranean—but not France—as well as a book of maps and charts, held open with a surveying instrument. Significantly, he does not look at the map; he does not have to gaze upon his own territory, because it is an extension of himself. Instead, viewers of the portrait, in the role of his subjects, are meant to look at the territory, and through their act of looking affirm their status as subjects.

That royal portraiture would often picture maps and other cartographic objects is not due merely to their symbolic power. The correspondences between the sitter and

FIGURE 1.2 Louis Tocqué, *Louis, Dauphin de France*, 1739. Oil on canvas, 77 × 57¾ in. (195.6 × 146.7 cm). Musée du Louvre.

his portrait and between the map and the terrain are based on the possibility of visual comparison by a viewer. In Denis Diderot and Jean le Rond d'Alembert's *Encyclopédie ou dictionnaire raisonné des sciences, des arts, et des métiers* (1751–80), a portrait is defined as a work that depicts the appearance of its sitter, accurately representing physiognomy and expression.[15] This definition was further articulated in the *Encyclopédie méthodique par ordre de matières* (Methodical encyclopedia by order of subject matter) (1782–1832), which qualified the meaning of "accuracy," defined as being achieved when "the characteristic resemblance can be *easily recognized* for that of the person whose traits one proposes to render."[16] The *Dictionnaire portatif de peinture, sculpture et gravure* (Portable dictionary of painting, sculpture and engraving) (1781) provided an elaboration by requiring the recognition to be immediate, such that the viewer is "able, in one glance, to recognize the person as one may have known them."[17] Immediately recognizable resemblance was central to the success of the portrait, which relied on the

FIGURE 1.3 Henri Testelin, *The Establishment of the Academy of Sciences and the Foundation of the Observatory, 1667, Based on a Drawing by Charles Le Brun*, 1673–81. Oil on canvas, 137 × 232¼ in. (348 × 590 cm). Musée national des châteaux de Versailles et de Trianon. © RMN-Grand Palais/Art Resource, NY.

viewer's capacity to compare represented and real figures. Yet if comparisons between the pictured dauphin and the dauphin himself could be made, they were not stable. The young dauphin, painted at the age of nine, would soon mature, and in time, the portrait would become a reference to the past. Moreover, comparisons between image and reality, though implied, were not possible for the territories drawn on the map, as the accuracy of those coastlines and landmasses could not be confirmed by visual comparison. Without the possibility of verification by the painting's viewer, the pictured map's graphic claim to be an accurate description was given authority primarily through the resemblance of the sitter. Moreover, the hand gesture of the dauphin pointing toward the globe and looking out at the viewer secures the correspondence not only between the terrain and map, but also between ruler and subject, between pictorial and lived spaces.

Henri Testelin's monumental painting of the Académie royale des sciences displays similar compositional strategies to Tocqué's (fig. 1.3).[18] Commissioned for the *Histoire du roi*, the tapestries that would depict the illustrated

series of battles and events of Louis XIV's reign, the scene holds Louis at the center, surrounded by the founding members of the new academy (who are being presented to the king by Jean-Baptiste Colbert), as well as books, maps, plans of military structures, and a globe. By placing a representation of the newly funded royal observatory in perspective, the group portrait signals the different modalities of representation linked to the projection, knowledge, and the building of spaces.[19] The objects are turned not toward the king for his viewing, but toward the tapestry's viewers. The globe is turned to reveal France, highlighted for the viewer's benefit, while plans hang over the table turned outward. And although the figures are secondary, they are nonetheless critical in legitimating the king's rule. What is a king without subjects, to say nothing of a king without a domain, without institutions, or without a history? Through their attention to the grand map being hung, to the king, and to other spectators, these woven figures—and the tapestry's viewers—become networked spatially and temporally as the king's subjects through the shared act of looking at the maps.

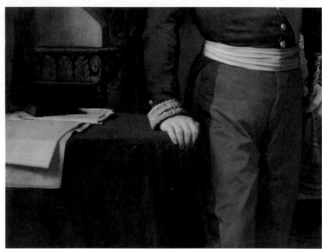

FIGURE 1.4 (left) Jean-Hippolyte Flandrin, *Napoléon III, in Uniform as Division General in His Official Office at the Tuileries*, 1862. Oil on canvas, 83½ × 57½ in. (212 × 146 cm). Musée national des châteaux de Versailles et de Trianon.

FIGURE 1.5 (below) Detail of Jean-Hippolyte Flandrin, *Napoléon III, in Uniform as Division General in His Official Office at the Tuileries* (fig. 1.4).

Jean-Hippolyte Flandrin's portrait of Napoléon III, displayed at the 1867 Exposition universelle, references this iconography (fig. 1.4).[20] As in Testelin's composition, the leader's hand rests on the table where maps are scattered casually, subtly connecting themes of rule, direction, and control. In his office at the Tuileries, appointed with national icons and colors, the emperor stands against the corner of a desk where two large maps are laid on top of one another. On top is a map of France, and below it, a plan of Paris. Their text is oriented so to be read by the viewer (fig. 1.5). They do not show the cartographic content itself, as in Testelin's or Tocqué's paintings, but rather serve enunciative and discursive functions within the painting by revealing the titles: this is Paris; this is France. The emperor looks directly out at the viewer, and while wearing the uniform of brigadier-general, grasps the handle of the sword at his side. His groin is positioned level to map, hand, and sword, further signaling his masculine authority. Unlike in the dauphin's portrait, he is not in the midst of studying the plans, but rather he is in full possession of them, ready and alert to defend the French

territories. He does not point to them as the young prince so eagerly does; he assumes the authority over the terrain represented by the maps.

The scattered papers on the desk and Napoléon III's posture by the table recall an earlier portrait by Jacques-Louis David of Napoléon Bonaparte in his study at the Tuileries palace (fig. 1.6). The papers on the desk also partially reveal words; here, "COD" alludes to the Napoléonic Code, established in 1804 as a legal code to govern all territories under French control. A map is among these documents, but it appears to have rolled off the table and fallen on the floor beside the desk, where it becomes a pictorial device for the artist to include his signature and date on its underside. The map reveals little, except for the word "tableau," hinting that if it were unrolled, it would provide an overview of France. As Philippe Bordes has argued, the significance of the ruler's portrayal at a desk rather than in royal garb was a reimagining of the emperor "as the busy head of government."[21] Unlike the 1690–91 portrait by Nicolas-René Jollain, le Vieux, of Louis XIV with a plan for the Maison Royale

FIGURE 1.6 (left) Jacques-Louis David, *The Emperor Napoléon in His Study at the Tuileries*, 1812. Oil on canvas, 80¼ × 49¼ in. (203.9 × 125.1 cm). National Gallery of Art, Samuel H. Kress Collection, Washington, D.C.

FIGURE 1.7 (above) Nicolas-René Jollain, le Vieux, *Louis XIV Holding a Plan of the Maison Royale de Saint-Cyr*, 1690–91. Oil on canvas, 87 × 65 in. (221 × 165 cm). Musée national des châteaux de Versailles et de Trianon.

de Saint-Louis at Saint-Cyr, in which the king casually gestures at the building that is to be constructed by others (fig. 1.7), here, Napoléon Bonaparte is a skilled administrator who himself consults maps and plans and composes rules and laws. Paralleling the portrayal of Louis XIV in Testelin's painting, in the painting by Jollain, Louis's aloof affect demonstrates possession, but not the personal investment demonstrated in David's portrait of Napoléon Bonaparte and the later paintings of Napoléon III.

Within the context of this cartographic iconography, Napoléon III's gesture with the colored map, as recounted in numerous histories, performs a legible and symbolic command. Like these portraits, the scene of a sovereign gesturing at a map frames how geographic representations are inextricably tied to claiming territories under a leader who is simultaneously upheld by those very maps. The authority of the map is guaranteed by the ruler, as

much as the ruler is legitimized by the map. And in this case, for the "modern" ruler, the map is a visible symbol of a new bureaucratic order with which that ruler is actively involved.

MAP AS INSTRUMENT

In "Deconstructing the Map," J. B. Harley uses the term "mimetic bondage" to describe how maps are often seen as a mirror of nature, and that "by the application of science ever more precise representations of reality can be produced."[22] In this model, nature is privileged over the representation as the origin of and basis for visual knowledge, and the goal of mapmaking is to come ever closer to that reality. His critique considers the map not as mere description but as a value-laden image wherein "social structures are often disguised beneath an abstract, instrumental space," thereby challenging the definition of the map from being a scientifically objective image to a socially constructed object.[23] Harley's project attempts to move away from evaluations of maps based on empiricist and nomothetic claims.[24] By denaturalizing the link between terrain and map, the map then becomes a site where social, political, and ideological performances are mediated, negotiated, and figured.

If maps become sites of investigation, we can begin to understand their modality of representation as constituted not just in terms of technique but also within the context of larger epistemological and cultural questions about how cities were seen and defined. During the mid-eighteenth century, French maps of the country's cities gradually shifted from representing the city in linear perspective to projecting the city orthographically. Since the Renaissance, perspectival views had dominated urban maps. They depicted an approach into town through the surrounding countryside, providing a picture and profile of an urban landscape (fig. 1.8). With a fixed horizon line, a map aimed to simulate visual and physical movement through the city, where the landscape would unfold incrementally with each imagined step. In this context, introducing the bird's-eye view into perspectival images in the sixteenth century was radical: it offered a comprehensive view of the city that had not been possible in earlier cartographic representations.[25] According to Hilary Ballon and David Friedman, bird's-eye maps employed techniques from perspectival representations in order to show depth, but the image was now contextualized as an aerial view of the city's totality. The plane of paper was still a window, but one that made the city illusionistically visible in a manner that was unavailable to direct experience.[26]

In the case of Paris, many of these aerial views drew the Seine vertically, revealing the façades of the city's monuments to a viewer observing ships heading to and from the ocean. Like strict perspectival projections, these views emphasized building façades and the city as an agglomeration of individual masses. The city revealed its face and its character through its specific buildings.[27] Often themselves referred to as portraits (*pourtraict*), all of these types of maps represented the appearance of the city in its monumental singularity, and they were intended as expressions of a ruler's power and domain.[28] While these bird's-eye views represented a profound departure from ground views by elevating the viewer's position and negating the possibility of direct visual comparison between map and city, they nonetheless continued to

FIGURE 1.8 Mathieu Mérian, *Plan de Paris sous Louis XIII* (Plan of Paris under Louis XIII), 1615. Engraving, 21$^{11}/_{16}$ × 39¾ in. (55 × 101 cm). Bibliothèque nationale de France.

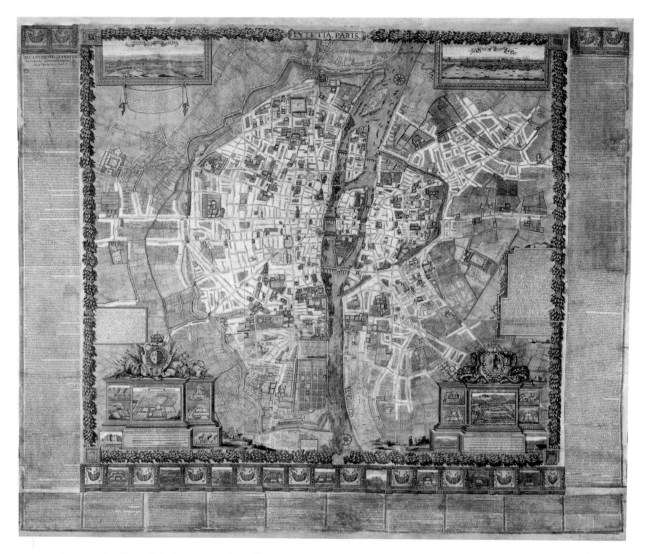

FIGURE 1.9 Jacques Gomboust, *Lutetia Paris,* 1652. Map sheet, 59¾ × 71¼ in. (152 × 181 cm). Bibliothèque nationale de France.

presuppose a viewer, allowing one to "move through" the representation in the air.

Because visual comparisons to confirm a map's verisimilitude were no longer possible in this modality, new rhetorical strategies were required to sustain the belief that this abstract image corresponded to the terrain. Previously, royal decrees unfolding at the margins or allegorical figures pointing into the landscape vouchsafed the map's veracity, aligning its authority as an accurate representation with pictorial traditions and iconography.[29] However, the linear qualities of orthogonal images were tied to an aim to no longer reproduce the direct visual experience of a built environment; rather, these images aimed to represent the city's structure, with its graphic validity laying precisely in its inability to be compared to direct experience. Planar representations threatened the rhetorical tradition of the urban map as a portrait, in which the city's character was understood through descriptions of its monuments. The horizon as the viewer's orientation and drawing's reference disappeared, and there was neither a fixed view nor a specific viewer.

The gradual shift from bird's-eye to planimetric views in the seventeenth century was not, however, without some experimentation, and these two categories were often not separable. For example, when bird's-eye views dominated cartographic representations, *Lutetia Paris* by Jacques Gomboust (1652) represented streets and many lots orthogonally, but large edifices and monuments were shown at an angled aerial view, with the Seine oriented vertically (fig. 1.9). Conversely, almost a century later,

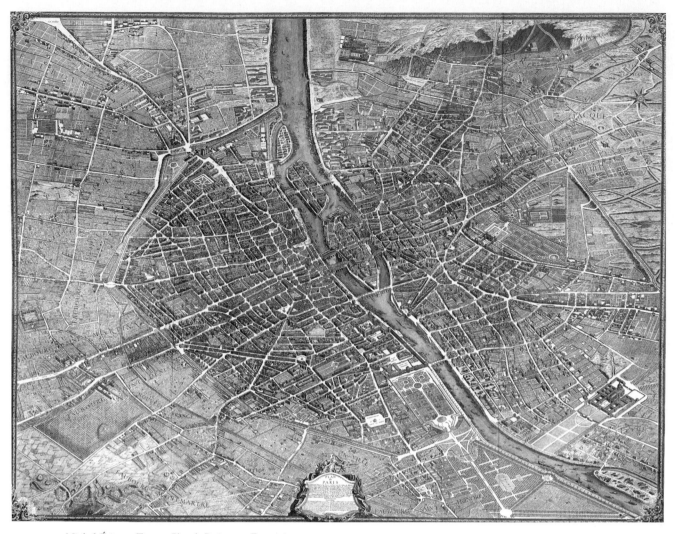

FIGURE 1.10 Michel-Étienne Turgot, *Plan de Paris,* 1739. Engraving, 96½ × 128¼ in. (245 × 326 cm, in 20 plates). Bibliothèque nationale de France.

when bird's-eye views were less common, the exceptional *Plan de Paris* by Michel-Étienne Turgot (1739) took a perspectival view oriented to the northwest with depictions of the buildings' façades, while simultaneously attempting to meet the synoptic advantages of planimetric modes by gradually angling the surface upward as the eye looked toward the horizon line (fig. 1.10). The persistence of different types of representational modalities did not contradict the general tendency, and my aim is not to argue for one form of representation supplanting another, but rather to claim that the role of the map expanded its descriptive and symbolic functions.

The orthographic mode signaled an important shift from "the map as spectacle" to "the map as instrument."[30] Quantitative methods of its production and its graphic composition changed the status of a map from an object to describe the city based on pictorial verisimilitude to an instrument to project urban designs based on geometrical correspondence. Orthographic composition constructs a view of an object at right angles in a single drawing. It is not a mimetic representation insofar as it is not a copy of how an object is seen. An orthographic map is no longer a portrait of a terrain. Instead, it produces graphic spatial knowledge of an object based on quantitative correspondence. The shift from bird's-eye to orthogonal views of cities represented a change from representing what could be seen and visually compared to what must be drawn in order to be seen.

The primary feature of this cartographic paradigm was the appearance of blank spaces within the composition, one that Michael Marrinan and John Bender have referred to as "material whiteness," to emphasize that such spaces are not to be understood as voids.[31] Analyzing the plates from the *Encyclopédie*, they argue that the appearance of these absences denaturalizes the image from that which it references, denying the fixity of a particular viewpoint while simultaneously facilitating projective analysis. Marrinan and Bender understand this blank space as allowing for correlations to be made across the different parts of the *Encyclopédie*, supporting "the composite play of imagery and cognition that is the motor-engine of diagram."[32] The appearance of the white spaces allowed for operations, such as comparison, analysis, and critique, based on reduction and isolation.[33] The white space functioned not as a passive ground onto which the figures were placed, but as an active element in its own right, whose very opacity allowed for the specific temporality of the graphic elements to be suspended, and for users to project their own thoughts onto its surface.

Standing in front of a large plan that had reduced the city to orthographic lines and blank spaces, Haussmann recalled neither looking out the window nor walking the streets of Paris to generate new planning ideas. He said: "I could, at any minute, by turning around, look in detail, check certain indications, and recognize the topographical correlations of the arrondissements and neighborhoods of Paris between them. Very often, I found myself before this accurate tableau with fruitful thoughts."[34] Here, he described in rich language all of the epistemological operations that the orthographic map afforded. Trust in the map's accuracy allowed Haussmann to make apparent connections within the composition and in relation to the urban terrain. For the prefect, the plane of paper was no longer a window that offered an aerial view of the city, but a medium for the linked practices of description and composition mediated through planar projection.[35] His capacity to use the map as a means "to see" the city depended on geometric correspondence.

A MAP'S COMMAND

If shifting epistemological values of description had supported the act of surveillance through this orthographic map, Haussmann's acts of projection recalled a history of military practices. Scenes of generals determining battle strategies over large maps are well established in the collective imagination; however, the totalizing view of a military map was a relatively recent development. In the early modern period, manuscript maps of the frontiers and of specific conflict areas were produced by the *ingénieurs du roi* (engineers of the king), but a map of an entire region was not commonplace. Large area maps were reserved for depicting the coastlines for sea navigation, called portolan charts, in which vast swaths of the ocean space between coastlines were left empty. Most often, military mapping in France consisted of surveyors drawing maps on-site to be used where they were drawn. The representations provided by these maps were contingent upon and distinct from the event and the place. They were used for outlining strategies for a specific battle at a particular moment. They were not reproduced and not part of a comprehensive cartographic program.

The exception were fortification plans, many of them constructed in relief to represent topographic information. Under Louis XIV, these fortification plans were compiled in 1683–88 as the *Recueil des plans des places fortes* (Collection of fortification plans), which contained over three hundred plates of all the fortresses in the French territories. These relief maps were both descriptions of the terrain and architectural scale models. Most of the first maps were of fortifications constructed or reconstructed by the military engineer Sébastien Le Prestre de Vauban. At a scale of about 1:600, they were built from stacked wooden plates and assembled with iron bars, each layer carved away to create the varied terrain, fortifications, and surrounding buildings. They were kept secret as part of the private property of Louis XIV, who used them for war games or to follow the progress of an actual siege from the safety of his palace. Some of these map-models were extensive: the relief map of Namur is fourteen plates high; Saint-Omer is about fifty square meters; Strasbourg is 10.86 meters long and 6.66 meters wide. Yet, however large, these representations compensated for the lack of more comprehensive topographical maps and drawings.[36]

Military mapping became a crucial operational tool for Napoléon Bonaparte: he reestablished the state title and privileges of *ingénieurs-géographes* (engineer-surveyors) (originally founded in 1769, but disbanded by the Directory in 1791), and charged them with systematically drawing maps of the French territories. In 1802, with a revision in 1807, he also decreed that a systematic cadastral survey be executed on the level of the commune.[37] Without a standard surveying method already established at any of the existing schools, a school of geography and topography was decreed in 1795, and the techniques developed there represented an attempt to standardize mapmaking within France. Meanwhile, French-occupied Egypt was a particularly important place for discovering and testing mapping techniques; one engineer, Gilbert-Joseph-Gaspard, comte de Chabrol de Volvic, would eventually take those methods and apply them in Paris when he became prefect of the Seine in 1812.

By the late eighteenth century, maps became integral to every part of the military's organization, and Napoléon Bonaparte was described as relying on them for all matters. In 1813 one of his generals, Ernst Otto Innocenz von Odeleben, recounted the centrality of a map in Napoléon's office:

> In the middle of the room was placed a large table, on which was spread the best map that could be obtained of the seat of the war. . . . This was placed conformably with the points of the compass . . . pins with various colored heads were thrust into it to point out the situation of the different corps d'armée of the French or those of the enemy. This was the business of the director of the *bureau topographique* [topographical office] who constantly laboured with him, and who possessed a perfect knowledge of the different positions. If this map was not ready, it was to be brought immediately on the arrival of Napoléon for he attached more importance to this than any want of his life.[38]

Accordingly, paintings depicted Napoléon Bonaparte on the battlefront, and these heroic representations were supported by descriptions of his military preparations using maps. Indeed, Napoléon's cartographic performances were part of a larger program of war propaganda that sought to demonstrate his superior planning and rational judgment. He was not a bloodthirsty brute, but a general capable of careful assessment and endowed with intellectual skill. As Susan Locke Siegfried has demonstrated, the authority of historical truth in paintings was based on an emergent appreciation and new use of empirical evidence.[39] These representations of contemporary events, in contrast to the idealizing and classicizing pretense of earlier practices of history painting, were valued for their documentary qualities based on authoritative firsthand accounts. Von Odeleben's direct recollection gives credibility to Napoléon's capacities as a leader, and it becomes part of the historical record, explaining the role of maps as the necessary instruments of any sovereign and government.

Whether it was Haussmann imagining new spaces and designing new forms, Louis XIV playing out battles on his wooden relief maps, or Napoléon moving pins to devise attacks and defenses, these men used cartographic representations to plan. Due to its large size, the map became a terrain unto itself, and its high resolution supported a belief in the accuracy of its description. This was a privileged and dominating view, supported by an institutional infrastructure and epistemological shifts that allowed for the plans to project back onto the terrain—as a battlefield or as a city.

MAPPED LINES

Even before Haussmann's transfer to the French capital, there had been many urban development projects underway. Claude-Philibert Barthelot, comte de Rambuteau, prefect of the Seine from 1833 to 1848, recognized the necessity of a comprehensive building program. Demographic and infrastructural problems were evident throughout the 1840s, fomenting the political dissent that would result in the revolution of 1848. The infrastructural conditions of the city, and particularly its poor housing, were decried by members of the Conseil général de la Seine, and in 1850, Louis Lazare, the director of the important journal *Revue municipale,* called on the government to convene a commission to study and propose a new plan of the city. Napoléon III eventually addressed the matter directly on August 2, 1853, in a letter to his minister of the interior, Jean-Gilbert-Victor Fialin, duc de Persigny, who convened the Commission des embellissements de Paris under the direction of Henri Siméon.

When Siméon's commission was formed, de Persigny conveyed the priorities of the emperor in a letter addressed to the head of the commission. The demands were mainly concerned with creating new circulation arteries up to the fortifications and especially around the train stations. Significantly, the letter assumed drawing as the basis for building. It recommended "that in the alignment of the main streets, the architects make as many angles as necessary, in order not to knock down either monuments or beautiful houses, while keeping the same width to the streets, and so that one does not become a slave to the alignment exclusively in a straight line."[40] De Persigny warned against designing a Paris "enslaved" by straight lines. The instructions specified a particular urban effect to be created from architectural designs, and yet those very designs—those drawings and projections of the future city—were composed exactly with straight lines. The straight lines of a triangulation survey. The straight lines of the orthographic grid. The straight lines of a ruler laid on a plan. The straight lines of Haussmann's *percements*, or openings.

The question of straight lines was precisely at the center of the controversy during and after the transformation of Paris. At the end of the nineteenth century, Austrian architect, planner, and writer Camillo Sitte organized his ideas on urbanism into a text titled *Der Städte-Bau nach seinen künstlerischen Grundsätzen* (commonly translated as *The Art of Building Cities*). As the third network of urban projects was under construction, Sitte's writings presented a theoretical attack on the "inflexible, geometric regularity" that had become the paradigm of modern urban planning.[41] In his chapter "Moderne Systeme," Paris was conspicuously absent in name, yet unmistakably present in allusion when Sitte vigorously condemned the technological approach to urban planning that had created a city dominated by a mechanical rectilinearity. Through a study of the interrelationships of the street, block, and building, Sitte insisted that modern city planning had reduced the experience of urban space to a plan. Its subsequent abstract forms were projects of development and profit instead of beauty: "Successively the parcel plan of the new city quarter can be made by even the lowest ranking administrative employee or courier."[42] His text represented a call to rescue the city from the clutches of the plan and its bureaucrats. But, by arguing for a redefinition of space as three-dimensional, Sitte had already conceded that urban space was regarded as, at present, flat. The peculiar character of these claims, opinions shared internationally by many architects, writers, and artists at the end of the nineteenth century, raised a historical question: How had space become two-dimensional?

Part of the answer was the formation of the real estate and private property market that became part and parcel of the imperial building policies and practices in Paris. Although only focusing on the Third Republic, Alexia Yates demonstrates that the major development projects already underway at the beginning of the Second Empire privileged financial transactions that translated stones into paper. Large public development projects facilitated the integration of the urban property market to the stock market. This market "turned land, the construction it supports and the revenues it generates, into paper shares that can be folded and placed in a wallet."[43] This process

of transforming physical materials into paper shares was aptly coined *"papier-pierre"* (paper-stone). Thus, Parisian space itself became a commodity whose definition and circulation was mediated by documents. Maps and plans were key instruments in the commodification of space, in which urban lots were quantified, measured, and objectified, bought and sold in the frenzy of speculation, and abstracted from particular contexts.

Due to the divergent and growing interests (both public and private) in the development of the city, the administration began to manage the method and means of shared communication among the growing number of agents in the shaping of the city. Laws and regulations were passed to determine not only the kinds of documents that had legal standing, but also the ways in which these documents were drawn up. The increasingly popular conviction that diagrammatic and graphic forms were more scientific than text dovetailed with the necessity to regulate the multiple interests involved in any discussion and land transaction. The outcome was a bureaucratic culture and a growing paper culture that included plans.

It is into this context that we must place Adolphe Yvon's painting of Napoléon III handing Haussmann a paper document annexing the suburban communes (fig. 1.11). It does not include but implies a map. Principally a history painter, Yvon was commissioned by the Conseil municipal to produce a tableau for the council's meeting room at the Hôtel de Ville.[44] Unlike his more well-known battle scenes, his subject here is the monumentalization of bureaucracy. Still holding a quill in his right hand, Napoléon III gives a freshly signed paper to Haussmann, while members of the council enter the room in the background. The prefect eagerly leans forward to receive the paper, but his gaze is not met by the emperor. None of the figures look at the document, and instead Yvon turns the paper toward the viewer to reveal the word *"Décret"* (Decree). Since the painting was completed and presented in 1865, a map including the suburbs would not have been available. In 1859 Haussmann had anticipated the annexation and ordered a new ground survey of the city, but the results would have yet to be printed.[45] Yvon worked around the problem of depicting the annexed territories by instead representing the annexation order. The administrative scene plays out under Jacques-Louis David's portrait, which similarly represented Napoléon I in his role as an administrator surrounded by legal documents. The unique subject of Yvon's painting made visible

FIGURE 1.11 Adolphe Yvon, *Napoléon III Handing to Haussmann the Decree for the Annexation Plan of the Peripheral Districts (16 February 1859)*, 1865. Oil on canvas, 128 ¹¹/₁₆ × 90 ⁵/₈ in. (327 × 230 cm). Musée Carnavalet.

the often-obscure practices of shaping the city. Here was a record of the less spectacular but essential bureaucratic steps of the city's modernization.

These projections about the urban process, the practice of translation between thinking, drawing, and building, were also the subject of an 1865 painting by Jean-Baptiste-Ange Tissier (fig. 1.12). Tissier depicted an unusual scene from 1852 of the architect Ludovico Visconti displaying his final plans for reuniting the Louvre and the Tuileries to Napoléon III and Eugénie. Leaning forward in their chairs, the imperial couple studies the large plan attentively. With other drawings scattered on the floor, Visconti clearly takes time to explain the architectural resolution rendered by dividing the space of the Louvre longitudinally, forming a parallel wing to the

Grande Galerie and creating a new *Cour* Napoléon. This would mask the difference in angles between the Tuileries and the Louvre, and the heights between the Place du Carrousel and the Quai du Louvre. Depicted at an angle in the painting, the plans—and there are only ground plans—reveal the building's massing to demonstrate the formal problems.[46] As the scene takes place in a room of the Tuileries palace, the architectural solution could have been indicated by looking out the window located behind the court's audience. However, perhaps then their gaze would have fallen on the cluster of houses between the two palaces that Haussmann would so proudly demolish when he wrote, "It gave me great satisfaction to raze all that during my debut in Paris. . . . Since my youth, the dilapidated state of the Place du Carrousel, in front of the Tuileries seemed

FIGURE 1.12 Jean-Baptiste-Ange Tissier, *Napoléon III and Eugénie Approving the Plans Presented by Visconti*, 1865. Oil on canvas, 70 × 91 in. (178 × 231 cm). Musée national des châteaux de Versailles et de Trianon. © RMN-Grand Palais/Art Resource, NY.

to me a disgrace for France, an admission of the government's impotence, and I had no desire for it."[47] However, unlike in Jollain's painting, where Louis XIV merely gestures to the plan of the Maison Royale (see fig. 1.7), here all attention is focused on the plan, making visible the axes that could only be known through drawing on paper.

By the time Tissier's painting was exhibited in the Salon of 1866, it would have been fourteen years since the first stone had been laid under Visconti's direction. After Visconti's sudden death in December 1853, his successor, Hector-Martin Lefuel, pictured to the right in the painting, finished the major construction of the Pavillon de Flore and developed its overall ornamentation from 1857 to 1865. Significantly, the scene would have predated the actual union of Louis Napoléon and Eugénie, who were

married in civil and religious ceremonies on January 29 and 30, 1853. As Alison McQueen discusses, marriage was a necessity when Napoléon crowned himself emperor, in order to ensure the continuity of the empire.[48] Thus, while historically inaccurate, the inclusion of the empress acknowledges her status as a prominent public figure who patronized important architectural projects during the Second Empire. Her presence adds to the painting's broader political message around the process of monumental building during Napoléon's reign.[49] Thus, Tissier's chosen subject is not the monumental building itself, but the authoritative act of looking at plans and the negotiation of its construction through imperial patronage. The architectural plans represented in this painting function simultaneously as description, projection, and

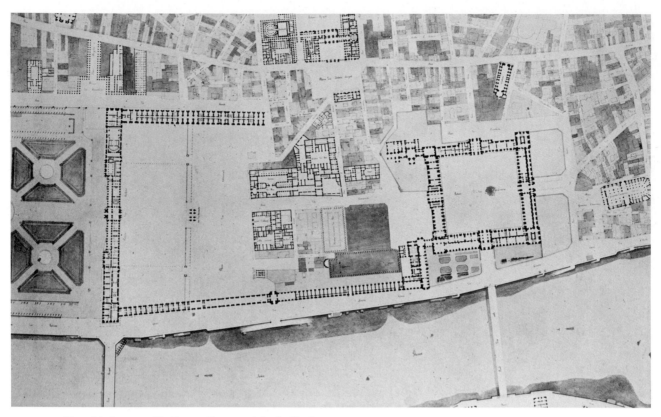

FIGURE 1.13 *Plan of the Louvre and Its Surrounding around 1830 by Charles Vasserot.* Salted paper print, 7⅛ × 11¼ in. (18.1 × 28.6 cm). From Édouard Baldus, *Du Louvre et des Tuileries (d'après les plans officiels)* (Paris, 1853). J. Paul Getty Museum.

confirmation. They describe the operations required in building the Louvre; project the future building through its synoptic and diagrammatic composition; and confirm that the final building relates to the plans as pictured.

This section of the Louvre that Lefuel would complete also became the key reference for a new avenue leading to the new opera building by Charles Garnier, to be discussed in Chapter 3. This long gallery linked the old palace to the Tuileries and would anchor the awkwardly angled approach to Garnier's building. But what is shown in Tissier's painting is neither Visconti's nor Lefuel's plan, but a version of Charles Vasserot's plan from 1830 showing the northern side of the palace yet unconnected (fig. 1.13). In the painting, Visconti gestures precisely at the specific section that would entail the most significant demolition and eventually create an entirely new district between the Louvre and the Palais-Royale. His proposal was not merely an architectural composition, but an entire urban system that would regularize the surrounding streets and blocks and connect the Louvre to the extension of the rue de Rivoli (fig. 1.14).[50]

What is not pictured in this unique scene is the political, social, and urban context of those plans. Mixed with *hôtels particuliers,* artist studios, and small workshops, the eclectic neighborhood of the Doyenné first emerged around Henri IV's expansion of the Louvre and then grew when the royal court moved to Versailles. It was known for its central location and bohemian character. Honoré de Balzac would house his fictional character of the old maid Bette in this district, and both Gérard de Nerval and Théophile Gautier would find cheap accommodation and low rents in the neighborhood.[51] The symbolic consolidation of urban power through the architectural intervention of the Louvre, beginning already during the Second Republic, led to the expropriation and eventual demolition of the neighborhood during the Second Empire.[52] Thus, the area would undergo demolition and development, and its residents would eventually succumb to drawn lines on the plans of the palace and the city.

Few scenes hold a more secure standing in histories of the modern city and the history of Paris than Haussmann's

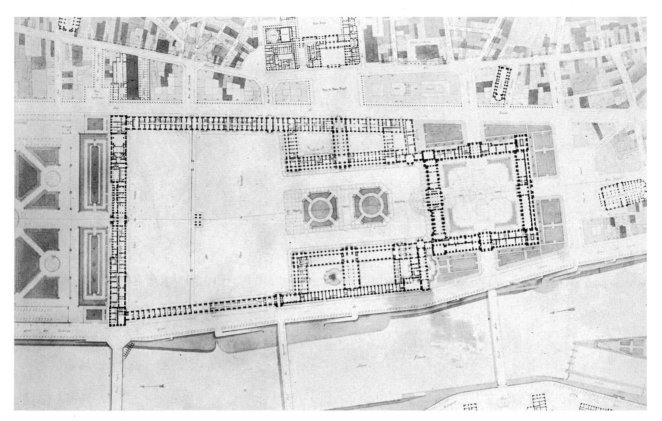

FIGURE 1.14 *Plan for the Nouveau Louvre by L. Visconti.* Salted paper print, 7³/₁₆ × 11⁵/₁₆ in. (18.2 × 28.8 cm). From Édouard Baldus, *Du Louvre et des Tuileries (d'après les plans officiels)* (Paris, 1853). J. Paul Getty Museum.

appointment to the post of the prefect of the Seine. Eight maps were produced by the Commission des embellissements before Haussmann took office, but the particular story of Napoléon's colored map continues to maintain a strong grip. Its endurance in the historical imagination of Paris relates to a long iconographic tradition of rulers pictured with maps and plans. Maps legitimized the sovereign's dominion. Yet as the value of maps changed during the nineteenth century, their representational meaning subtly shifted, as became evident in the visual culture of cartographic representations. They came to validate a leader's governance capacities, associated with the skillful management of paperwork.

The growing complexities of city building fostered a desire for the clarity of a singular master plan. Given the number of agencies, institutions, and roles created to build new monuments, roads, sewers, parks, and squares, the idea of one map that could capture the extent and intention of all such works was alluring. It would present a physical object to direct attention and, significantly, to redirect any accountability for the social consequences of

building from its geometry. It was an abstract spatial representation that conformed to the medium of the paper on which it was drawn. Two-dimensional, gridded, and reduced to linear forms, the renderings offered neither an identifiable author nor any specific temporal or qualitative context. This very decontextualization became the basis of the map's standing as objective, and, accordingly, it became a blank slate on which both spaces and histories were created.

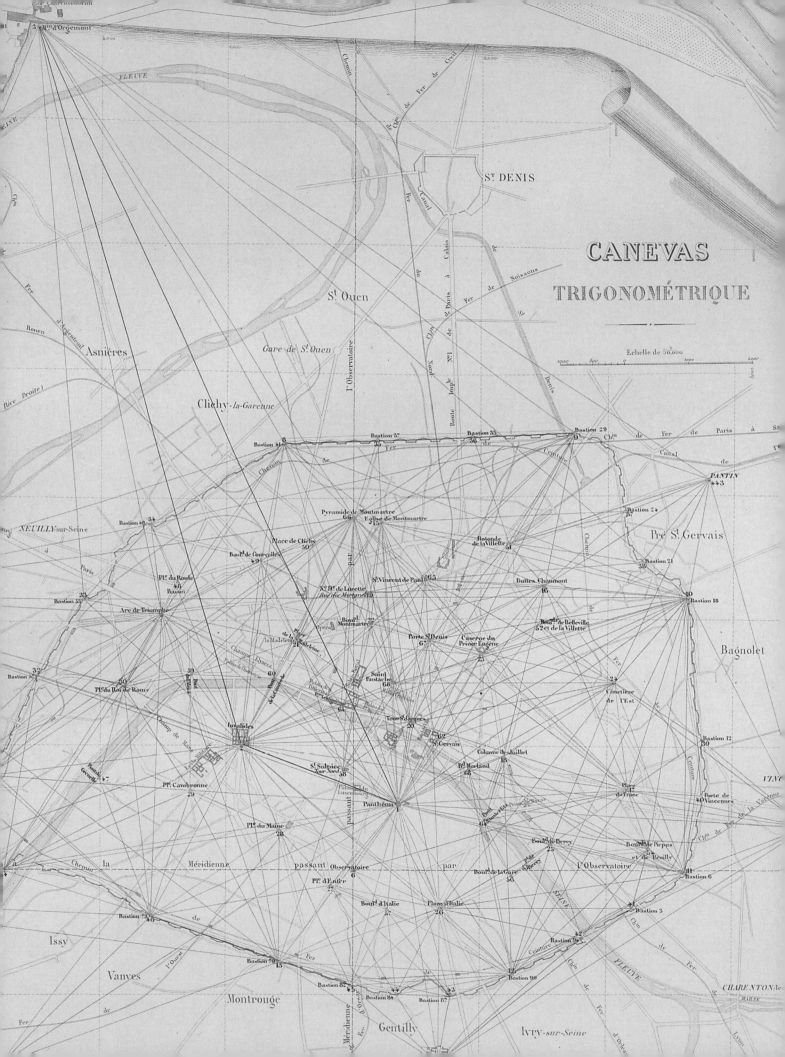

2 Triangulating the City

Everything you measure must be measured by triangles.
—Sebastian Münster, *Cosmographey; oder, Beschreibung aller Länder* (Cosmography; or, the description of all countries) (1550)

The triangle is always assumed to be flat.
—Lancelot, *Nouveau traité d'arpentage et de toisé* (New treatise on surveying and measuring) (1833)

The topic of cartographic fidelity has been fertile ground for many writers. Lewis Carroll wrote about a fictional map with an ever-reducing scale, until at last, a ratio of 1:1 was established between map and terrain. "'Have you used it much?'" the narrator inquires. "'It has never been spread out, yet,' said Mein Herr, 'the farmers objected: they said that it would cover the whole country, and cut out the sunlight! So we now use the country itself, as its own map, and I assure you, it does nearly as well.'"[1] The scene takes place in a topsy-turvy world where the terrain is used as a map and where even feudal relations are reversed, with multiple kings ruling over one subject. Here, fidelity is linked to governance, and Carroll's implicit claim is that maps are embedded in and maintain sociopolitical relations. Jorge Luis Borges's "On Exactitude in Science" (1946), a one-paragraph story of an empire, mirrored these themes.[2] In this work, cartographers were called upon to draw a map the same size as its terrain, such that land and map were synonymous, until one day they were not. No longer equivalent, the map lost its usefulness and was left to fray. As with Carroll's piece, Borges pinpoints the central tension in the map's capacity to represent something as complex as a country's geography. On the one hand, a map is too abstract and too reductive to describe sufficiently the complexity of the world; on the other, a map is itself a complex system of relationships, epistemologies, and symbols. In both cases, however, the stories highlight more than the difficulties in establishing a correspondence between a map and the world; they also examine the capacity to trust a map to represent that world, whether it is through the ruling authority in Carroll's case, or an academic institution in Borges's example.

TRUST IN MAPS

Although abstract and bearing no semblance to the real world, orthographic maps and plans became trusted as accurate images. Historically, maps generally offered a privileged view regardless of modality, and thus were often controlled by a ruling authority or maintained within a monastic setting. Drawn in the margins, royal decrees, seals, and symbols were a constitutive part of the map in guaranteeing its fidelity to the terrain as well as the command of royal control.[3] Moreover, with the perspectival modality used to depict the city of Paris in three dimensions, many of the monuments depicted were recognizable even if they were distorted. As Hilary Ballon has shown, Benedit Vassallieu dit Nicolay's 1609 map of Paris exaggerated the comparative size of the Pavillon du roi and described a much more ordered city than its dense reality.[4] In this case, accuracy was not necessarily defined by mathematical calculations, but by pictorial correspondence within the contexts of patronage and governance and their specific representational objectives.

However, beginning with the establishment of the Observatoire de Paris in 1667 and the ascendancy of French academics in the practice of mapping and surveying during the eighteenth century, the meaning of cartographic fidelity shifted away from the royal body as guarantor. Trust was placed in extrinsic entities just as cartographic aesthetics shifted from pictorial to diagrammatic forms. The triangulation survey was still the basis for these new orthographic and two-dimensional compositions. However, the material and epistemological meaning of accuracy changed, and royal cartouches no longer guaranteed the exactitude of the map. New strategies had to be communicated to authenticate the graphic image that depended less on resemblance and aligned with emergent quantitative practices. Geometric correspondence over pictorial verisimilitude became the basis for cartographic fidelity, and by the nineteenth century, maps had a new symbolic system firmly in place to communicate assurance.

In the case of Paris, fidelity was first established through two comprehensive triangulation surveys—one during the late eighteenth century, and another in the second half of the nineteenth century. However, more than simply establishing fidelity, these surveys made Paris's space graphic, and the explicit representation of the correspondence between terrain and map became essential to the map's classification as scientific. Once that correspondence was created, once the map was thought to accurately represent the terrain, it became the privileged site of observation, planning, and many other epistemological functions, for administrators, engineers, and architects.[5] As Haussmann explained, he no longer had to look outside in order to make decisions about the city's development.[6] The map could be substituted for that which was represented; by being regarded as an accurate description of the terrain, it

became autonomous from the very terrain it purported to describe.

Triangulation is a geometric technique to measure land area, based on creating a network of triangles across a given surface. However, the surveying procedures that generated the map's correspondence and autonomy involved more than just technical skills—they also required political actions and social conditions. The geometric forms drawn from these triangulation surveys constituted the historically specific merger of speculative, administrative, and spatial practices, entrenched in the imperial context of their materialization. Moreover, without a fixed community of surveyors or an established iconography of these geometric maps, the norms of surveying and mapping had to be negotiated and themselves represented. In focusing on the triangulation surveys, then, this chapter aims to describe the process of their execution, as well as to understand how the relationship between cartographic images and the terrain conditioned a new graphic conception of space in post-Enlightenment France. Ultimately, triangulation was tied to assertions of methodological transparency—an ideological claim in and of itself—that obscured the map's capacity to condition social relations through its graphic abstraction. For French administrators and engineers, space was putatively neutralized through these surveying practices as the map came to be understood as an objective image of science.

SURVEYING PARIS

A map's utility is found in its reduction of the material and experiential qualities of the world. It makes the multiple dimensions of a territory visually manageable. Its abstraction of that terrain, whether composed orthographically or pictorially, is an essential basis for its function as an instrument of governance, military action, and knowledge. Yet what amount of abstraction is acceptable? How are the conventions and norms of that abstraction determined? How, in other words, is the connection between a map and territory established and maintained?

First, correspondence was often established through surveying. Based on archaeological evidence, measured ground surveying in Europe dates back to at least the Greeks and Romans, who worked from a system of squares to create a regular grid of perpendicular lines.[7] Greek planners applied this system to many cities in their colonies, and the Romans extended it through their vast territories into the countryside.[8] For triangulation, there is debate over its origins.[9] Although Euclidean geometry provided a method for constructing triangles, creating a chain of triangles requires trigonometry, whose invention is dated to the second century.[10] When triangulation became used in surveying during the early modern period in Europe, trigonometric knowledge had arrived from India via the Islamic Empire.[11] In France, the cartographer Jean-Félix Picard used methods based on Willebrord Snellius's triangulation system to create a chain of thirteen triangles from Paris to the clock tower of Sourdon, close to Amiens.[12] His survey was significant for establishing one degree of latitude along the Paris meridian, which would become the primary reference for subsequent surveys of Paris and of France. By the mid-eighteenth century, ground surveys based on triangulation—as opposed to astronomical observations—became the primary method for establishing correspondence between maps and terrains, especially cities.

However, earlier, during the *ancien régime*, mapmaking was tied to the expertise of a particular *géographe de cabinet* (scholarly geographer), and fidelity was guaranteed through royal patronage. The authority of the monarch was transferred to that of the map, and maps were accordingly invested in presenting an image of a unified kingdom. Taking a top-down approach, the resulting map was based on the ability of geographers to discern the validity of geographical writings and to construct a coherent image based on heterogenous textual sources.[13] The primary method of verifying a location consisted of textual analysis of authoritative—often ancient—writings, comparing them with calculations from older maps and recent surveying information. Because there were often gaps between and within these various sources, the method consisted of first establishing the coastlines and waterways, filling in specific locations after a general outline was determined. Each landmark was established independently.

Triangulation, on the contrary, is a relational system of defining space through the geometric rules of three-sided forms. As Jean Delagrive, renowned mapmaker with the official government rank of *géographe ordinaire de la Ville de Paris,* explained in his 1754 manual on triangulation, "Geometry is a science that deals with quantities and the relationships they have with each other."[14] Unlike other polygons, the elements that define a triangle—the lengths of three sides and the degrees of three angles—are not independent. Given two angles and one side, or two sides and one angle, the area contained within its boundaries

FIGURE 2.1 The Paris meridian, executed by Jean-Dominique Cassini and Jacques Cassini. From Jacques Cassini, *De la grandeur et de la figure de la terre* (Of the size and shape of the earth) (Paris, 1720). Bibliothèque nationale de France.

can be determined. This relationship between angles and sides is unique to triangles, and thus it offers surveyors the greatest possible number of combinations for calculating surface area, with fewer opportunities for miscalculations when confronted with topological challenges. Triangles' advantage over other polygons for surveying was that the measure of a triangular area required less information than, for example, a square. Creating chains of linked triangles provided flexibility for surveyors and allowed them to make calculations with minimal data. Moreover, by linking triangles so that two triangles shared one side, the system included a self-correcting mechanism, and errors did not propagate across the network. The length of one triangle's side and its angles were measured twice and verified for the next triangle.

The method required a baseline to be established, on which the first triangle could be constructed and from which a chain was built. For the first comprehensively triangulated plan of Paris, the meridian line originating from the Observatoire de Paris served as that foundational line (fig. 2.1). A meridian is a longitudinal line of Earth's circumference, passing through the terrestrial poles, and, in this case, the observatory. It is a means of division and, together with the equator and its parallels, of imposing a grid onto Earth's surface, conceiving the

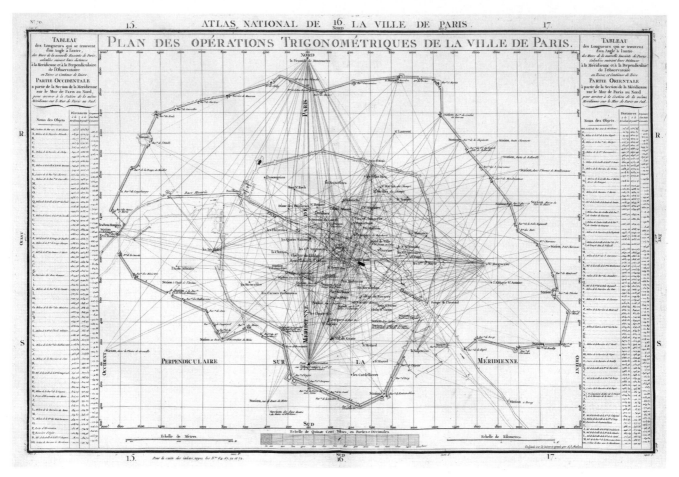

FIGURE 2.2 *Plan des opérations trigonométriques de la Ville de Paris* (Plan of the trigonometric operations of the city of Paris). From Edme Verniquet, *Atlas du plan général de la Ville de Paris* (Paris, 1795). Bibliothèque nationale de France.

planet as a gridded sphere. For these lines to correspond to topographic reality, a ground survey is necessary to establish the link. Defining the measure of the meridian line established a fictional and absolute reference that allowed a representational anchor and geographic measure to be created. It also served as the basis for the cartographic grid that controls the scale and measures across the entire paper surface. Using the meridian as its base, a national cartographic survey began in 1683, but it was half a century until a rigorous geodetic survey with triangulation chains across the whole of France started in earnest. This undertaking was carried out by the astronomer Nicolas-Louis de Lacaille, and his work was eventually published in 1744 as an eighteen-sheet map at the behest of engineers who aimed to use the data for maps of the canals and road network within the French interior. In 1789 the first topographical map of France based on those extensive triangulations, the *Carte de France*, was produced by César-François Cassini de Thury and his son Jacques Dominique.[15]

If the meridian line and its perpendicular formed the basis of a cartographic grid, a base measure had to be established for the triangulation system. For the Paris survey, this measure runs diagonally across the table and is indicated as the *ligne de vérification* and marked as a thick, black line on Edme Verniquet's map (fig. 2.2). It is anchored to the southern tower of Notre-Dame de Paris, indicated with a small circle, with endpoints at the Pompe de l'Arsenal to the east and the marked pavilion at the southwest corner of the Place Louis XV. It is a fixed distance to which all other measures relate. With this baseline and the meridian intersecting at the Observatoire de Paris, the first triangular area was determined. From there, a chain of triangles could be built to extend in any direction through the city, only limited by towers that could allow for distances to be seen unobstructed.

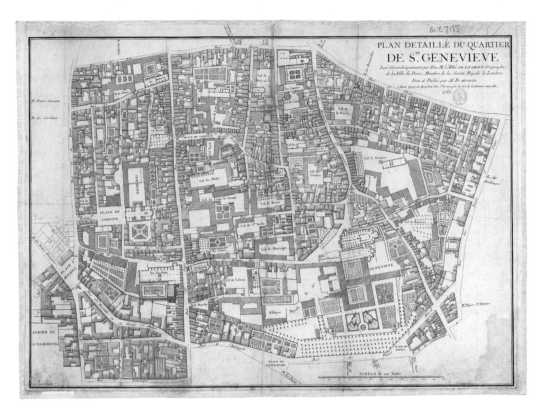

FIGURE 2.3 Jean Delagrive and Alexandre-François Hugnin, *Plan détaillé du quartier de Sainte-Geneviève* (Plan Detailing the Neighborhood of Sainte-Geneviève), 1757. Map sheet, 19¼ × 27³/₁₆ in. (49 × 69 cm). Bibliothèque nationale de France.

After studying mathematics and architecture in Dijon and Strasbourg, Verniquet was employed by several noble families as a surveyor and *terrier,* a land manager in charge of ensuring that an estate met its legal obligations. Before moving to Paris in 1772, he also worked in the atelier of his father, who was a carpenter and managed the royal forests of Châtillon-sur-Seine. In Paris, he worked under Georges-Louis Leclerc, comte de Buffon, as the architect of the Jardin du roi (now Jardin des Plantes), designing its amphitheater. Eventually, he secured a position as one of four *commissaires de la voirie* (road commissioners), a post charged with overseeing all urban operations related to the roadways.[16] This work included the opening and closing of roads, alignments, and the construction of closures and borders.[17] In addition to the city's public functions, he also oversaw the construction of many private residences. In his capacity as a road commissioner, he worked from maps drawn by and relied upon procedures established by Delagrive.

While Verniquet's survey was considered to be the most precise to date, it relied on already established procedures. In his surveying manual, Delagrive explained that "a good plan must encompass the totality of the neighborhood and roads, which should be represented accurately in their width and length."[18] He participated in measuring the latitude of Paris, perpendicular to the meridian at the Observatoire de Paris, and established the standard for geometric precision in representing the topography of Paris. He was prolific, and stories of his exacting reputation were well published and circulated. One source claimed that he was so dissatisfied with the engraving of his first plan of Paris in 1728 that he broke the plates and vowed to engrave any future works himself.[19] Beginning in 1731, Delagrive produced a series of manuscript maps that detailed every neighborhood of Paris, including the outlines of individual buildings and ground plans of the major monuments. As an early cadastral map—a map used for determining taxation based on land—this *plan parcellaire* (plan of land plots) traced the borders of each district and its individual properties. The sheets provided information about spatial relationships in proportion, scale, and position. He died before completing the project, and his student Alexandre-François Hugnin continued the work. Hugnin published a plan of Sainte-Geneviève (fig. 2.3) and engraved the ones for the Île Saint-Louis and Île Louviers (fig. 2.4), but others were never completed.[20]

These maps constituted the most detailed ones available at the time, but they proved difficult to use by the road commissioners. The utility of cadastral mapping was its demarcation of the dimensions and shape of an

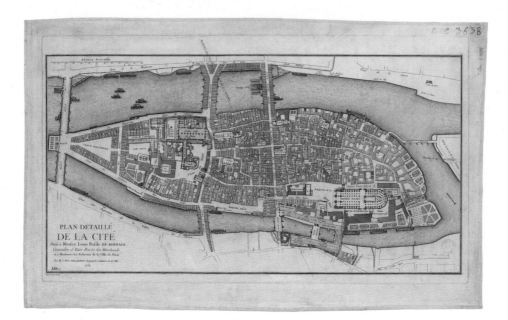

FIGURE 2.4 Jean Delagrive and Alexandre-François Hugnin, *Plan détaillé de la Cité* (Plan detailing the City), 1754. Map sheet, 17 × 32 in. 43 × 81 cm). Bibliothèque nationale de France.

individual lot and its relation to other properties within a given administrative area. The road commission's use of cadastral maps was an inevitable mismatch of aims and instruments. Delagrive depicted Paris in fragments. The identity of each neighborhood was prioritized over a general view of the city, an approach that had already earned Delagrive criticism for his maps of the Seine and its tributaries.[21] While he aimed to describe both lot and road, the maps' ultimate purpose was to serve as a cadastre, and accordingly, they emphasized the block allotments and their interior agglomeration and organization over the street systems.[22] A useful map of the roads needed to show circulation across the city, and these property maps did not fulfill all of the necessities of managing and improving the road network across all of the quarters.

Verniquet began work on a new map of Paris from 1774 to 1783, surveying alone at night without government approval. He carried his surveying equipment through the streets and up towers, using torchlight to identify the distances between different points across the city. The *Journal des géomètres* explained that his clandestine labor was simply due to the poor conditions of the roads ("inextricable difficulties associated with most of the streets of old Paris").[23] There was also the general chaos of the many small and winding streets, with noise and traffic filling the city during the day.[24] However, historians have noted that there were also suspicions of any form of oversight in the prerevolutionary years.[25] There were even reports of a surveyor being killed by villagers south of Paris; many peasants saw surveyors as the handmaidens of a feudal system that impoverished them to servitude.[26] House numbering was also met with resistance, not only from residents but also from government officials.[27] In 1779 Marin Kreenfelt de Storcks, the publisher of the *Almanach de Paris,* hired a group of men to number houses along Parisian streets, but these efforts were stopped by François-Louis Joly de Fleury, a Parisian magistrate, who ordered the lieutenant-general of police to shut down all work.[28] Surveying and house numbering were parallel practices of rationalizing space through a uniform system of numbers and geometric lines. Accordingly, these practices were forms of surveillance, and it was understood that whoever controlled them would gain a privileged view.[29]

Verniquet's surveying calculations were eventually accessed by government administrators when the royal declaration of April 10, 1783, established for the first time a regulation between the width of roads and the height of buildings. This regulation obligated a new survey of the existing built environment for future building projects, and in September 1785, Louis XVI agreed to establish and fund a new surveyed plan of the city and its surroundings. Based on his position and the work he had already been carrying out, Verniquet was charged with this immense task. He and his team of surveyors took their initial measurements on the ground and then drew an initial plan based on those calculations.[30] This plan was verified against trigonometric calculations at Verniquet's atelier in the Couvent des Cordeliers and then redrawn scrupulously to the measure of a *pouce,* a premodern and imperial unit.[31]

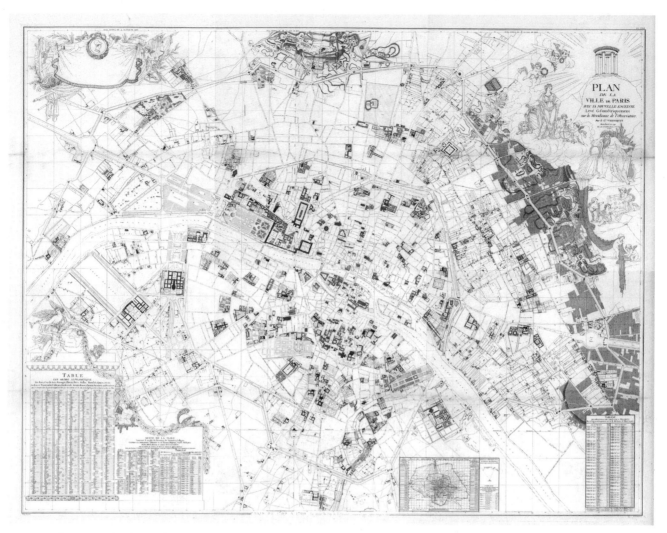

FIGURE 2.5 Composite plan from Edme Verniquet, *Atlas du plan général de la Ville de Paris* (Paris, 1795). Bibliothèque nationale de France.

The scale of the map's production made it impossible to point to an individual author, and the collective work was aggregated into a seamless graphic image. About sixty engineers and more than eighty assistants labored on this project that lasted from 1791 through the French Revolution.[32] Their work included triangulating, surveying, charting, calculating, and drawing the plan, which Michael Lynch has called "normalizing observations," such that differences between materials, dimensions, heights, and locations—all spatial qualities—were translated into standardized linear notations.[33] Each particular space was made comparable to another, creating a common graphic denominator, such that years of labor by multiple men were condensed into a single orthographic composition.

THE IMAGE OF TRIANGULATION

In the final print, Verniquet's triangulation measurements were marked in a table and diagram in the margins of the map (fig. 2.5). This cartouche was the only place where the labor involved in determining distances could be glimpsed. It listed the exact locations where the surveying measurements had been taken. It recorded the distances across the city that the surveyors and engineers had to carry their heavy equipment, and the calculations that they repeated and reproduced in the drawing office. Yet, while referring to the work involved in the map's production, the diagram also was an expression of the map's scientific objectivity that relied on obscuring the physical traces of its making.

From its place on the margins, the triangulation measures were central to establishing the map's authority as

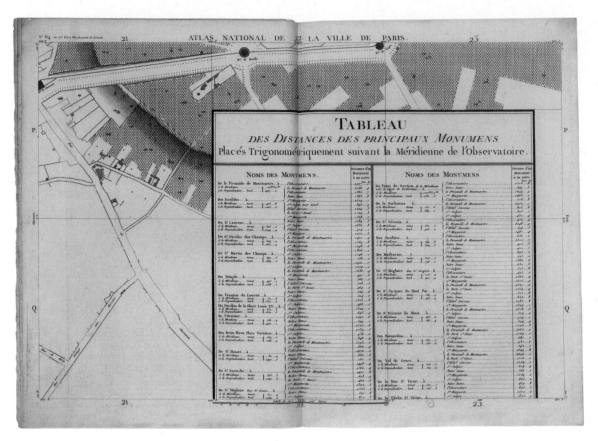

FIGURE 2.6 *Tableau des distances des principaux monuments* (Table of distances from principal monuments). Plate 64 from Edme Verniquet, *Atlas du plan général de la Ville de Paris* (Paris, 1795). Bibliothèque nationale de France.

accurate. The diagram of the survey was titled *Plan des opérations trigonométriques de la Ville de Paris* (see fig. 2.2). It is an abstract cartouche that confirmed the city as a measured object. On this sheet, a grid of lines forms a coordinate table, oriented along cardinal directions with north at the top of the page and south at the bottom. These directions are defined by longitudinal and latitudinal lines intersecting at the Observatoire de Paris, indicated as the Paris meridian and its perpendicular. Layered on top of the coordinate grid is another network of lines, this one made up of triangles. These triangles record and represent the surveying work involved in drawing this new map. This strange diagram was a new element to be included in a map, and one that signaled a shift in the epistemological foundations that supported the map's descriptive authority. The diagram recorded the labor of triangulation, and while this surveying technique had long existed, the act of publishing it aligned mapmaking with the Enlightenment values of methodological transparency and objectivity.

Delagrive had published his trigonometric calculations separately as part of a textual manual, dividing the presentation of the maps from the explanation of their geometric production.[34] By contrast, Verniquet's inclusion of his triangulation methods within the *Atlas du plan général de la Ville de Paris* was novel: it was an implicit argument that the city of Paris itself was the consequence of such scientific and technical activities.

The dense web of triangles pinpoints a few reference points along the perimeter of the city, and then forms increasingly tighter clusters toward the center of the city. The measures are made and remade into a matrix of triangles, whose points correspond to the sixty-seven *points de station.* The observatory served as the center of this map, and the logic of its gridded layout was determined by the intersection of the meridian line and its perpendicular. Directly on the meridian line, the northern station was determined at the Piramide de Montmartre. Several western and eastern points were chosen to create large

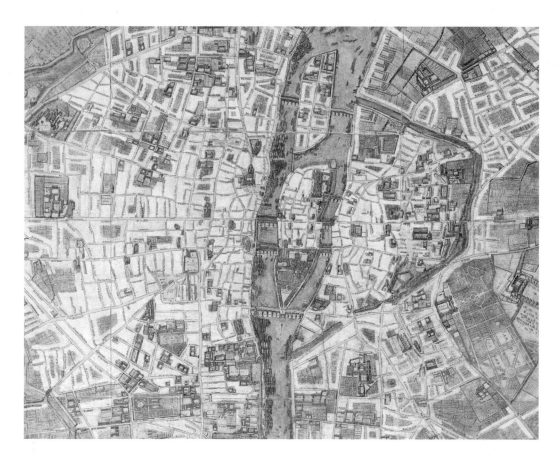

FIGURE 2.7 Jacques
Gomboust, *Lutetia Paris*
(detail), 1652. Map sheet,
59¾ × 71¼ in. (152 × 181 cm).
Bibliothèque nationale
de France.

triangles, which was shown in the triangulation diagram, but also in another table where distances to the meridian and the perpendicular were listed (fig. 2.6). The primary layer of triangles was based on the points: (1) the Piramide de Montmartre; (2) the Observatoire de Paris; (3) Notre-Dame; (4) Sainte-Marguerite; (5) Saint-Sulpice; (6) the Hôtel de Soissons; (7) the Porte Saint-Denis; and (8) the Pompe de l'Arsenal. Structures that had tall towers for observation, as well as other stations established in areas where no tall buildings stood, served as secondary points; these determined increasingly smaller triangles constructed in the city center. Each of these points of reference, listed along the side margins of the triangulation table, were identified by their individual distances from the meridian and perpendicular lines, as well as by a third point, creating a relational system of representational space.

There were "geometric plans" that preceded Verniquet's triangulated map, including a plan of Paris from 1652 by Jacques Gomboust that remains one of the earliest orthographic representations of the city (fig. 2.7, see fig. 1.9).[35] Gomboust, who was the royal engineer for the fortifications in Normandy and Picardy and compiled many plans of France's northern cities, maintained that the map was based on a survey using the most up-to-date tools. With the assistance of Pierre Petit, the inspector-general of fortifications, Gomboust claimed to have employed strictly mathematical methods for the calculation of distances.[36] His plan, however, is not a true orthogonal plan and is in fact more of a mixed map: while a diagrammatic view of the city as a network of roads embracing the Seine is clear, it is equally counterbalanced by the pictorial elements of the major buildings' façades, the boats and bridges on the river, and the figures depicted at the bottom of the map, as if the sheet of paper rolls out into a landscape picture. The widths and direction of the streets were rendered proportionately and in detail, whereas most of the interior spaces of a block were left blank. Prominent buildings were privileged with three-dimension representations, with perspectival views of Montmartre and the Louvre highlighted in cartouches at the top of the map.[37] For Gomboust, the façades of religious and royal buildings at the margins provided comparative verification of the map's fidelity and testified to the city's royal patronage.

FIGURE 2.8 François Blondel and Pierre Bullet, *Plan de Paris* (sheets I–XII), 1710 (originally published in 1675, with several updates in 1710 and then reprinted in 1900). Engraved composite map of twelve sheets, 37 × 37⅞ in. (94 × 95 cm). David Rumsey Historical Map Collection.

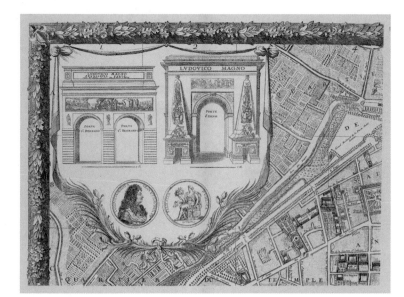

Even if earlier maps were surveyed, the guarantee of their correspondence was still found in cartouches along the margins showing either figures or texts that claimed the royal authority of the plans. In response to the demolition of the fortifications ringing Paris, a new plan of the city was executed in 1675 by Pierre Bullet, architect of the Bâtiments du roi, draftsman of the Académie royale d'architecture, and, from 1674 to 1675, architect of the city of Paris. He made this plan under the direction of François Blondel, professor of mathematics to the dauphin and naval engineer, who in 1671 became professor of mathematics at the Collège royale and director of the Académie royale d'architecture, and then in 1672 director of public works for the city of Paris. (Their titles and responsibilities indicate the fluidity of the roles of architect, engineer, mathematician, and administrator.) The map was drawn from triangulated calculations based on new surveying instruments (fig. 2.8).[38] Like the *Plan Gomboust*, this map showed the network of streets in plan and significant buildings and bridges in three-dimensions as well as the gardens and Seine. However, unlike it, the edges of the cartographic picture functioned differently, with no attempt to link the cartographic image to a pictorial one. The orthographic map continues all the way to the border of the page, and the cartouches on the corners seem to be placed atop the surface of the map, so that there is a graphic acknowledgment that these are two separate modalities: one that takes Paris as an object of representation and the other that acknowledges itself as a cartographic representation.

While the plan is planimetric in its composition, the cartouches at the top corners of the page appear conventional in their depiction of significant monuments of the city: façade elevations of the city's gates, textual descriptions, an outline of its royal commission, and a circular profile portrait of Louis XIV in the upper-left corner of the map sheet (fig. 2.9). However, the lower cartouches are novel in what they depict, although they are still framed traditionally. In the lower-left corner of the plan, there is a linear diagram of the water network of fountains in the city illusionistically drawn onto a textile and then pinned, with a classic ground view of the city from the Seine. In the lower-right corner, a map of the route of the Seine within the region is held up by cherubs. These cartouches offer a new conception of the city as a system, describing a structure of the city invisible to the eye and embedded in the imagination. By simultaneously describing and creating the spatial relations of the roads, the water, and the river, Bullet and Blondel provide a cartographic response to the royal order that specified a new, accurate plan of Paris.[39] It was a response that was concerned with circulation, in terms of roads, water, commerce, and the king's glory in the pictured triumphal arches. In its totality, the map demonstrated a tense balance between Enlightenment objectivity and royal tradition in terms of who decided what truth was.[40] The measured survey might have assured the map's accuracy, but the royal cartouches guaranteed its authority.

Verification of accuracy had often been based on a comparison between the image and the object that was represented.[41] Yet in the case of these geometric maps, comparison was difficult. Cartographic lines did not necessarily correspond to any physical line in the lived environment, nor was the view afforded by these maps verifiable by the experience of walking through the city. Verniquet's *Plan des opérations trigonométriques* thus was an attempt to offer a form of confirmation (see fig. 2.2). The triangulation diagram was a claim of authority based not on an individual, but on the surveying process. As part of a new survey of Paris, it attested to the cartographic image as a product of geometric methods of measurement derived from numerical measures, and as such, it was to be trusted.

In this period, with the appreciation of empiricism, the mapmaker was obliged to make the map's objectivity visible. Since the scientific revolution, truth was increasingly understood to be found in what was not immediately visible. Through inductive reasoning, general laws and principles governing nature were drawn from discrete and specific phenomena.[42] The observable world did not offer the fundamental explanation for a given event, which was instead derived through reason. When Delagrive purposefully aligned surveying with geometry and trigonometry in his *Manuel de trigonométrie pratique* (1754), he was making an argument for mapmaking as a science. Maps, in their orthographic and geometric expression, were products of mathematical functions, their accuracy guaranteed not by their pictorial verisimilitude but by the rational process of their production. As such, the inclusion of the trigonometric diagram in Verniquet's map of the city represented a new cartographic practice, not only in terms of map production and publication, but also in the shifting epistemological and political values of maps to represent the built environment.

The exposition of those surveying techniques through the *Plan des opérations trigonométriques* was aligned with a basic tenet of the scientific method: reproduction.[43] The orthographic mode assumed two kinds of reproduction: first, the reproduction of surveying practices; and second, the reproduction of those practices through print (to be addressed in Chapter 4). Both were in part codified by the triangulation table incorporated into Verniquet's *Atlas*. This table was meant to demonstrate the scientific methods used to produce the map, as well as to justify its accuracy by making transparent the processes used in the map's production through quantitative means. Since the image of Paris was not comparable to the lived city

as an object of experience, its legitimacy as an accurate image was instead based on a claim that the methods of the map's graphic production, if repeated, would result in the same image.[44]

The reproducibility of the measures resulting in the same image was, on the one hand, a central scientific value of the Enlightenment that suggested the absence of a renderer's influence. On the other hand, as Ken Alder has shown regarding mechanical drawing, the rigor of this process and the capacity to reproduce the same results were contingent upon the renderer's self-discipline and judgment.[45] Moreover, in the attempt to efface the trace of the maker and erase a particular viewpoint, this diagram reasserted a relationship between the mapmaker and the viewer. It represented a proposal from the maker to the viewer that, if the procedures were reproduced, the viewer would arrive at the same conclusions. However, it was also clear from how schematic the triangulation diagram was that there was little possibility of a viewer being able to reproduce the steps taken. The surveying stations were imprecisely indicated, and there was generally a lack of contextual detail to understand how those lines were formed and measured. Thus, in the end, the triangulation diagram functioned symbolically. It performed a rhetorical and ideological function, much like the cartouches of royal decrees and crests symbolized the king's authority. Only in this instance, the ideology was scientific objectivity scrubbed of subjective influence. Thus, not only was Paris measured and drawn based on consistent mathematical calculations; the map also had to present itself as a result of those scientific values.

THE LABOR OF SURVEYING

The appearance of triangulation cartouches made visible the procedures of surveying, and in doing so represented a major change in the value of labor involved in a map's production. During the revolutionary years, manual labor was no longer regarded as lowly and corrupting; moreover, the repetitiveness of the scientific labors required to ascertain exact knowledge came to be recognized not only as necessary, but valuable.[46] The efforts to standardize surveying practices were concerned with creating a criterion for how measurements were to be taken and retaken. Yet, as Alvaro Santana-Acuña has argued, among the effects of this standardization was the disassociation of the physical body from the practice of surveying, in order to attain "impersonal" universality.[47] There was a

FIGURE 2.10 A. Teyssèdre, *Nouveau manuel de l'arpenteur, contenant toutes les instructions nécessaires sur cet art, le lever des plans, le toisé, etc.* (New manual for the surveyor, containing all the necessary instructions on this art, the surveying of plans, measuring, etc.) (Paris: Dépôt des nouveaux manuels, 1836). Bibliothèque nationale de France.

accurate means of establishing land distances, social distinctions within the field were increasingly untenable. The boundaries of expertise in mapmaking expanded beyond the official institution of the Académie royale, and there was increasing recognition of the work of technically trained men.

A. Teyssèdre's *Nouveau manuel de l'arpenteur* (New manual for the surveyor) (1836) included an image of two well-dressed men in long coats, hats, and dress shoes (fig. 2.10). They are pictured assembling a surveying instrument along the banks of a river. It is an unusual composition for many reasons. First, the men seem dressed more for a leisurely stroll than for the strenuous activity of surveying. They are also missing essential materials, notably a way to record their findings. Second, it is rare, especially for a surveying manual, to narrate a scene of an instrument being assembled, as opposed to the actual act of surveying. If the instrument is the focus, it is often depicted with its components separated and then fully assembled, but rarely accompanied by figures. Additionally, the river behind them is spanned by an arched bridge, suggesting that this generic terrain would have already been well surveyed by civil engineers of the Corps des ponts et chaussées (Corps of bridges and roads). In sum, the engraving is less about indicating the actual techniques or instrumentation of surveying than it is about the ascendant social status of the surveyor.

This image in Teyssèdre's manual was part of a series of basic technical guides on several crafts and scientific hobbies, from the entertaining ("*de physique et de chimie amusantes,*" or "fun physics and chemistry") to the potentially hazardous ("*de médecine et de chirurgie domestique,*" or "medicine and domestic surgery") to the mundane ("*economie domestique,*" or "home economics"). A significant portion of the manual on surveying addressed topics related to measuring, maintaining, and improving land. The text's overall simple language and basic organization addressed landlords, who might have taken a direct interest in the management of their properties. Indeed, the lack of technical details, along with the inclusion of

constant tension in the creation of the norms of mapmaking between appreciating ground surveys over theoretical methods in principle, while simultaneously trying to erase the evidence that this surveying had been undertaken by individual bodies. The value of exactitude had the price of making invisible the labor invested in its production. Thus, as surveying came to be accepted as a scientific practice, upheld by new norms of objectivity, empiricism, and quantification, the means of guaranteeing those values were obscured.

As the director of the Observatoire de Paris, Joseph-Jérôme Lalande, wrote regarding Verniquet's surveyed plan, "This plan, whose work [*travaux*] I have followed and whose accuracy, I admire, appears to me as a most perfect work [*ouvrage*] of its kind that has ever been executed."[48] Lalande's comment emphasized "work" as the guarantor of exactitude. Yet the specific meaning of surveying in his assessment was contingent upon the moral economy of scientific labor, whose normative values were in flux. Until the mid-eighteenth century, the practice of mapmaking had been split between mathematicians and practitioners, but as triangulation became seen as the most

FIGURE 2.11 Frontispiece from Louis-Charles Dupain de Montesson, *La science de l'arpenteur dans toute son étendue* (The science of surveying in all of its scope) (1766; repr., Paris: Goeury, 1800). ETH-Bibliothek Zürich, Rar 1099.

the least technical excerpt from Louis-Charles Dupain de Montesson's well-known *La science de l'arpenteur dans toute son étendue* (The science of surveying in all of its scope) (1766), positioned this text as a manual for leisure rather than serious study.[49] The engravings included in the respected volume of Dupain de Montesson, a military cartographer, contrast with those of Teyssèdre. In the 1766 volume, surveyors are pictured diligently determining sight lines on planchettes, tracing lines with compasses, and measuring with levels. In addition to their instruments, some of them are pictured armed, as surveying often provoked hostility. The volume's frontispiece features men who, instead of standing in a picturesque

landscape, are found alongside farmworkers in the midst of planting or harvesting fields, or with the landed elite in the large gardens of their country estates (fig. 2.11). These men are not surveying for leisure but employed to ensure the borders and limits of properties.

Teyssèdre's self-help guide and the genteel surveyors pictured can be seen as the end of a long trajectory of the land surveyor's attempts to gain status and authority, and the full acceptance of their methods as normative in scientific communities and the literate mainstream. Surveying and its associated mathematical skills became an important part of an education for affluent men, whose families were responsible for managing their properties, including gardens. The shifting social values of these gardens during the eighteenth century created a broad need to understand surveying. Gardens shed their previous associations with agriculture and horticulture and gained new connections to architecture, in which their complex designs served as "a practice of world-making."[50] Accordingly, many of these texts were also significant symbols of erudition and dominion. They were written by those who had military experience and had trained in mathematics.[51] One 1702 example entitled *La géométrie pratique*—by Alain Manesson-Mallet, "*maitre de mathématiques*" at the court of Louis XIV—was based on the courses of Philippe Mallet, a professor of mathematics and military engineer.[52] Essentially creating a military textbook, Manesson-Mallet included engravings that provided many views of the grounds at the palace of Versailles, making an important link between military engineering and garden design.[53] Owning tomes such as these also gave men pretensions to territorial control and intellectual mastery.[54]

The increasing publication of these manuals paralleled the popularity of Parisian guidebooks beginning in the mid-eighteenth century, which represented an expanded understanding of urban space by the reading public. These guides were intended for the growing number of travelers as well as urban residents whose conceptions of the city extended beyond their particular neighborhoods.[55] Unlike geometry manuals, guidebooks were textual portrayals that offered itineraries through the city. While many used literary devices of emotion, narrative, and drama to compel readers to imagine and seek out urban sites, Victoria Thompson remarks that many of these guidebooks also emphasized "abstract knowledge," providing descriptions of spaces that were based in measures. For example, in Edme Béguillet's 1779 *Description historique de Paris,* readers learned that the Place Louis XI "forms a parallelogram 130 *toises* in length by 105 in width, the angles of the parallelogram form four outer corners [*pans coupés*], of twenty-two *toises* each."[56] These were spatial accounts that cultivated a broad audience for geometry, offering a descriptive modality that transformed each reader into a surveyor.

The new ways in which surveying and geometry were made accessible to the public contrasted with the closed nature of mapmaking during the ancien régime. At that time, drawing maps had been left to *géographes de cabinet*, (office geographer), whose tasks consisted of collecting and compiling information from different surveyors and texts from the confines of their offices, guided by Ptolemy's theories in *Geography*.[57] The map was an expression of the mapmaker's unique education in mathematics and singular erudition, and signaled his individual expertise. Having carried out the gathering of sources, the compilation and assessment of data, the determination of mathematical calculations, the drawing of geometric lines, and sometimes even the engraving and printing, the géographe de cabinet controlled the entire process. Still, without a centralized school or curriculum that defined standards for mapmaking or a corporate body that protected mapmaking as a profession, maps remained distinct and particular objects that did not necessarily relate to each other.

The intellectual work that marked the géographe de cabinet contrasted with the manual labor of *arpenteurs* (land surveyors). While the former generally produced large-scale cartographic representations of entire regions, kingdoms, or countries, the work of arpenteurs consisted of more local assignments. It entailed resolving local border disputes, verifying property titles, and surveying small tracts of land for property owners. It did not necessarily require an academic education; knowledge was disseminated mainly through apprenticeships. There was no standard body of knowledge that had to be mastered, and no exam to attain the title of surveyor. Beyond technical training, the physical requirements of surveying were considerable. These men had to trudge through often unforgiving environmental conditions and difficult terrain, carrying heavy but fragile equipment, only to be greeted with suspicious and even hostile locals.[58] Thus, the work of the arpenteur was particular: it required basic theoretical skills in mathematics and geometry as well as manual labor, which was not highly valued within the moral economy of labor during the ancien régime.[59]

The result of this combination of arduous duties was that surveying was an unattractive career. With little social status and no community, it was seen as "inferior to the objects of the administration."[60] Moreover, when the office of land surveyors was put up for sale under Louis XIV, feudal lords took control by purchasing the rights to it and placing their preferred surveyors. After this, the position became inheritable, and thus, without any checks on the occupation from the monarchy or scientific organizations such as the Académie royale, surveying practices became further subject to wide divergences in skill, knowledge, technique, and results.[61] As one important surveying manual admitted, "land surveyors rarely agree on the measurement of land."[62] This was not merely a glib statement. Without standardization in training, procedures, and measures, each calculation and each map was determined by the unique practices of a particular surveyor, which meant that no map could relate to or be compared to another. Each map stood as its own unique representation.

A law of October 22, 1795, attempted to systematize and professionalize land surveying, and served to bridge the division between géographes de cabinet and arpenteurs. The law reestablished the professional title of ingenieurs-géographes with the École des ingenieurs-géographes, along with many of the schools of public services.[63] Parallel to these establishments was also the École des géographes under the Dépôt de la Guerre, a branch of the military dedicated to preserving and managing its archive and all documentation, including maps and plans. The goal was to develop a practical curriculum aimed at training armies of surveyors in field and office work, and trainees were to spend time at the newly created Bureau de cadastre to reinforce their skills. However, very few men actually enrolled in the École des ingenieurs-géographes. Initially admission was open, yet because of the technical complexities of land surveying, a competition was established for one of fifty positions a year.[64] The school never reached this yearly quota: in the first year, there were eleven students; in the second, thirty. Over the course of its existence, the school only graduated sixteen ingenieurs-géographes, and in 1802 it closed.[65]

While the École des ingenieurs-géographes was short-lived, the value of surveying did grow through these institutions, and its inclusion in school curricula demonstrated an increasing acceptance of surveying methods and results. Yet because the schools did not actually train

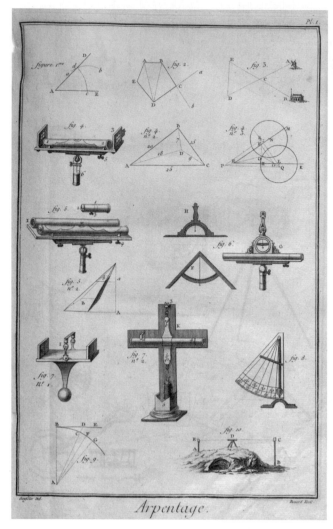

FIGURE 2.12 "Arpentage" (Surveying). Plate I from Denis Diderot and Jean le Rond d'Alembert, eds., *Encyclopédie, ou dictionnaire raisonné des sciences, des arts et des métiers, etc.* (Encyclopedia, or a systematic dictionary of the sciences, arts, and crafts) (Paris, 1751–80). University of Chicago: ARTFL Encyclopédie Project (Autumn 2022 edition), edited by Robert Morrissey and Glenn Roe.

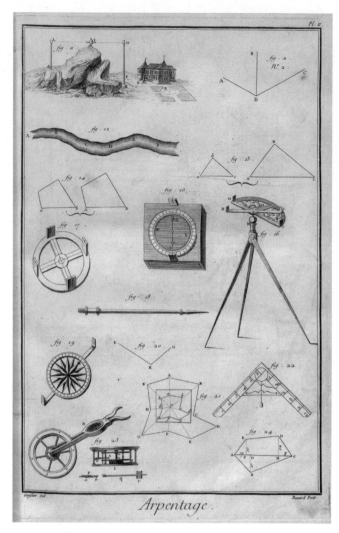

FIGURE 2.13 "Arpentage" (Surveying). Plate II from Denis Diderot and Jean le Rond d'Alembert, eds., *Encyclopédie, ou dictionnaire raisonné des sciences, des arts et des métiers, etc.* (Encyclopedia, or a systematic dictionary of the sciences, arts, and crafts) (Paris, 1751–80). University of Chicago: ARTFL Encyclopédie Project (Autumn 2022 edition), edited by Robert Morrissey and Glenn Roe.

many people, surveying continued to be largely dispersed in practice. Precisely due to this diversity, geometry became a necessary way to communicate cartographic information consistently beyond a given locale. The topographic details that would have once been depicted through pictorial modalities were shed in response to the map's application to increasingly different purposes, such as demarcating private property and edifying the growing audiences of a literate public. Geometry, used in a standardized fashion, thus promised greater sociability and communication among its practitioners through its uniformity.[66] It gave surveyors the potential to compare their work, their measures, and their maps. However, it came at the cost of obscuring the very labor that guaranteed the accuracy of their results. The uniform straight lines did not reveal the toil of carrying heavy equipment along remote roads, the calculations drawn and corrected, or the various hands that went into engraving. The lines pictured a collective, but ultimately disembodied, work.

SURVEYING DIAGRAMS

Without a centralized school to train surveyors and without a corporate body to establish norms, surveying manuals became important sources for disseminating knowledge. Published since the sixteenth century, these texts, whose production and printing increased

dramatically during the eighteenth and nineteenth centuries, provided information regarding trigonometric
calculations, geometric principles, and guides to measuring instruments. The authors of these manuals shifted
from mathematical theoreticians and members of the
Académie royale des sciences to empiricists who were
experienced in fieldwork.[67] Subtle but significant changes
to the authority of expertise and the social value of labor,
as well as the encouragement of the scientific principles of
objectivity, empiricism, and quantification, were charted
in these technical manuals for their changing audiences.[68]

The images in both Dupain de Montesson's and
Teyssèdre's volumes are exceptional. The majority of
images in surveying manuals did not include figures.
Denis Diderot and Jean le Rond d'Alembert's *Encyclopédie*,
a paragon of Enlightenment values, included images
showing the spatial and social contexts of many mechanical arts; however, the entry on surveying ("Arpentage")
included only geometry diagrams and illustrations of
the surveying instruments (figs. 2.12, 2.13). In images that
contain actual workers, tools and instruments are given
priority, crowding out the human figures in a plate's composition.[69] This focus on the equipment of work rather
than the labor itself was informed by Diderot's aim "to
raise the mechanical arts from the debasement where
prejudice has held them for so long" and elevate them to
the level of the sciences.[70] The instruments are organized
typologically, with each drawn as a distinct and decontex-
tualized item. They are contorted to display the various
parts in three dimensions and, at least in the two plates
dedicated to surveying, they are not shown in use.

Yet even these geometric diagrams (understood to
be scientific and static) were variable and underwent a
process of standardization. In *La géométrie pratique* (1702),
Manesson-Mallet pictured geometric diagrams embedded within a composite landscape. In his first plate for
his section on trigonometry, three triangles are rendered,
with each point connected to a specific landmark or a
surveyor (fig. 2.14). A straight line connects the surveyor
with a planchette at G and the steeple at B and tower at
A; in between, workers carry supplies to riverbanks. The
distance between points C and D is marked by dashed
lines and triangulated by a surveyor, depicted with his
equipment. The lower figure is in the midst of measuring between E and F, near the depth of a well whose
opening is distorted upward so that its interior can be
better revealed. Framing the scene are two figures at the

FIGURE 2.14 Plate I from Alain Manesson-Mallet, *La géométrie pratique* (Paris: Chez Anisson directeur de l'Imprimerie royale, 1702). ETH-Bibliothek Zürich, Rar 1930.

lower-right corner of the plate—one seated, the other
standing—who carry on an animated conversation, pre-
sumably about triangulation. This plate, along with the
others in Manesson-Mallet's volume, contextualize the
geometric diagrams of triangles. The points and lines have
substance and are related to specific landmarks. The land-
scapes through which these lines cut are also consistently
productive and active (figs. 2.15–2.17). Fields are plowed;
goods are carried down rivers; and people occupy the
streets. The implication is that surveying is embedded in
the daily activities of the world.

The illustrations in Jacques Ozanam's 1716 surveying manual *Méthode de lever les plans et les cartes de terre et
de mer* (Method of mapping plans and maps of the land
and the sea) took a different direction (fig. 2.18). Ozanam,

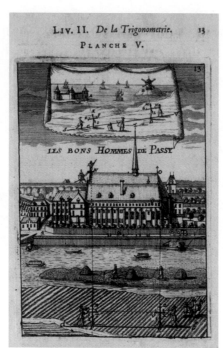

FIGURES 2.15, 2.16, AND 2.17 (left to right) Plates III, V, and IX from Alain Manesson-Mallet, *La géométrie pratique* (Paris: Chez Anisson directeur de l'Imprimerie royale, 1702). ETH-Bibliothek Zürich, Rar 1930.

an academician trained in mathematics instead of land surveying, embedded diagrams in the landmarks; unlike Manesson-Mallet, he omitted people. Instruments are represented in detail and in their totality, fixed to the ground and not floating on the page. To indicate the methods of triangulation, the surveying instruments are depicted, and dashed lines extend outward from them. References to the physical landscape are included in some illustrations, but they are schematic (fig. 2.19). Plate 9 represents a triangulation diagram, creating a web of triangles to determine the distances betwcen Paris, Gentilly, and Botouille (fig. 2.20). There is a radical reduction in this representation, with only lines, points, and letters. Sites are indicated not by a particular monument but solely by their names; there is no depiction of the topography, and the white space supports the drawn lines and circles. The plate that follows in the book shows the "geometrization" of a winding river, in which straight lines circumscribe its bends, such that the process of reducing variable topographic features to abstract lines and points is visualized.

A 1781 reprint of the same text includes a different set of plates, with the geometric figures composed into grids. Instruments are presented as assemblages, deconstructed with details of the individual parts. Representations of buildings and terrains are orthographic, and geometric figures are drawn without context. While there are no figures included in these plates, a floating eye does appear, indicating sight lines for perspectival renderings (fig. 2.21). This eye, while referencing a body, represents no particular body, proposing that the measures to be taken and the calculations to be made are not physical but mental efforts. With little sense of the actual process of surveying, the context, and the sequence, the floating eye signals surveying's value as scientific and disembodied.

These geometric diagrams informed the composition of the triangulation tables included in Verniquet's and subsequent maps. The technical images established norms to represent the practice of surveying. And as the surveying manuals indicate, Verniquet drew his map by first surveying landmarks in relation to each other, drawing not inward from coastlines or borders, but outward from triangulated, gridded points using a scale of *"une demi-ligne par toise"* (a half-line by toise) that had never been achieved before. For Verniquet and his team of surveyors, the work involved in triangulation and drawing the map entailed physical and intellectual labor, from carrying the cumbersome instruments to determining the trigonometric calculations—and all were reduced to straight lines.

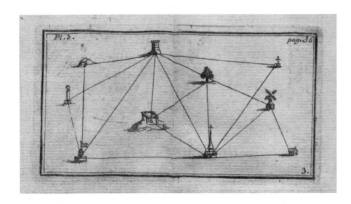

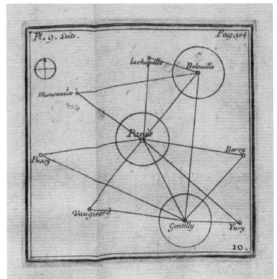

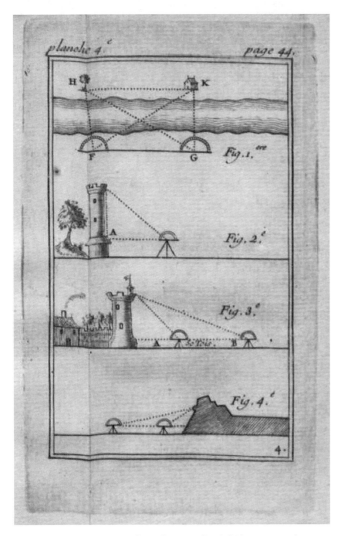

FIGURES 2.18, 2.19, AND 2.20 (top, above, and right) Plates 3, 4, and 9 from Jacques Ozanam, *Méthode de lever les plans et les cartes de terre et de mer avec toutes sortes d'instrumens, et sans instrumens* (Method of mapping plans and maps of the land and the sea with all sorts of instruments and without instruments) (Paris: Chez Claude Jombert, 1716). ETH-Bibliothek Zürich, Rar 4102.

IMPERIAL DESCRIPTIONS

Colonial settings were often the testing ground for new methods of mapping and surveying based in imperial pursuits. For those connected to the administration of Paris, their military service in the French colonies helped to develop and hone skills that were then applied to France.[71] Both Edme-François Jomard and Gilbert-Joseph-Gaspard, comte de Chabrol de Volvic, served in Egypt as ingénieurs-géographes during Napoléon Bonaparte's invasion in 1798. This conquest attempted to thwart British control of the region, as well as to protect French commercial and military interests in Egypt.[72] Significantly, the campaign mixed military and scholarly interests, when more than 150 engineers and academicians accompanied the soldiers: the conquest was executed explicitly under both military and civilian auspices. These men, often fresh out of school, learned surveying skills in the context of the military invasion and later brought them to bear within the French administration.[73] Jomard, who began as a field surveyor, became the director of the *Description de l'Égypte* (a monumental study of Egypt's geography, history, and culture that was published between 1809 and 1822) and eventually the director of the geography and maps division of first the Bibliothèque du roi and later the Bibliothèque nationale. He would write and lecture specifically about the scientific qualities of the cartographic medium in justifying the creation of the cartographic collection. Chabrol de Volvic, as his fellow classmate at the École polytechnique, would institute the first census of Paris, among other urban development activities, as prefect of the Seine from 1812 to 1830.

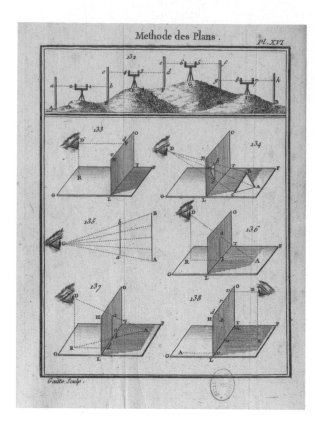

FIGURE 2.21 Plate 16 from Jacques Ozanam, *Méthode de lever les plans et les cartes de terre et de mer avec toutes sortes d'instrumens, et sans instrumens* (Method of mapping plans and maps of the land and the sea with all sorts of instruments and without instruments) (1716; repr., Paris: Jombert le jeune, 1781). Bibliothèque nationale de France.

During France's occupation of Egypt, mapping initially concerned establishing Egypt's geography and understanding its resources through a topographic survey based on triangulation. The goal was to measure the arc of the meridian in latitudes south of France, as well as to cover Egypt in a chain of primary and secondary triangles that would link first to Corsica and then ultimately back to France.[74] In the end, due to the resistant local conditions—both climatic and political—as well as the sinking of the ship *Patriote* with its surveying equipment, the survey was never fully conducted.[75] Only specific local areas were triangulated and then compiled from diverse manuscript maps using astronomical control points. Surveying concentrated on the Nile and the extent of its seasonal floods, the location of wells and canals and other waterways, and coastlines of defensive concern.

However, in the last year of the expedition, from June 1800 to August 1801, General Jacques-François Menou shifted priority to cadastral mapping in order to undermine and usurp the Coptic monopoly of taxation, and to appropriate well-needed funds by replacing the local *iltizām* system with a direct tax on land.[76] In order to gain control of the land revenues for their own coffers, the French were interested in eliminating Mamluks, intermediaries who collected taxes from the peasantry on behalf of the Ottomans.[77] On March 1, 1801, a cadastral commission of Egypt was convened, and the ways in which the cadastre should be conducted were discussed.[78] However, with military defeat imminent, the French surveyors were never able to actualize any of their objectives. Thus, although the survey was diligent, it was not thorough, reaching only places that soldiers could defend.[79]

While Napoléon's full military ambitions failed, the conquest, however, produced the twenty-two-volume *Description de l'Égypte*.[80] A section devoted to Egypt's topography was intended to be the study's chief contribution, representing the major work that was undertaken by the conquest's scientific mission. The cartographic section included a *carte topographique*, comprising forty-seven sheets at a scale of 1:100,000 that covered Egypt and Syria; a *carte géographique*, made up of three sheets at 1:1,000,000; and a single-sheet *tableau d'assemblage* that provided a general overview.[81] Because of their detail, production and publication were deemed too militarily sensitive. Thus, control over the maps was maintained by the Dépôt de la Guerre, and control over the text was handed to the largely civilian Commission de la Description de l'Égypte.[82] Jomard, as the director of the commission, vigorously argued for the joint publication of maps and text, as he saw the maps as integral to understanding the text. His arguments failed to persuade, and only years later, in 1828, were the maps finally printed.

The maps of the *Description de l'Égypte* were not merely geographic descriptions, as the title might suggest. As Anne Marie Claire Godlewska has demonstrated, they were used as a primary tool for a complete restructuring of the country and its alignment to France.[83] This information was compiled by the surveyor through a set of instructions elaborated by the director of the Egyptian survey, Pierre Jacotin. These instructions included one section related to surveying the terrain, a second section for observations related to population, family groups, type of agriculture, commerce, and industry, and a third section for general remarks about arts and culture, animals, sanitary conditions, and natural resources, including speculations about the reasons for particular patterns or habits.[84] The data was then compiled into maps—although

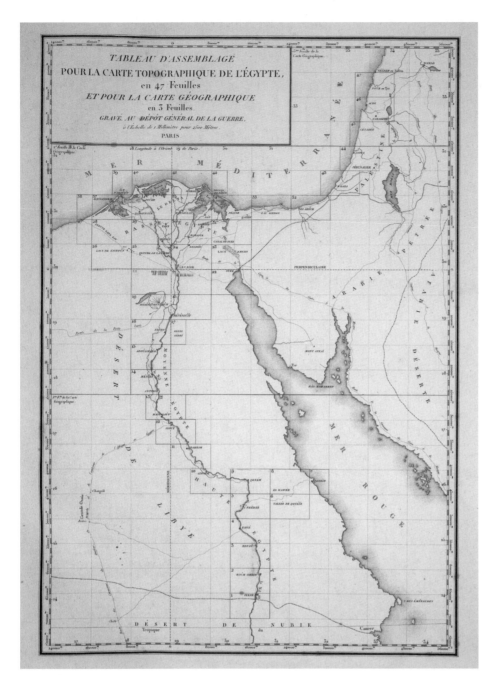

FIGURE 2.22 "Tableau d'assemblage pour carte topographique de l'Égypte" (Index for topographic map of Egypt). From Edme-François Jomard, *Description de l'Égypte: Ou, recueil des observations et des recherches qui ont été faites en Égypte pendant l'expédition de l'armée française* (Description of Egypt: Or, summary of the observations and research that was executed in Egypt during the expedition of the French army) (Paris: Imprimerie impériale, 1809–28). Rare Book Division, New York Public Library.

not all regions were surveyed with equal detail—and anticipated the spatialized data of thematic cartography, such as maps of population, transportation routes, and resource distribution, that would become part of the French administrative apparatus during the late nineteenth century.[85]

These unfulfilled plans led to a highly mixed atlas—including the *Carte de l'Égypte*—that did not always correspond to the actual terrain, and that was instead created for the Orientalist imagination of its French audience. As Godlewska has argued, the images relegated Egypt to an

ancient past in both explicit and implicit ways through cartographic codes. First, the same scale of 1:100,000 was used for these maps and the *Carte de France*, the triangulation project executed by the Cassini family, some of whose surveying methods were outdated.[86] Moreover, in the *Description de l'Égypte*, distances between sites in Egypt and the Paris meridian were marked on the corner of every map sheet as well as on a chart, making a clear case for Egypt as part of French territorial claims (fig. 2.22). The Observatoire de Paris, a symbol of scientific progress, served as the anchor for the meridian, but its equivalent

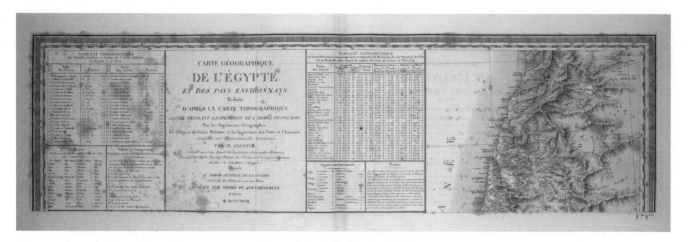

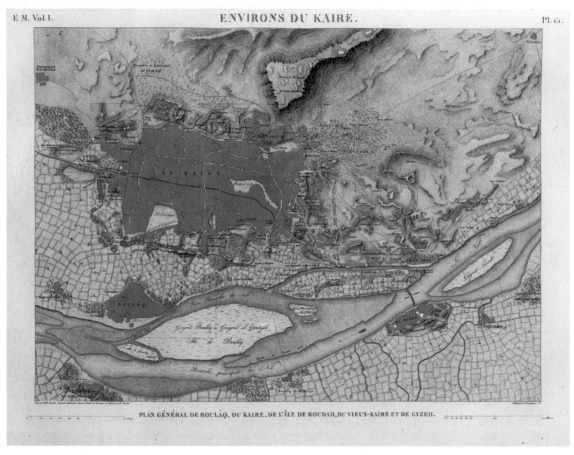

PLAN GÉNÉRAL DE BOULÂQ, DU KAIRE, DE L'ÎLE DE ROUDAH, DU VIEUX-KAIRE ET DE GYZEH.

in Egypt was the pyramid of Giza, an asynchronous choice associating the country with the past (fig. 2.23). Accordingly, many of the maps rendered in meticulous detail the ancient structures and sites, while present-day cities and towns were sketchy and sometimes left blank (fig. 2.24). Even in the section dedicated to "The Modern State" (*État moderne*), there was continued emphasis on ancient monuments, dissected into elevations and plans, rather than on the contemporary lived conditions and built forms (fig. 2.25).

The colonial practice of fixing Egypt in an imagined past was enabled by a cartographic expression that was all-encompassing and putatively scientific. Timothy Mitchell has argued that the colonization of Egypt was based in a fundamental cultural mechanism of dominance that turns the world into an exhibition for the pleasure and fantasy of the European.[87] This worldview, which positioned Europe in the present and froze the Other in history, was precisely based on mixing reality and fantasy. Mitchell offers the example of the 1889 Exposition

FIGURE 2.25 "Environs du Kaire. 1. Vue de la plaine de la Qoubbeh; 2–4. Prise d'eau de l'aqueduc du Kaire; 5–8. Pont de la plaine des pyramides" (Surroundings of Cairo. 1. View of the plain of Qoubbeh; 2–4. Water intake from the aqueduct of Cairo; 5–8. Bridge of the plain of pyramids).

universelle in Paris that constructed a typical street in Cairo in which the buildings were painted intentionally dirty—even as the meticulously rendered façade of the mosque opened into a café filled with young Egyptian women dancing.[88] There was just enough accurate detail to certify the whole experience as authentic. Temporal differences were discounted, with ancient sites pictured in detail while contemporary ones were left vague. Sites that were not surveyed yet were drawn using the graphic language of triangulation. For maps, this amalgam was concealed by the continuity of geometric forms.

As with French surveying manuals, the *Description de l'Égypte* included affirming depictions of surveyors exploring, drawing, and measuring the monuments and sites. The authority of these maps was located in their claim to geometric measurement and methods of quantification. In his *Mémoire sur le système métrique* (1817), Jomard wrote, "Geometry, more than any other branch of knowledge, offers the means of achieving truth. In effect, the theorems of geometry do not allow vague interpretations to hold."[89] Jomard's three years of surveying in Egypt formed the basis of his professional career during the early nineteenth century and his significant influence on the development of cartography and geography in France.[90] In addition to his position as director of the *Description de l'Égypte*, he also became a head of the Comité des géomètres-conservateurs du cadastre and was one of the founders of the Société de géographie, posts that provided a platform to promulgate mapping as a scientific practice based on a graphic language. According to him, geometry, and its graphic expression, was understood as an unmediated path. It reduced the world to lines, incorruptible by human errors and fallacies, and offered absolute clarity where there was none.

However, a line on the *Carte de France* did not equal a line on the *Carte de l'Égypte*, even if the same methods of measurement, the same techniques of observation, and the same resolutions were established in both.[91] The *Carte de France* represented three generations of triangulation work, whereas the *Carte de l'Égypte* was conducted in three years and could not have achieved the same detailed level of surveying, of which the final images disclose none of these differences. In his *Mémoire sur la construction de la Carte de l'Égypte*, Jacotin admits that all the procedures necessary for exact measures, such as measuring baselines, determining the length of an arc of the meridian, and triangulating the entire country, were not achieved.[92]

Ultimately, this was the power of the orthographic map: to conceal variability as well as the epistemological and social practices of power that informed its production, such that it appears given. Jomard argued for the medium specificity of the map, claiming that "a map permits one to embrace the whole at once."[93] He made the same case for the medium specificity of the map in order

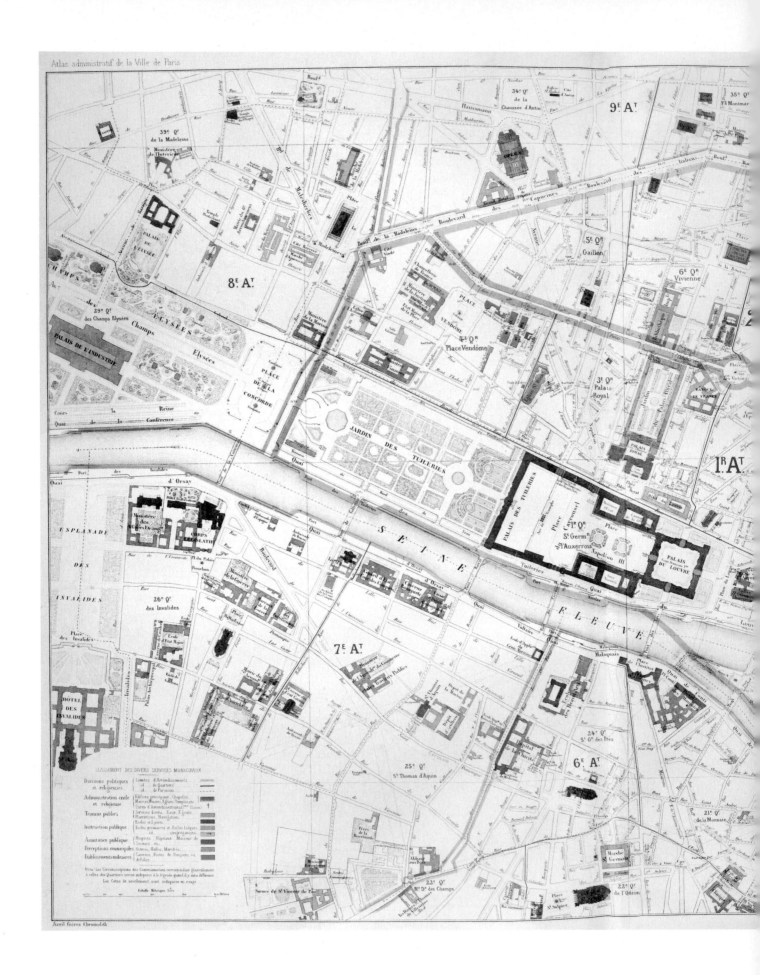

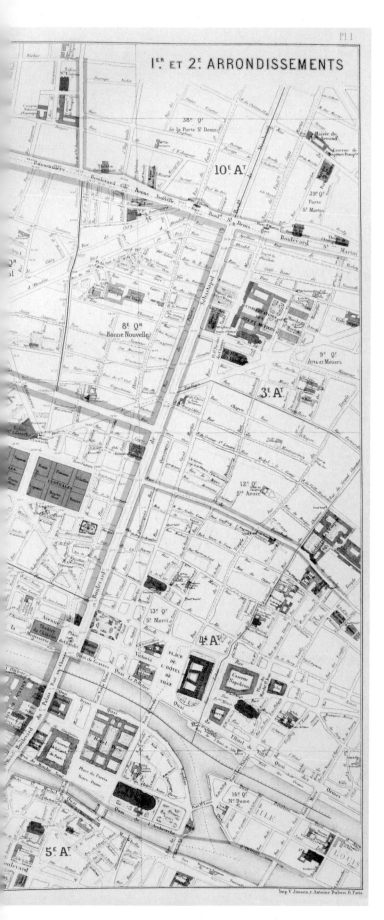

1ᵉʳ ET 2ᵉ ARRONDISSEMENTS

FIGURE 2.26 Plate 1 ("1ᵉʳ et 2ᵉ arrondissements") from Eugène Deschamps, *Atlas administratif des 20 arrondissements de la Ville de Paris* (Paris, 1868). Archives de la Musée Carnavalet.

to establish a separate department dedicated to cartographic objects at the state library. While maps are now accepted in a variety of representational modalities, it was during this period that a specific definition of a map was argued for and developed—one that sought unequivocal uniformity in its graphic modality. Moreover, this claim of the map's capacity to capture the whole at once was based in imperial power. It sutured geometric lines together in the name of measurement to construct an imaginary coherence and correspondence to a place. The map became its own cartographic site that was neither entirely fictive nor entirely accurate: a hybrid, projecting the values of its makers and users onto the world.

A CARTOGRAPHIC SITE

With no survey since the 1780s, new efforts to triangulate Paris started on June 16, 1859, compelled by the annexation of the suburbs five months earlier. The appropriation and the organization of the new urban territories required an update. The survey results were eventually compiled and published in 1868 as the *Atlas administratif des 20 arrondissements de la Ville de Paris,* which provided the first view of the newly incorporated districts and expanded city.[94] It was the only map that Haussmann commissioned during his tenure during the Second Empire, and likely the map he kept mounted on a folding screen in his office. Until this point, however, Haussmann had used plates based on Verniquet's survey from the 1780s to conceive and plan Paris's modernization.

In histories of Paris's transformation, the 1868 map is often used to describe the modern city: an illustration of modernization accomplished. Yet a basic chronology raises questions about what the map described. The survey occurred in the middle of major upheaval, when the major boulevards of Sébastopol, Rivoli, Saint-Germain, and Saint-Michel were under construction, and the districts of Châtelet, L'Opéra, and Les Halles Centrales were all unrealized. How could one then map the blocks, the alignments, and the streets that were yet to be completed (fig. 2.26)? The answer argued for here is that the map was as much projection as it was description, and in this

FIGURE 2.27 Illustration of a surveying tower. From "Triangulation de Paris," *L'illustration* (December 17, 1859): 432.

PYRAMIDE en charpente dressée (barrière du Tróne) pour les opérations de la triangulation de Paris.

hybridity, the map served as its own site, functioning as an instrument to build and to imagine a modern Paris.

The triangulation work was coordinated by the Service du plan, a new organ in Haussmann's administration that centralized the production of all official maps and plans.[95] At its head was Eugène Deschamps, who was not trained as an engineer like the other directors but rather as an architect, having studied at the École des Beaux-Arts under the architect Henri Labrouste.[96] Deschamps began his career as an *architecte-voyer*, responsible for ensuring that building conformed to government plans (especially roadways), in the Service des travaux

(Public works service), where Haussmann said that "geometry and graphic drawing play[ed] a more important role than architecture."[97] Before he became head of the Service du plan, he held the post of *conservateur* (curator or custodian) of the *Plan de Paris,* during which time he became intimately aware of the previous surveys and government maps of Paris.[98]

Deschamps's intimate knowledge of the maps of Paris helped him to conceive of a strategy to eliminate any potential surveying errors. Unlike in Verniquet's survey, extant buildings were not used for measurements; instead, surveying towers were constructed throughout the city.

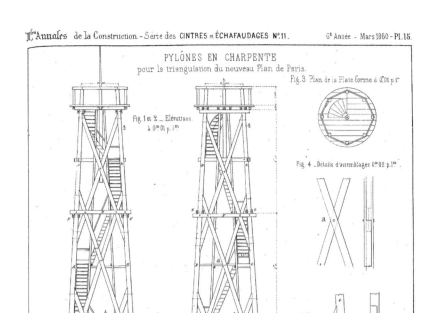

FIGURE 2.28 Elevation and plan of a surveying tower. From *Nouvelles annales de la construction*, no. 11 (March 1860): pl. 15.

These towers represented a method to control the triangulation measures in terms of distance and elevation. The objective was to translate the multidimensional spaces and buildings of the city into a planar representation. Sixty-five pine towers were constructed across the city, onto which tricolor flags of blue, white, and red were attached to aid in signaling to the other towers. Their heights were varied by design, to make up for the uneven terrain. The largest, measuring thirty-seven meters, was located at the Madeleine, then followed by Saint-Vincent de Paul and the barracks at boulevard du Prince-Eugène (now boulevard Voltaire), which both measured thirty-four meters high; all the others reached twenty-eight meters (figs. 2.27, 2.28).[99] At the center was a set of stairs that reached the top platform where the measurements were taken.

These towers were left in place for two years after their initial erection, in case corrections to the survey had to be made.[100] Others were built in the midst of the city under major construction, such as the tower at the church of Notre-Dame-de-Lorette, when on January 31, 1860, it was reported that the last surveying tower was erected in the midst of sewage pipes being placed behind the apse.[101] The caricaturist Amédée de Noé, who went by the pseudonym "Cham," captured the spectacle of

FIGURES 2.29–2.35 From Amédée de Noé, *Croquis contemporains* (Contemporary sketches) (Paris, 1853), 6:331, 6:333. Bibliothèque nationale de France.

FIGURE 2.29 "The surveyors charged with the triangulation survey all go to the Jardin des Plantes to see how they should go about climbing into their office."

FIGURE 2.30 "Madame Saqui was charged with putting the surveyors in touch with one another during the triangulation survey."

FIGURE 2.33 "The giraffe's neck used for the triangulation survey in the quarter of the Jardin des Plantes."

FIGURE 2.34 "The surveyors charged with the triangulation survey had the cautious wisdom of attaching a line to their legs in case there was a strong wind during their working hours."

FIGURE 2.31 "The artillery service was at the disposal of the surveyors to send them to their offices."

FIGURE 2.32 "The postal service added chimney sweeping to the delivery of letters addressed to the triangulation surveyors in the execution of their tasks."

FIGURE 2.35 "—Come down! We won't climb up there! —My wife and I rented these spots from a gentleman to see to his business in Sicily."

the towers in *Le charivari*, providing a rare record of the surveyors (figs. 2.29–2.35). His humorous illustrations highlighted their dizzying height, and how these surveying towers became fixtures of popular culture in a city under construction. In one picture, a famous tightrope walker, Madame Saqui, is shown traversing a line between the numerous towers, rising well above the skyline (see fig. 2.30).

The towers were constructed not only to measure surface area, but also to establish a level plane that could translate the hilly terrain of the city to the flat surface of a printed map. In his *Atlas,* Verniquet had used tall buildings to triangulate the distances within the city, and thus had to compensate for the various heights at which he took his measures. Each reference tower acted as a leg holding up a flat surface, in which all legs reached the same given height, regardless of topography. Verniquet's map shows none of these leveling calculations; in the tables, he provided only horizontal distances, no vertical elevations. In the map, however, the transformation of the city into a flattened space is well marked: the center of Paris

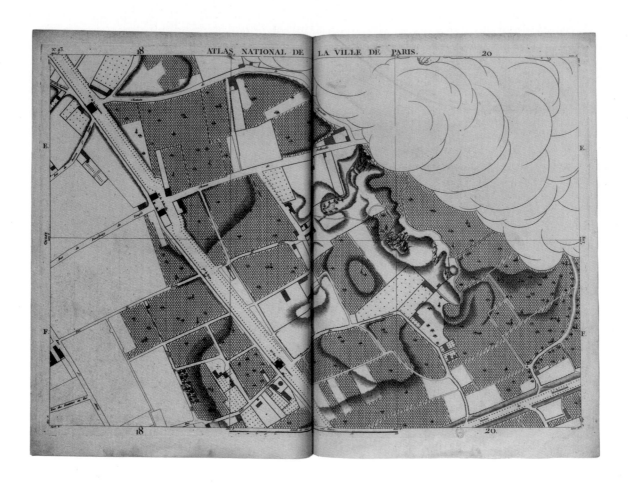

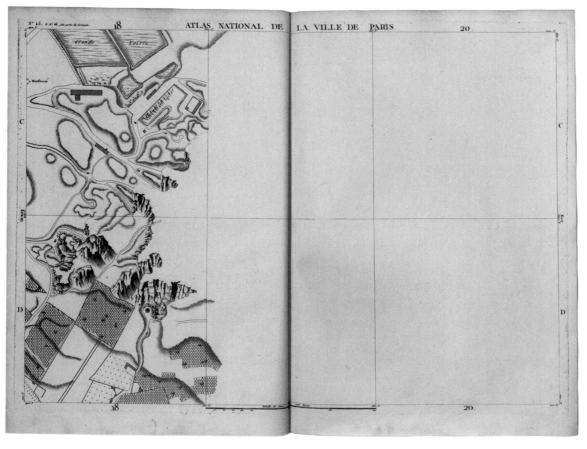

is described in an abstract and orthographic manner, and, as the urban territory reaches the margins, topographic details appear (figs. 2.36, 2.37). Hills and valleys are marked by hatchings to convey shape, slope, and height. These naturalistic features seem to be pushed out from the geometric core of the city, the straight lines expanding into the hinterlands. The symbolic effect is of a city as a flat artifice produced through geometric calculations, separate from the pictorial and even allegorical space populated by gods and goddesses floating on clouds at the margins of the map sheet.

Deschamps eliminated all these distinctions in his maps, and a planar geometry dominates the entire sheet. But what was the topographical basis of this flatness? There was no standard level of measures used until the second half of the nineteenth century, and previously, each urban surveyor typically had to develop his own system of levels or none at all. This was often based on the instruments taking measurements at ground level, calculating divergences between the flat surface of a map and the curved surface of Earth. For the relatively small surface area of Paris, these divergences are not great; however, surveyors still had to contend with the uneven and hilly terrain of the city's topography, and ultimately had to render the city flat.

Eventually, a law of March 25, 1852, mandated that all plans include a leveling measure, determined by the Préfecture de la Seine in relation to the mean sea level (*niveau moyen de la mer*), which was marked as zero.[102] A vertical reference was created and given a base quantitative value. Elevations were to be determined relative to this level that was determined by an average, a statistical method newly applied by the Paris administration.[103] Because seas change constantly and differently depending on the specific location of the measurement and body of water, the definition of a sea level is an abstraction. Height measures are then determined from this abstract line, creating a fictive level to cartographic space. By 1856 a level was marked at the Pont de la Tournelle and La Villette, and leveling became incorporated into the curriculum at the École polytechnique, such that a planar standard was established to control for topographic variance.[104]

Deschamps's final published atlas of 1868 did not include a triangulation table. Nor did it include tables indicating the location of the surveying measures. Even the frontispiece contained only a short explanation that Haussmann had directed the atlas's production, and that it was a scaled reduction of a larger general plan executed by the *geomètres*, or surveyors, of the Service du plan. The legend that was featured at the corner of the sheets only indicated the color codes for the different types of public buildings and administrative boundaries. Here, the validity and authority of the map's correspondence to the urban terrain were fully assumed in its graphic composition. Geometry had conquered the terrain. If Verniquet's map had distinguished between the geometry of the city and the natural topography of its hinterlands, then in Deschamps's, the comparison emphasized differences only among the maps themselves. The primary juxtaposition shows the extension of Paris before and after the annexation of the suburbs, in which the shaded area expands smoothly across the same base map (figs. 2.38, 2.39).

In Haussmann's personal atlas from 1866, however, a triangulation table was included on plate 20, "Canevas trigonométrique" (fig. 2.40). At a scale of 1:5,000, it presents the city as a gray shaded form delimited by a green boundary and the Seine as a blue ribbon twisting through the city. In contrast, the straight triangulation lines weave in and out of the ovular form, connecting to points within and beyond the shaded area. The observatory is the center point of the grid; the Panthéon, Invalides, and Moulin d'Orgemont complete the base triangle to the entire system. The illusionistic composition of the frame around the "Canevas" acknowledged the cartouche's dual role as part of and distinct from the map of the city. Along the bottom, its outline conformed to the frame of the plate with a dark line. However, when it confronts the image of the terrain, the "Canevas" appears to peel away using a *trompe l'oeil* technique.

The map's reflexivity as a paper surface refers to the artifice of the cartographic system. In Haussmann's case, the belief in the accuracy of the map could not rely on visual comparison to the physical environment, because it described a city that did not yet exist. The cartographic image of Paris produced from Deschamps's survey was of a city in the midst of major building projects. Even if

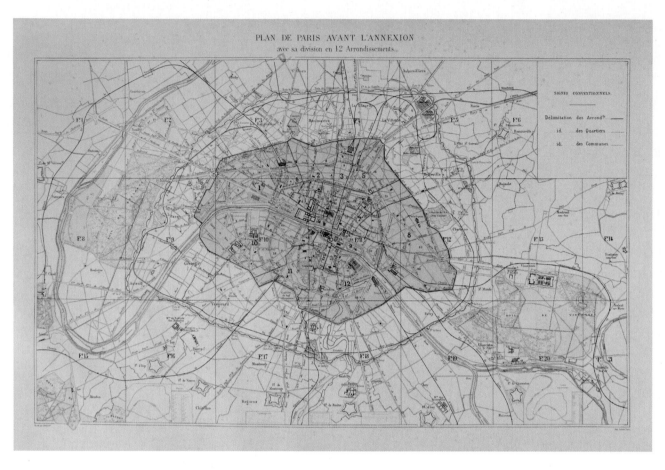

PLAN DE PARIS AVANT L'ANNEXION
avec sa division en 12 Arrondissements

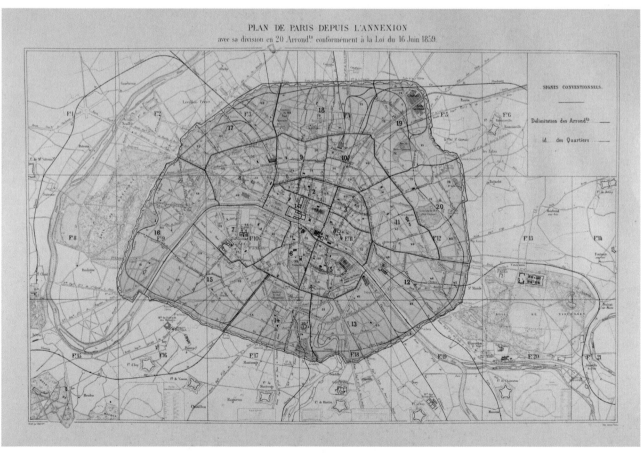

PLAN DE PARIS DEPUIS L'ANNEXION
avec sa division en 20 Arrond.ts conformément à la Loi du 16 Juin 1859.

Haussmann had wanted to look out his office window, the view would have offered a forest of scaffolds, stone, and dust. For constructions that had not yet begun, the map functioned projectively, outlining the form of the city and the projects in gray; for constructions underway or completed, the map functioned descriptively, as a means of placing those actions in context.

Recalling Carroll and Borges, the map became its own site, and its graphic form and geometric composition produced through the triangulation survey defined Paris's terrain as regular. The surveyed orthographic map thus purported to offer a view from nowhere to no particular individual. The varied topography of the city was not simply absent from the cartographic representation; urban space itself became frictionless. Particular sites lost their discrete value over the uniformity of the ensemble that ultimately afforded straight lines to be drawn. In visual terms, these surveyed geometric maps marked the adoption of a concept of absolute space by administrators, engineers, and architects, not inflected by any relations or processes that might determine the site in the first place.[105]

Paris was made to conform to a map. The process of triangulation and its geometric expression created a cartographic terrain. Surveyors and engineers (through their instruments and towers), as well as administrators (through their discourse and regulations), neutralized the variable qualities of the topography. Its composition into printed lines and blank spaces was part of a regime of description, promoted through an ever-increasing number of surveying manuals that privileged diagrammatic forms and promulgated a conception of space as gridded and flat. Grounded in imperial pursuits and military actions, geometry conquered the terrain.

By the end of the eighteenth century, the map's connection to a given territory was not based on pictorial comparison for verifying its accuracy. Instead, it relied

FIGURE 2.38 (opposite, top) "Plan de Paris avant l'annexion" (Plan of Paris before annexation). From Eugène Deschamps, *Atlas administratif des 20 arrondissements de la Ville de Paris* (Paris, 1869). Archives de la Musée Carnavalet.

FIGURE 2.39 (opposite, bottom) "Plan de Paris depuis l'annexion" (Plan of Paris after annexation). From Eugène Deschamps, *Atlas administratif des 20 arrondissements de la Ville de Paris* (Paris, 1869). Archives de la Musée Carnavalet.

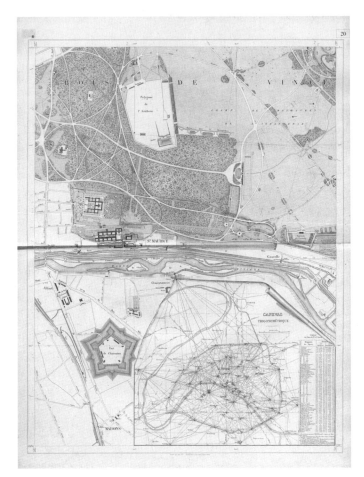

FIGURE 2.40 "Canevas trigonométrique" (Trigonometric canvas). Plate 20 from Eugène Deschamps, *Plan général de la Ville de Paris et de ses environs comprenant les bois de Boulogne et de Vincennes*, 1866. Map sheet, 35 11/16 × 26 11/16 in. (90.6 × 67.8 cm). ETH-Bibliothek Zürich, Rar KA 95.

both on ground surveys carried out by teams of surveyors and on rhetorical codes, such as triangulation tables, to account for its geometric correspondence. These figures symbolized the map's status as accurate; maps never become *more* accurate, only accurate in different ways.[106] For example, pictorial maps of Paris were not less accurate than orthographic; rather, they provided information about material, spatial, and physical qualities that geometric maps could not. As Borges recognized in his short story, the exactitude of the map never eliminates the difference between a representation and its object. These graphic symbols acted as compensation for the map's differences from the terrain. Thus, inherent to a map's value as a valid and accurate representation of a terrain is the idea of difference; and, paradoxically, in this difference, the map can make graspable an urban terrain that is ungraspable in and of itself.

3 Drawing Grids

The French language gives a double meaning to the word *plan*. It may be the recording of a de facto topographical state (hence, urban maps); it may be a projection into the future.

 Pierre Lavedan, *Histoire de l'urbanisme à Paris* (1975)

Grids have long been used in maps. They functioned for users as a locating device; specific places could be identified by their position in a particular graticule. Coordinate grids were used by mapmakers to plot points on the surface of the paper and to preserve the proportions of a drawn figure across different media. This functionality is exemplified in Albrecht Dürer's well-known woodcut, which shows a draftsman making a drawing of a woman with the aid of a "perspectival frame" (fig. 3.1).[1] A gridded screen stands between the model and artist, maintaining the fidelity of proportions between the real and drawn figures. This translational capacity was always appreciated in various arts, and grids are found as the armature of paintings, frescoes, sculptures, drawings, and maps. Yet often, even if the grid was integral a map's making, it was often omitted from the finished print.

By the late eighteenth century and throughout the nineteenth century, however, the grid began to appear consistently on the final map as a printed graphic element in its own right. As the meaning of the map's accuracy came to be tied to new values of precision based on geometric correspondence, rather than pictorial verisimilitude, the grid became an important symbol of objectivity and the map's adherence to scientific principles. More than a means to establish the correspondence between the urban terrain and the print, this grid also functioned as a compositional reference for drawing projections (fig. 3.2). The repeating and regular squares of the printed cartographic grid oriented new marked lines, such that it allowed both the printed description and the drawn projections to cohere onto a single map sheet.

A MAPPED GRID

The symbolic and functional significance of the cartographic grid in urban plans was directly connected to its integration with the standardized measure of the meter. The adoption in France of a single system of measurement had rendered all distances relatable. Conceptually,

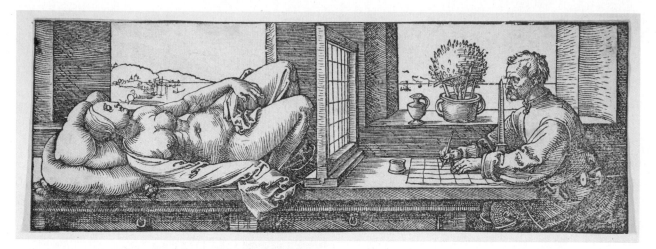

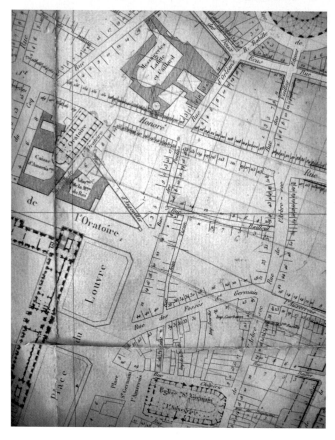

FIGURE 3.1 (above) Albrecht Dürer, *Draughtsman Making a Perspective Drawing of a Reclining Woman*, ca. 1600. Woodcut, 3¹/₁₆ × 8⁷/₁₆ in. (7.7 × 21.4 cm). Metropolitan Museum of Art.

FIGURE 3.2 (left) Detail showing printed horizontal and vertical lines as well as a grid that is drawn in pencil on top of the printed plan. From Théodore Jacoubet, *Atlas général de la ville, des faubourgs et des monuments de Paris* (General atlas of the city, suburbs, and monuments of Paris) (Paris, 1827–39). Archives nationales de France.

This chapter discusses the ways in which the drawn line on the plan was linked to the terrain, and the specific values that the grid acquired in the late eighteenth and first half of the nineteenth century. Synthesizing intellectual values of objectivity, economic goals of transferability, and political aims of standardization, the grid became a device that permeated all aspects of spatial design. It was both instrument and representation. At the same time, the grid's ubiquity also made it difficult to grasp in material and physical terms. Its abstract form could represent both part and whole, order and disorder, concentration and dissipation, detail and outline, container and contained, space and line, and boundedness and endlessness, and thus was easily adaptable for varying political and social advantages. Its expansive logic pervaded modern life, but it was nearly impossible to trace how and where.

This intangibility was precisely the quality that made the grid integral to a definition of modernity, a phenomenon that during the nineteenth century was marked by a tension between disenchantment from and nostalgia for explanatory totalities.[2] In his writings on avant-garde painting in the second half of the nineteenth century, T. J. Clark describes the attempts and failures to fix an image

the metric system defined space as regular and neutral: a meter was the measured distance regardless of who was measuring it, the particular site or media, or who was traveling that span. Through a shared mathematical reference, distances were understood quantitatively rather than experientially. And once the cartographic grid of a map was linked to a standard measure, the paper surface of a plan could also relate to this network; it allowed for the plan's communicability while remaining immutable across translations between terrain and image.

FIGURE 3.3 Title page from Théodore Jacoubet, *Atlas général de la ville, des faubourgs et des monuments de Paris* (General atlas of the city, suburbs and monuments of Paris) (Paris, 1836). Bibliothèque historique de la Ville de Paris.

of Paris.[3] As artists such as Édouard Manet, Edgar Degas, and Georges Seurat responded to a city under construction, government architects, engineers, and administrators employed by the Préfecture de la Seine and its various agencies also endeavored to fix and stabilize their own image of the city—one that created its own myths of modernity. For much of the nineteenth century, Paris's terrain was marked by stone hills caused by demolitions, craters of excavated soil, and rising wooden scaffolds and towers. The lived experience of the city did not necessarily express itself as a stable image, which fed into a visual culture that was invested in projecting images of Paris and in which Haussmann himself was deeply committed. State architects, engineers, and administrators, unlike the avant-garde artists of this era, however, were involved in different kinds of alignments and political values—not critique, but entrenchment. The images that they produced and used to build were understood to be thorough and objective, precisely describing the world around them. This belief defined modernity as much as the distortions that became apparent through the artistic descriptions in oil on canvas.

Regardless of whether the modernization program was successful (and Louis Lazare, reporting in *La revue municipale*, argued that it failed), there was a feeling that Paris was being shaped into a spectacular image that would eventually come to be understood as modern, based on another image, in which those long, straight lines of the boulevards were prefigured on a map.[4] In this context, the grid became modernity's cipher, so mundane

that its profound effects were understood as causes. When printed, the grid allowed for a drawn line to tether what was to what could be, transforming the map into a site outside of any specific time and, as a consequence, even outside of history. It gave the illusion that the plan was not a contingent artifact, and that its use by architects, engineers, and administrators was totalizing.

THE MEASURED LINE

As prefect of the Seine from 1812 to 1836, Gilbert-Joseph-Gaspard, comte de Chabrol de Volvic, directed Théodore Jacoubet to produce new plans of the city. Urban developments in Paris since 1785, when the first triangulation survey had been executed, through to the Bourbon Restoration necessitated an update. The municipal administration had used Edme Verniquet's surveyed plans until 1827, when the new *Atlas général de la Ville, des faubourgs et des monuments de Paris* (General atlas of the city, neighborhoods, and monuments of Paris) was directed and published. Because Paris's limits had not expanded beyond the border of the Mur des Fermiers généraux (Wall of the Farmers general) from the time of Verniquet's survey, and because of the high cost of surveying, a new survey was not justified.[5] Instead, a new engraving was commissioned based on copying and assembling extant maps, in which Jacoubet adroitly avoided a reference to the French Revolution by naming neither the Commission des artistes nor Verniquet (fig. 3.3).

As the head of the Bureau des plans d'alignement, Jacoubet had access not only to Verniquet's plans, but

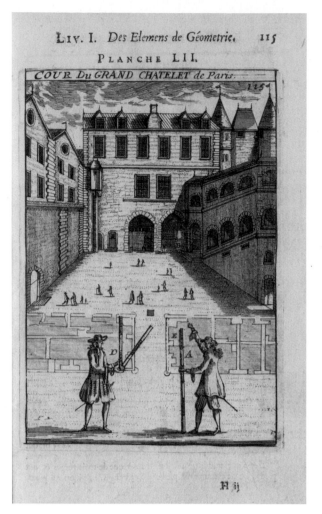

FIGURE 3.4 Plate LII from Alain Manesson-Mallet, *La géométrie pratique* (Paris: Chez Anisson directeur de l'Imprimerie royale, 1702). ETH-Bibliothek Zürich, Rar 1930.

also those kept at the Direction des contributions du Département de la Seine, which held records of the plans parcellaires of Paris as well as ground plans of buildings that property owners were required to submit for building.[6] Cadastral maps drawn by Philibert Vasserot and J. S. Bellanger would also have been accessible resources.[7] In addition, Jacoubet had use of the depository of plans held by the Préfecture de police, which relied on various *plans topographiques* from the eighteenth century. In aggregating these diverse maps, it was necessary for Jacoubet to find a common measure so that the maps could correspond to one another. His efforts thus went into updating and producing a current map that conformed to the new metric standard.

If Verniquet had used geometric triangulation to transform Paris into an orthographic matrix of lines, Jacoubet

used the metric measure to translate them into a scalable cartographic system. While there had been royal edicts of official standard weights and measures employed in Paris since François I in the sixteenth century, their consistent employment was haphazard, much to the frustration of the Ministère de l'intérieur. Several figures, including the military engineer Sébastien Le Prestre de Vauban, had long called for a standardized measure. Since the eighteenth century, a standard measure was formed into an iron bar, the *toise du Châtelet*, embedded into the corner wall at the foot of the Grand Châtelet's staircase; however, over the years, this rod rusted and gradually bent, providing an inconsistent reference (fig. 3.4). Moreover, outside of Paris, this measure was not referenced, forcing conversion tables between Paris and several major towns to be drawn up.[8]

This changed on August 1, 1793, when the meter, defined as one-ten-millionth of the distance from a pole to the equator, was mandated.[9] In Europe, until the eighteenth century, how land was measured was based on the cultural and political values of each particular site. Cultivated land was measured by the input of labor or the amount of seed necessary, and then differences were based on the kind of cultivation—that is, for plowing, for meadows, for gardens, or for vineyards. As Witold Kula describes, mensuration emphasized human relations to the land and the land's fertility.[10] During the ancien régime, different objects were measured with different units, and different trades employed different systems, creating a dense and intricate web of anthropometric values that relied not on a shared and universal quantity, but that suffused particular objects, personal negotiations, artisanal practices, local regions, and even specific rulers. Units of measure were qualitatively determined, and in this way, almost all premodern measures were defined locally.

However, this did not mean that standardization was not attempted. The French crown identified three itinerary measures for road distances that consisted of the *pas militaire* (military march) of two *pieds* (feet), the *pas commun* (common march) of two and a half pieds, and the *pas géométrique* (geometric march) of five pieds. Paris determined the standard for the *mille* at five thousand pieds, but these hardly resulted in any regularity.[11] The degree of variability in these measures was so great that geographers were required to create or copy conversions when using them.[12] Another example of premodern standardization was the public display of the toise bar, and its manifest accessibility at the Grand Châtelet to resolve disputes in a public space was just such an attempt: the bar's materiality was designed to provide a sense of immutability and to make an authoritative claim to a common measure.[13] Measures from the ancien régime were contingent upon the particular social system that they represented and constituted.

The ground survey by Jean-Baptiste-Joseph Delambre and Pierre-François-André Méchain, who set out from the Observatoire de Paris to measure the meter, altered that value of space from the qualitative to the quantitative. Using the surveying methods of triangulation, the aim was to define the baseline of the meridian to which all other maps would refer and to create a standardized and relational system of distances and measures contingent upon neither human relations nor the human body. Instead, this was a measure defined by a planetary reference: Earth's meridian.[14] Its survey became subject to turbulent political events, from reversals of the meter's acceptance and the suppression of the academies that directed the surveys to the execution of the king and war. During this period, surveyors traversing the entire French territory as well as crossing borders to neighboring countries worked under dangerous conditions, with the most minor example being arrest and confiscation of supplies, and the gravest being execution.[15] Once the survey was completed, the troubles continued with controversies due to mistakes in the triangulation that had not been made transparent. Méchain had incorrectly measured his chain of triangles between Barcelona and Perpignan due to the irregularity of Earth's curvature.[16] Despite the error, in June 1799 the government was presented with the official platinum meter, valued at 443.296 *lignes.*

The ground survey had been instigated by legislation presented in 1791 by Charles-Maurice de Talleyrand and Marie-Jean-Antoine-Nicolas de Caritat, marquis de Condorcet, for a national standard for measures, which was eventually adopted by the revolutionary government.[17] The change in units of measurements from various local systems to a single metric system had two goals: first, to facilitate the free exchange of goods by introducing a uniform means of comparing and defining them; and second, to enable the government to generate and collect accurate information about its resources. The adoption and use of a universal unit of measure held the ideological promise of equality. It presented a radically new logic that was based not on social hierarchies, particularities of physical labor, and temporal variations, but rather, as Condorcet and Talleyrand claimed, on the facilitation of circulation and exchange.[18] In his definition of surveying in the supplement to Denis Diderot and Jean le Rond d'Alembert's *Encyclopédie,* Condorcet understated the political and economic consequences of measuring: "Surveying is even more the art of recognizing, of sharing and of evaluating a field than of marking positions, of measuring and dividing. And it's in this civil and economic part where one finds some difficulties that can be solved easily in all cases with the help of the following principles."[19] Instead of relying on site-specific and personal relationships, exchange was based on the acceptance of shared principles. And as space became conceptualized as regular and fixed, objects and their movement through that neutral space was foregrounded.

The effect of this flattening of space was that the objects and their spatial contexts were separated, such that the immediate presence of the object—especially in its commodity form—was valued over the ways in which it could be tied to particular terrains, bodies, communities, and places.[20] In making possible exchange between distant places and strangers, the meter represented a new spatial order that denied local and particular meaning, a constitutive element that, according to Antoine Picon, defines the "modern terrain."[21] It did not matter if the context or terrain was city or countryside, inhabited or uninhabited, desert or forest, Egypt or Louisiana. The facility of communication and trade among strangers was more possible through this universal measure, which proposed a new sociopolitical body. Josef Konvitz has argued that this quality of the metric system encouraged a "fundamentally democratic" culture, allowing individuals to become less dependent on an elite that defined the terms of economic relations—an argument echoing Condorcet and Talleyrand.[22] Yet, as Witold Kula has suggested, every moment of metrological standardization has been associated with absolutism.[23] Napoléon Bonaparte's mission to Egypt was, for example, a means to extend France's interests as universal, and as described in the first chapter, the orientation of the maps demonstrated that Paris served as the cartographic reference for the survey of Egyptian cities. The standard measure was still set by elites. Moreover, the meter's definition and adoption gave cover for France's extractive and colonial interests in the name of equality and efficiency.

FIGURE 3.5 Title page from Edme Verniquet, *Atlas du plan général de la Ville de Paris* (Paris, 1795). Bibliothèque nationale de France.

With the uncoupling of measure and place, maps employing a universal measure and orthographic composition were concerned with neither a viewer nor its maker, but rather with how they related with other maps that shared the same reference. When used to determine a given area, the measured line permitted its representation to correspond proportionally to a metric grid that now marked the surface of Earth. Unlike the physicality of a single metal rod, this was a fixed but abstract definition that supported a belief in a universal measure, offering consistency at all scales, regardless of the place or user.[24] In this sense, by constructing a standard cartographic description around the meridian and a meter—both defined in relation to one another—the values of precision and accuracy merged, insofar as the correspondence of the representation to reality became equal to the representation's internal logic.[25]

Verniquet's *Atlas général de la Ville de Paris,* although contemporaneous with the determination and adoption of the meter, did not use the metric measure (fig. 3.5). Originally published in 1795, it employed the imperial measure of the toise for its triangulation calculations and

printed maps. These imperial measures, however, did not make the plans less exact. In fact, the ground survey guaranteed the map's accurate correspondence to the terrain. Nevertheless, the measure did make the map a unique and discrete object—one that could relate only to other plans that used the same measure in the same way.

While the metric system offered a conception of space as flat, there was still a question of making space as such. Verniquet had used existing towers to triangulate distances, and thus he had to compensate for the various heights from which he took his measures. Measuring surface area was one part of the surveying; the other was constructing a level plane. He devised an imaginary level plane, in which each reference tower acted like a leg holding up a flat topographic plane at the same given height. His surveying table, however, reveals none of these calculations; he only provides horizontal distances and no vertical elevations on the final printed map. Yet contours and topographic features appear along the margins of the complete map of Paris (figs. 3.6, 3.7). Hills and valleys are marked by hatchings to convey shape and slope. This naturalistic description pushes against the geometric one; the straight lines of the city expand outward into the countryside. The symbolic effect is of a city as a flat artifice produced through geometric calculations distinct from pictorial and even allegorical spaces inhabited by gods drawn in the cartouches.

Jacoubet's updated 1836 *Atlas* eliminated all allegorical features. Eventually, a grid colonized the entire paper surface and held the map together (fig. 3.8). It used a metric scale, which had been officially adopted in 1799, though before the meter was made compulsory in 1840.[26] The simplicity of its graphic composition elided the complicated and laborious process of its determination and adoption. The actual measures when surveyed diverged from the provisional values, creating the need for a process of verification that was hampered by revolution and war. These discrepancies, as well as those between the provisional metric bars produced first in brass and then in platinum, and the subsequent politics of its international adoption, made the metric system less constant than what its graphic definition denoted. Several publications provided the necessary conversions between the Parisian and the metric measures. In the *Instruction sur les mesures déduites de la grandeur de la terre, uniformes pour toute la République, et sur les calculs relatifs à leur division décimale* (Instruction on measures deduced from the size

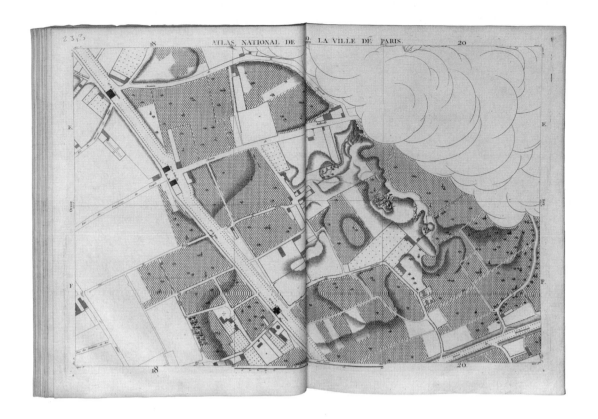

FIGURE 3.6 (top) Plate 23 from Edme Verniquet, *Atlas du plan général de la Ville de Paris* (Paris, 1795). Bibliothèque nationale de France.

FIGURE 3.7 (above) Plate 32 from Edme Verniquet, *Atlas du plan général de la Ville de Paris* (Paris, 1795). Bibliothèque nationale de France.

FOLLOWING SPREAD

FIGURE 3.8 *Plan d'assemblage* (overview). From Théodore Jacoubet, *Atlas général de la ville, des faubourgs et des monuments de Paris* (General atlas of the city, suburbs and monuments of Paris) (Paris, 1836). Bibliothèque historique de la Ville de Paris.

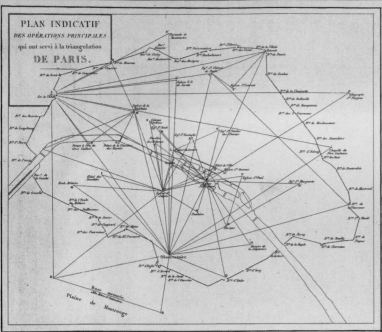

PLAN INDICATIF
DES OPÉRATIONS PRINCIPALES
qui ont servi à la triangulation
DE PARIS.

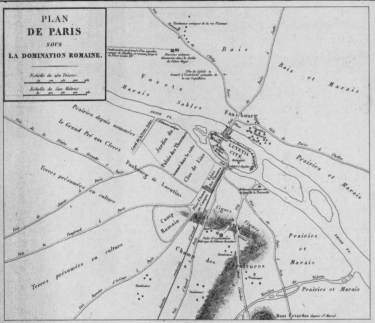

PLAN
DE PARIS
SOUS
LA DOMINATION ROMAINE.

Echelle de 250 Toises.

Echelle de 500 Mètres.

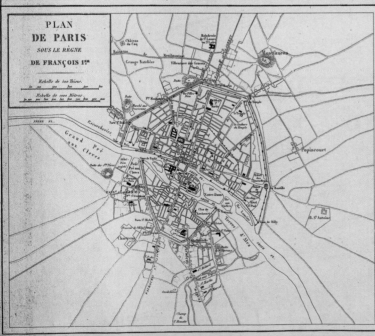

PLAN
DE PARIS
SOUS LE RÈGNE
DE FRANÇOIS Iᵉʳ.

Echelle de 500 Toises.

Echelle de 1000 Mètres.

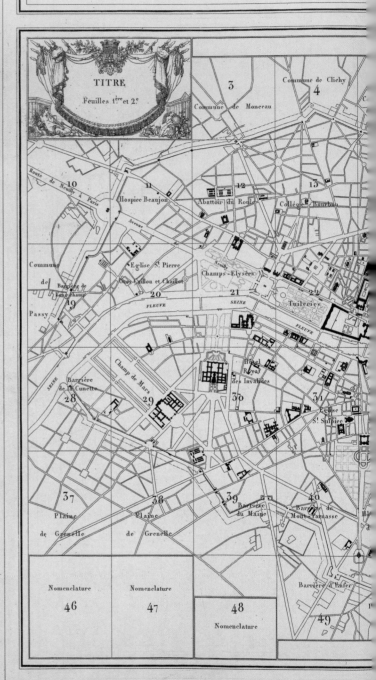

PLAN I
DE L'ATLAS GÉNÉRAL DE LA VILLE
LEVÉ
Rapporté et dessiné sur une éc
Par TH. J.

TITRE
Feuilles 1ᵉʳᵉ et 2ᵉ

Commune de Cliehy

Commune de Monceau

3

4

Route de Neuilly à Paris

10   11   12   13

Hospice Beaujon   Abattoir du Roule   Collège Bourbon

Eglise St Pierre   Champs-Elysées

Commune   Gros Caillou et Chaillot

de   Barrière de   20   21   22

Passy   Longchamp   19   Tuileries

SEINE   FLEUVE   SEINE   FLEUVE

Champ de Mars

Barrière   Royal
de la Cunette   des Invalides

28   29   30   31

Eglise
St Sulpice

37   38   39   40

Plaine   Plaine   Barrière   Barrière du
du Maine   Mont-Parnasse

de Grenelle   de Grenelle

Barrière d'Enfer

Nomenclature   Nomenclature

46   47   48
Nomenclature   49

1ᵉ LIVRAISON contenant les feuilles 1. 2. 3. 10. 11. 12.	2ᵉ LIVRAISON contenant les feuilles 4. 5. 6. 13. 14. 15.	3ᵉ LIVRAISON contenant les feuilles 7. 8. 9. 16. 17. 18.	4ᵉ LIVRAISON contenant les feuilles 19. 20. 21. 28. 29. 30.
NOTICE INDICATIVE pour chaque feuille.	NOTICE INDICATIVE pour chaque feuille.	NOTICE INDICATIVE pour chaque feuille.	NOTICE INDICATIVE pour chaque feuille.
1 Titre	4 Partie de la Commune de Cliehy et de ses environs.	7 Barrière de la Villette et ses environs.	19 Partie de la Commune de Passy, Barrière de Longchamp et ses environs.
2 Partie de la Commune de Monceau et de ses environs.	5 Partie de la Commune de Montmartre et ses environs.	8 Plan d'assemblage, Triangulation, Plan indicatif des nomemens et ses Plan de Paris à quatre époques différentes.	20 Eglise St Pierre Gros-Caillou et Chaillot et leurs environs.
10 Route de Neuilly à Paris et ses environs.	6 Partie de la Commune de la Chapelle et de ses environs.	9	21 Champs-Elysées et leurs environs.
11 Hospice Beaujon et ses environs.	13 Collège Bourbon et ses environs.	16 Hospice St Louis et ses environs.	28 Barrière de la Cunette et ses environs.
12 Abattoir du Roule et ses environs.	14 Académie Royale de Musique et ses environs.	17 Commune de Belleville et ses environs.	29 Champ de Mars et ses environs.
	15 Eglise St Laurent et ses environs.	18	30 Hôtel Rᵉ des Invalides et ses environs.

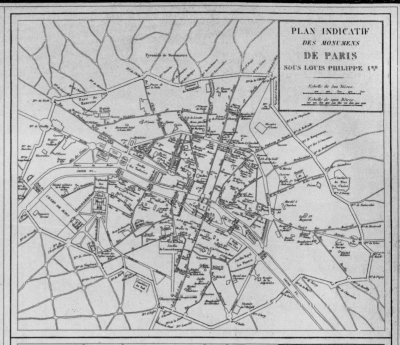

PLAN INDICATIF
DES MONUMENS
DE PARIS
SOUS LOUIS PHILIPPE 1ᵉʳ.

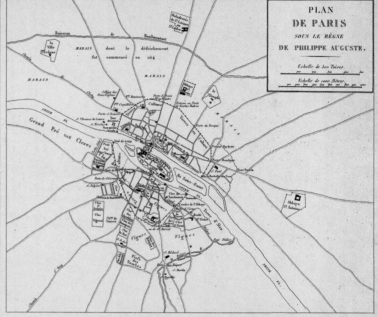

PLAN
DE PARIS
SOUS LE RÈGNE
DE PHILIPPE AUGUSTE.

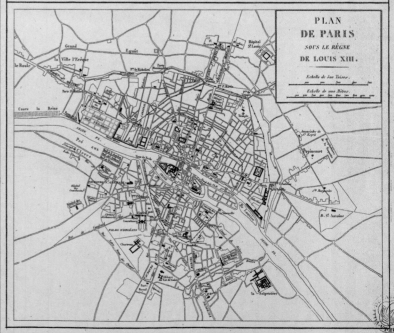

PLAN
DE PARIS
SOUS LE RÈGNE
DE LOUIS XIII.

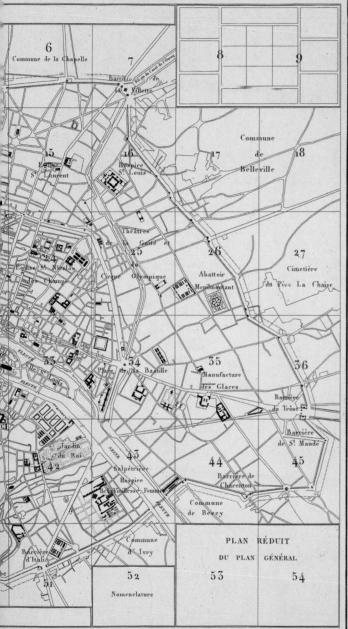

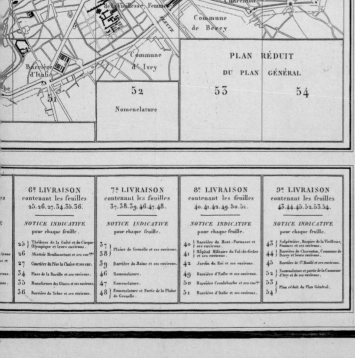

PLAN RÉDUIT
DU PLAN GÉNÉRAL

6ᵉ LIVRAISON contenant les feuilles 25.26.27.34.55.56.	7ᵉ LIVRAISON contenant les feuilles 37.38.39.46.47.48.	8ᵉ LIVRAISON contenant les feuilles 40.41.42.49.50.51.	9ᵉ LIVRAISON contenant les feuilles 43.44.45.52.53.54.
NOTICE INDICATIVE pour chaque feuille.	NOTICE INDICATIVE pour chaque feuille.	NOTICE INDICATIVE pour chaque feuille.	NOTICE INDICATIVE pour chaque feuille.

Écrit par Davy.

TABLE Ire. MESURES LINÉAIRES.

LIGNES.	MÈTRES.	TOISES.	MÈTRES.	TOISES.	MÈTRES.	AUNES DE PARIS.	MÈTRES.
1	0,0023	1	1,9484	1000	1948,3948	1/32	0,037127
2	0,0045	2	3,8968	2000	3896,7896	1/16	0,074253
3	0,0068	3	5,8452	3000	5845,1844	1/8	0,148507
4	0,0090	4	7,7936	4000	7793,5793	1/4	0,297014
5	0,0113	5	9,7420	5000	9741,9741	1/2	0,594027
6	0,0135	6	11,6904	6000	11690,3689	1/24	0,049503
7	0,0158	7	13,6388	7000	13638,7637	1/12	0,099005
8	0,0180	8	15,5872	8000	15587,1585	1/6	0,198009
9	0,0203	9	17,5356	9000	17535,5533	1/3	0,396018
10	0,0226	10	19,4839	10000	19483,9481	1	1,1881
11	0,0248	20	38,9679	20000	38967,8963	2	2,3761
Pouces.		30	58,4518	30000	58451,8444	3	3,5642
1	0,0271	40	77,9358	40000	77935,7926	4	4,7522
2	0,0541	50	97,4197	50000	97419,7407	5	5,9403
3	0,0812	60	116,9037	60000	116903,6889	6	7,1283
4	0,1082	70	136,3876	70000	136387,6370	7	8,3164
5	0,1353	80	155,8716	80000	155871,5852	8	9,5044
6	0,1624	90	175,3555	90000	175355,5333	9	10,6925
7	0,1894	100	194,8395	100000	194839,4815	10	11,8805
8	0,2165	200	389,6790	200000	389678,9630	20	23,7611
9	0,2435	300	584,5184	300000	584518,4445	30	35,6416
10	0,2706	400	779,3579	400000	779357,9260	40	47,5222
11	0,2977	500	974,1974	500000	974197,4075	50	59,4027
Pieds.		600	1169,0369	600000	1169036,8890	60	71,2833
1	0,3247	700	1363,8764	700000	1363876,3705	70	83,1638
2	0,6495	800	1558,7159	800000	1558715,8519	80	95,0444
3	0,9742	900	1753,5553	900000	1753555,3334	90	106,9249
4	1,2989			1000000	1948394,8149	100	118,8055
5	1,6237					1000	1188,0548
6	1,9484					10000	11880,5479

P 2

TABLE IV. MESURES DES SURFACES.

Toises-points.	MÈTRES CARRÉS	*Toises-pouc.*	MÈTRES CARRÉS.	TOISES CARRÉES.	MÈTRES CARRÉS	PIEDS CARRÉS.	MÈTRES CARRÉS.
1	0,000366	7	0,369079	100	379,6242	1	0,105451
2	0,000732	8	0,421805	200	759,2485	2	0,210902
3	0,001098	9	0,474530	300	1138,8727	3	0,316354
4	0,001465	10	0,527256	400	1518,4969	6	0,632707
5	0,001831	11	0,579981	500	1898,1212	12	1,265414
6	0,002197	*Toises-pieds*		600	2277,7454	18	1,898121
7	0,002563	1	0,632707	700	2657,3696	24	2,530828
8	0,002929	2	1,265414	800	3036,9939	30	3,163535
9	0,003295	3	1,898121	900	3416,6181	36	3,796242
10	0,003661	4	2,530828	1000	3796,2424	*Pouces carrés.*	
11	0,004028	5	3,163535	2000	7592,4847	1	0,000732
Toises lignes.		*Toises-carr.*		3000	11388,7271	2	0,001465
1	0,004394	1	3,796242	4000	15184,9694	3	0,002197
2	0,008788	2	7,592485	5000	18981,2118	6	0,004394
3	0,013181	3	11,388727	6000	22777,4541	12	0,008788
4	0,017575	4	15,184969	7000	26573,6965	18	0,013181
5	0,021969	5	18,981212	8000	30369,9388	36	0,026363
6	0,026363	6	22,777454	9000	34166,1812	72	0,052726
7	0,030757	7	26,573696	10000	37962,4235	144	0,105451
8	0,035150	8	30,369939	20000	75924,8471	*Lignes carrées.*	
9	0,039544	9	34,166181	30000	113887,2706	1	0,000005
10	0,043938	10	37,962424	40000	151849,6942	2	0,000010
11	0,048332	20	75,924847	50000	189812,1177	3	0,000015
Toises-pouc.		30	113,887271	60000	227774,5413	6	0,000031
1	0,052726	40	151,849694	70000	265736,9648	12	0,000061
2	0,105451	50	189,812118	80000	303699,3884	18	0,000092
3	0,158177	60	227,774541	90000	341661,8119	36	0,000183
4	0,210902	70	265,736965	100000	379624,2355	72	0,000366
5	0,263628	80	303,699388	1000000	3796242,3549	144	0,000732
6	0,316354	90	341,661812				

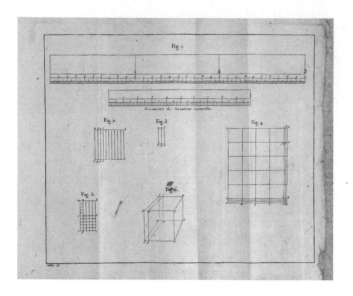

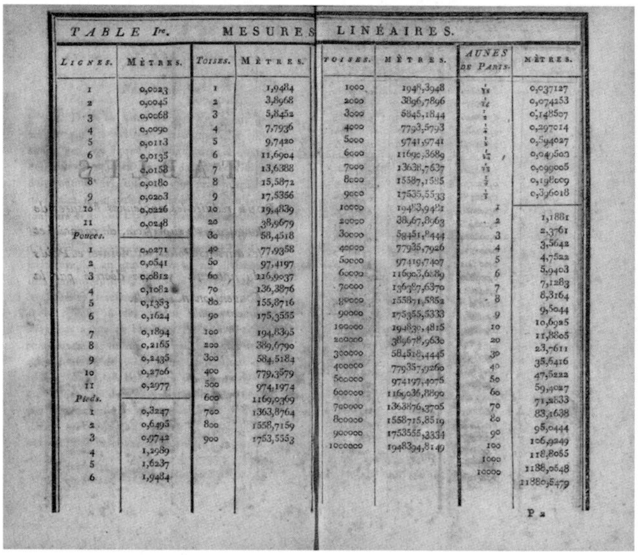

FIGURES 3.9, 3.10, AND 3.11 From René Just-Haüy, *Instruction sur les mesures déduites de la grandeur de la terre, uniformes pour toute la République, et sur les calculs relatifs à leur division décimale* (Instruction on measures deduced from the size of the earth, uniform for the entire Republic, and on calculations relative to the decimal point) (Paris, 1794). ETH-Bibliothek Zürich, Rar 4083.

FIGURE 3.9 Figures detailing grid compositions and providing mensuration references. Plate I.

FIGURE 3.10 Conversion charts for linear measures. Table I.

FIGURE 3.11 Conversion charts for surface area. Table IV.

of the earth, uniform for the entire Republic, and on calculations relative to the decimal point), simple diagrams of a ruler stood in stark contrast to the detailed textual descriptions and the conversion tables that provided metric conversions (figs. 3.9–3.11).[27] The procedures to make the toise conform to the new standards meant first identifying the measured object as linear, a solid, a surface, or a capacity, then as possessing one, two, or three dimensions, and finally by converting between fractions and decimals and translating between terminologies, before drawing lines.

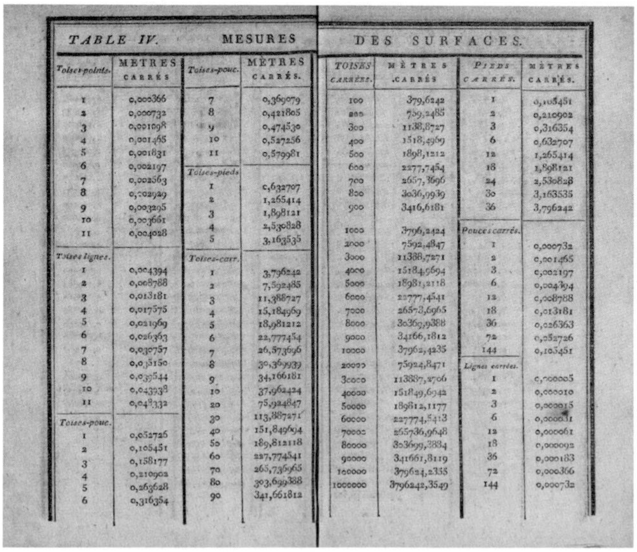

To recuperate the high costs of its engraving and printing, Jacoubet's *Atlas* was offered for sale to the public.[28] A subscription beginning in 1836 cost eighteen francs for each of its nine six-sheet installments, or one could purchase the whole thing for 163 francs unbound and 182 francs bound. Significantly, the ability to buy the *Atlas* in parts meant that a map was no longer produced as a singular object but rather conceived as a composite formed by numerous interlocking sheets.[29] In order for sheets to connect, shared references were necessary. The reference used was the Observatoire de Paris, whose ground plan was drawn at the corners of two different plates that were to be lined up (figs. 3.12a–b). The separate sheets then related to each other through a coordinate grid that remained consistent across the surface of the map and became a central landmark on this cartographic terrain.

Not only religious, royal, or public monuments but also graticules composed by orthogonal lines became reference points that shared the metric measure. Lines appeared within a common context, and their regularity and consistency through a geometric network unified the

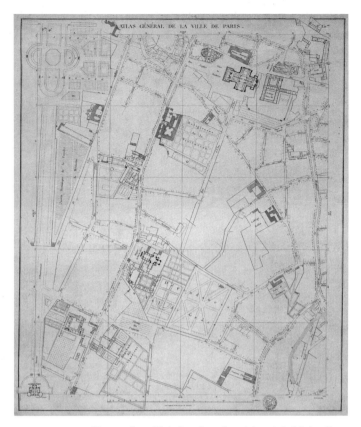

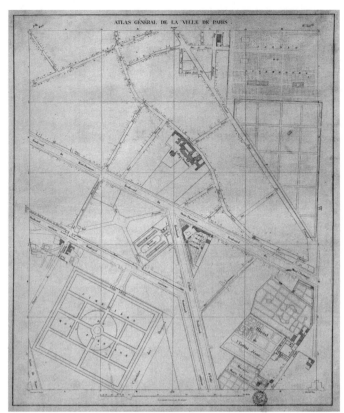

FIGURE 3.12A Plate 40 from Théodore Jacoubet, *Atlas général de la ville, des faubourgs et des monuments de Paris* (General atlas of the city, suburbs and monuments of Paris) (Paris, 1836). Bibliothèque historique de la Ville de Paris.

FIGURE 3.12B Plate 41 from Théodore Jacoubet, *Atlas général de la ville, des faubourgs et des monuments de Paris* (General atlas of the city, suburbs and monuments of Paris) (Paris, 1836). Bibliothèque historique de la Ville de Paris.

whole composition. Each line was simultaneously a part of a grid and its own coherent whole, and when combined through measure and mode found cartographic unity. The dominance of the line emptied the image of any details that could not be captured by its graphic limitations. If earlier maps had attempted to recreate the experience of physical movement through depth, volumes, and masses, Jacoubet's orthographic maps privileged movement across each sheet, whose divisions did not correspond to *quartiers* but to the dimensions of the printing plate and the paper. Its blank spaces replaced particular places and disguised the multiple cartographic sources used to construct the map. Moreover, the omission of certain kinds of representational details, such as materials, elevations, terrain, or vegetation, became the basis of the map's graphic universality.[30] However, this universality was not merely a result of a graphic modality. The cartographic grid of the map corresponded to a coordinate grid of the metric system that defined the surface of Earth; the measured lines of the plan linked to the measured lines of all distances,

abstracted within a standardized, homogenous, and flattened conception of space.

GRIDDED DRAWINGS

At the same time that the metric system merged with the map's coordinate grid, architecture in France was facing a crucial turning point in which the orthographic grid featured significantly in debates about the nature of design. By the end of the eighteenth century, classical traditions based on Vitruvius's formulations had been exhausted, yielding few solutions to the problems that architects now faced with respect to new political and social exigencies. As a basis for architectural composition, the strict ideals of the classical orders and their proportions seemed unadaptable to the changing tastes and forms of patronage that required new typologies, as well as unjustifiable in light of new evidence of the actual diversity of ancient buildings.

With this crisis of traditional models, new values such as utility and structure emerged that created fissures

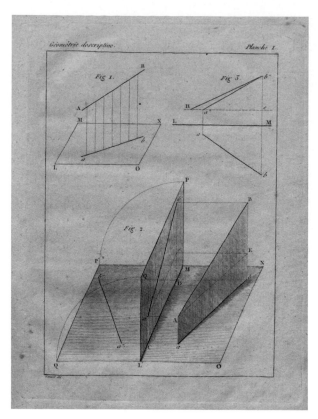

FIGURE 3.13 Plate 1 from Gaspard Monge, *Géométrie descriptive* (Paris, 1811). ETH-Bibliothek Zürich, Rar 21855.

between increasingly professionalized architects and engineers.[31] Aligning with the rise of empiricism, architecture oriented its practice to design, and the image gained value in architectural production. As Étienne-Louis Boullée evoked, "Our earliest forefathers did not build their own huts until they had first conceived the image of them. That production of the mind, that creation, constitutes architecture."[32] Meaning was situated in drawing, where composition followed analytical methods laid out by René Descartes and Étienne Bonnot de Condillac, among others. Diderot visually articulated this method most explicitly in the *Encyclopédie,* whose plates demonstrated an emphasis on analysis through visual decomposition and comparison.[33] Objects were represented in their composite parts, taken apart, and then put back together, often in the context of their use. They were drawn to scale and aligned on the sheet to create a gridded order to the plates and ultimately to the entire project.

The École polytechnique founded by Gaspard Monge in 1794 exemplified a pedagogy born out of these Enlightenment-era cultural and political values. The school's curriculum placed an emphasis on science,

technology, and practice, and made descriptive geometry a central subject. Formulated from projections used for stonecutting, descriptive geometry is a method to describe three-dimensional objects on a two-dimensional surface.[34] It is based on references, made along parallel lines, that consist of intersecting vertical and horizontal planes to form a ground line (fig. 3.13). By moving the object around this line, multiple facets of a solid could be represented on paper. One of the purposes of descriptive geometry is to define the position of a point in space in reference to objects of known position. It assumes a concept of space as relational and abstract. As Monge would explain, "Space is without limits; all parts are perfectly equal, they have nothing distinctive and none of them can serve as a comparison to indicate the position of a point."[35] Moreover, its historical development concerned architectural projections that were no longer based on complex numerical calculations for the construction of fortifications, but instead based on compositions through drawing. According to Alberto Pérez-Gómez, the method applied a "mechanism of transformation" from algebra to geometry that realized a "principle of continuity" between two and three dimensions, and "allowed for the first time a systematic reduction of three-dimensional objects to two dimensions."[36] Descriptive geometry was a means to generate and make visible an architectural object through lines, plans, and elevations on a paper surface. Explaining the relevance of descriptive geometry in cartography, Edme-François Jomard stated, "The products of geography depend uniquely on the exact sciences: maps are above all else a mathematical projection of the globe or its parts, an application of descriptive geometry."[37] In his account of the need for a distinct archive for cartographic materials, Jomard emphasized maps as exemplars of scientific practices such as descriptive geometry and geometric surveying.

As Monge's contemporary, Jean-Nicolas-Louis Durand was a strong advocate and proponent of descriptive geometry, and beginning in 1795, he integrated this geometric method into his lectures on design production, making the bridge between geometrical diagrams to architecture.[38] His *Précis des leçons d'architecture* (Summary of architecture lessons) (1821) outlined a mechanism of composition on a grid, eventually codifying and standardizing orthogonal drawings on the newly introduced medium of graph paper.[39] Instructions from a *Cahier classique sur le cours de construction à l'usage des élèves de*

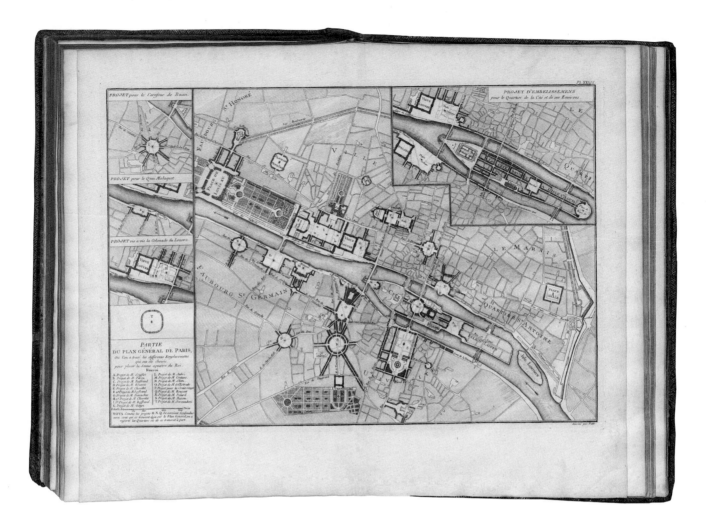

FIGURE 3.14 Pierre Patte, *Partie du plan générale de Paris où l'on a tracé les différents emplacemens qui ont été choisis pour placer la statue équestre du roi* (Part of the general plan of Paris where different locations for the equestrian statue of the king are drawn), 1765. ETH-Bibliothek Zürich, Rar 9876.

l'École royale de l'artillerie et du génie (Classic workbook on the course of construction for the students of the Royal School of Artillery and Engineering) (1819) read: "Gridded paper, which is used at the École polytechnique in the architecture course, facilitates grouping and sketching the arrangements of masses in a building project. We consider each square as forming an axis or one of its multiples."[40] In this methodology, the ground plan was disconnected from the elevation and, under Durand's system, became the privileged instrument for architectural composition. As he explained, "For the elevation, it is composed through the plan and the section."[41] The grid became a central reference and its intersecting lines organized both drawing and space.

Here was an architectural method that was in opposition to the architectural values of *embellissement* (improvement) promoted by Voltaire and Marc-Antoine Laugier, as well as François Blondel and Pierre Patte. Embellissement was a notion that appeared around the sixteenth century to compose space around a monumental edifice; it was a triumphal approach symbolizing royal authority. By the

eighteenth century, it became aligned with a syncretic idea of improving the circulation and moral health of a city and its inhabitants through architectural compositions. Under these earlier values, the elevation had priority, in which the character of an edifice and visual elements of its façade could be legible. Patte's 1765 *Monumens érigés en France à la gloire de Louis XV* is exemplary. It begins with the *Partie du plan générale de Paris où l'on a tracé les différents emplacemens qui ont été choisis pour placer la statue équestre du roi* (Part of the general plan of Paris where different locations for the equestrian statue of the king are drawn), an orthographic map that provides a single-image overview of all the possible locations for the new monument to Louis XV (fig. 3.14).[42] Nineteen different proposals for the 1748 competition for a *place* honoring this equestrian statue were drawn into a single plan, demonstrating the capacity for the orthogonal mode to depict different possibilities occupying the same space. Each architect's proposal is indicated by a letter, and the map situates the proposals' urban contexts. Then, in each chapter, a project is specifically discussed with a detailed elevation, delineating the monumental and balanced values of Beaux-Arts neoclassicism. Every project had its own discrete proposal, as in an example from the Pont Neuf (fig. 3.15). Thus, while Patte's plan presents an early example of an orthographic description of the city, providing an opportunity to compare and imagine the possibilities for different urban arrangements, it did not propose a systematic spatial understanding of the city, but was rather episodic, and spaces remained discrete.[43] There was no conception of a coherent and gridded space.

Parallel to Patte's study was the comparative *Plan général du projet des embellissements de Paris* (General plan of the improvement projects of Paris) by Charles de Wailly,

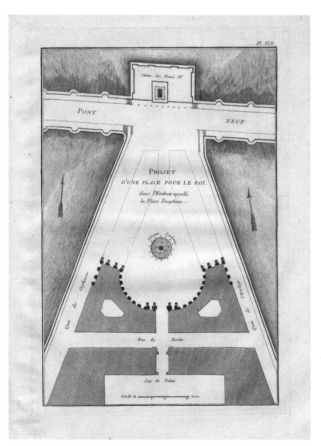

who also employed an orthographic format (fig. 3.16).[44] In 1777 Wailly had drawn a plan to restructure the neighborhoods on the Left Bank in specific relation to the Odéon. Two other plans of the areas around the École de médecine de Gondoin and the Jardin du Luxembourg succeeded it. These were distributed as engravings between 1778 and 1782 and exhibited at the Salon of 1789. The *Plan général* presented an overview, but it fell short of realizing a practice of holistic planning. Ultimately, de Wailly simply used his *Plan général* to present ideas for different quarters together and at once. The proposals were marked in red and were mainly located along the Seine and the Left Bank, in what was a finished image describing some potential embellissements based on monuments in the city. Patte's and de Wailly's images were both orthogonal, but neither orthogonal plans nor grids were necessarily used to draw on them. The orthogonality was a method to situate different discrete designs on a single image of the city. The projections were not based on the map as a terrain, but on the principles of embellissement. The map was then a descriptive tool, not yet a projective one.

It was precisely the value of projection that Durand's architectural system afforded to the grid (figs. 3.17–3.19). Under his system, the primary instrument for spatial composition was the ground plan. Moreover, as elevation became disconnected from plan in architectural composition, the practice of composing space was defined through (and limited to) a grid. The corollary effect was a shift in the representational value of architecture from what was visible in the expression of a façade to its invisible qualities in plan. Moreover, the printed grid, paired with the orthographic view, acted as a reference for projective composition. The layers of the descriptive map and the drawn projections could be graphically integrated, and could communicate with each other on that same printed surface via the grid's mediation. Concretely, a ruler could be laid down flat on the sheet of paper and aligned to the grid, and then a line could be drawn that could relate to both description and projection.

The members of the Commission des artistes did lay a ruler on Verniquet's plan. Antoine-Jean Amelot, the administrator of the Domaines nationaux, had the idea of assembling experts and men of the arts to study the value of the properties acquired by the state during the revolution. The Commission des artistes was made up of eleven members: the four general inspectors of the roads (Verniquet, Charles-François Callet, Pierre

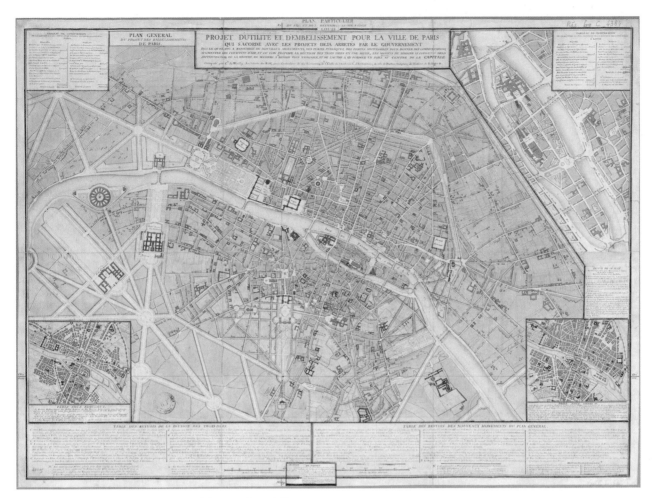

FIGURE 3.16 Charles de Wailly, *Plan général du projet des embellissements de Paris* (General plan of the improvement projects of Paris), 1785. Bibliothèque nationale de France.

Garrez, and George Galimard), and seven other architects (Louis-François Petit-Radel, M. Cambault, Nicolas Lenoir le Romain, Guy-Laurent Mouchelet, Jean-Baptiste Pasquiet, Marie-Jean Chabouillé, and Charles de Wailly).[45] Beginning in April 1794, they met irregularly at the Hôtel d'Uzès on the rue Montmartre, working from the printed sheets of Verniquet's engraved plans to develop a comprehensive plan of Paris to divide, sell, and incorporate properties within a new revolutionary order.[46]

The pressure to develop an accurate plan of the city led the commission to rely heavily on Verniquet's work, and in 1796, upon the demands of the Conseil des bâtiments civils and the Conseil général des ponts et chaussées, plans for the city were drawn. One of the commission's concerns focused on aligning the street network to the metric system, which was newly adopted.[47] These regulations and their drawn plans to make the city

conform uniformly to these spatial proportions were the main merits of the commission in its recommendations to the council.[48] However, it was clear from their reconstituted drawings that the reorganization of the capital city ideologically supported the principle of the new republic's unity—an ideological commitment taken up a century later by the municipal administration of the Third Republic.

The plan, however, remained unfulfilled; in 1797 the commission was dissolved, and the original plan was lost.[49] The plan available for current historical study is a reconstruction that was ordered by Adolphe Alphand for the 1878 Exposition universelle and for the *Atlas des travaux de Paris, 1789–1889* (Atlas of the works of Paris, 1789–1889), published on the occasion of the centennial celebrations of the French Revolution. This nineteenth-century version of the *Plan de la Commission des artistes*

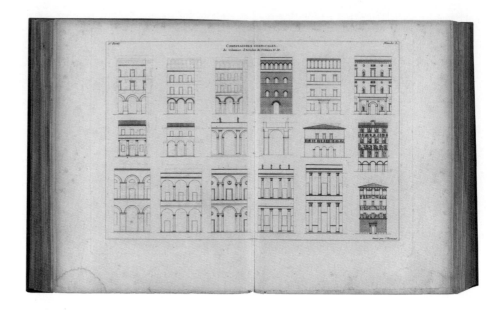

FIGURE 3.17 Part 2, plate 3, from Jean-Nicolas-Louis Durand, *Précis des leçons d'architecture données à l'École polytechnique* (Summary of the architectural lessons delivered at the École polytechnique) (Paris, 1802–5). ETH-Bibliothek Zürich, Rar 6766.

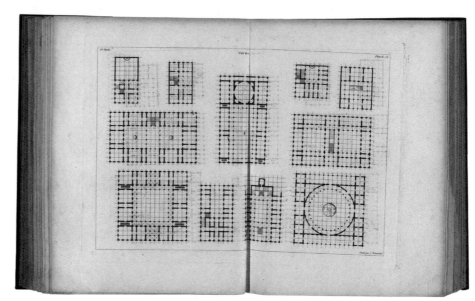

FIGURE 3.18 Part 2, plate 16, from Jean-Nicolas-Louis Durand, *Précis des leçons d'architecture données à l'École polytechnique* (Summary of the architectural lessons delivered at the École polytechnique) (Paris, 1802–5). ETH-Bibliothek Zürich, Rar 6766.

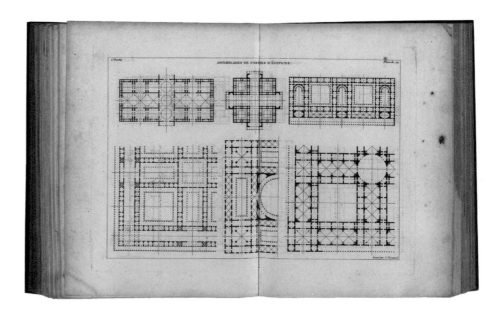

FIGURE 3.19 Part 2, plate 22, from Jean-Nicolas-Louis Durand, *Précis des leçons d'architecture données à l'École polytechnique* (Summary of the architectural lessons delivered at the École polytechnique) (Paris, 1802–5). ETH-Bibliothek Zürich, Rar 6766.

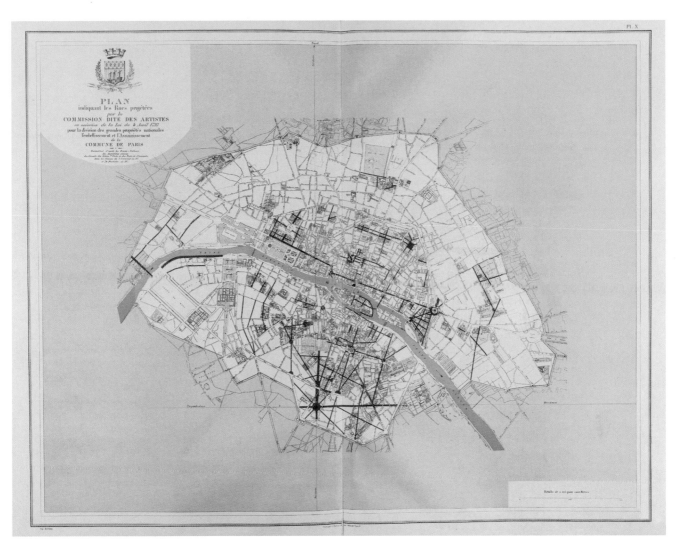

FIGURE 3.20 *Plan de la Commission des artistes,* showing the meridian and perpendicular lines intersecting at the Observatoire de Paris. From Adolphe Alphand, *Atlas des travaux de Paris, 1789–1889* (Atlas of the works of Paris, 1789–1889) (Paris, 1889). Bibliothèque historique de la Ville de Paris.

was compiled through archival reconstructions overseen by an anticlerical historian and writer, Léo Taxil, whose historical writings became popular during the Third Republic.[50] Therefore, it is important to regard this plan used to read late eighteenth-century planning concepts as a mediated view that passes through the eyes and hands of early Third Republic French administrators, who were keen to justify their own initiatives.

Tellingly, the reorganization of Paris by the commission, based on the Third Republic's reconstruction, focused on two symbolically significant areas: the site of the Place de la Bastille, symbolic of the French Revolution, and the Observatoire de Paris, the central landmark of the academic community (fig. 3.20). The Bastille is planned on

axis with Claude Perrault's façade of the Louvre palace; the observatory is situated on axis with the Luxembourg palace; and there is no communication between these two projects. Although this plan presented an orthogonal view of the whole of Paris on a single sheet, the planning still employed the logic of the quartier, presenting an idea of circulation that was nonetheless based on local embellissements.[51] The city was broken down into parts, new streets and blocks were aligned and widened, and their redirection created the space for the monuments upon which they were based. However, unlike Patte's and de Wailly's maps, the designs demonstrate the grid as a projective reference.

The red line drawn from the Place de la Bastille to the

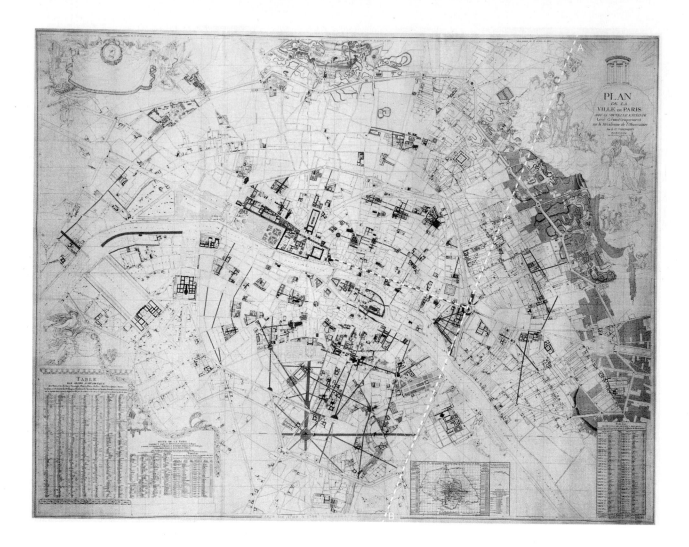

Louvre bisects the façade perfectly. White dashed lines drawn by this author emphasize the angles' intersections (fig. 3.21). Cutting through a series of blocks, including the immediate area around the church of Saint-Germain de l'Auxerrois, the orientation of the palace is shifted to anchor an east–west axis. New streets radiate from the circular place that was to be constructed on one of the corners of the former ramparts of Paris into a hierarchical organization. The angles of those radiating streets seem to be determined by the symbolism of Republican France, especially the wide street extending to the north of the proposed place, where the angle intersects with the cartographic grid at the top-right corner (A) and the bottom corner of the triangulation table (B). Additionally, the angle of the Louvre line is perpendicular to the northern red line, suggesting that the cartographic grid does more than serve as a tool to locate the monuments: the grid stabilizes a monument's proposed orientation.

The conflation of the grid and the design for the city is more evident around the observatory, colored in black and encircled in red with proposed radial avenues. The building also sits at the center of the cartographic grid, and the proposed horizontal avenue follows the Paris meridian (fig. 3.22). The angles of the four red secondary radials, narrower in width to the cardinal ones, are determined by lines connecting the far corners of the four square units to the center point of the observatory. The interventions follow Durand's functionalist approach, based on the duplication and division of gridded forms (fig. 3.23). These lines bisect the squares diagonally to form triangles along defined axes, creating new spatial relations around the building. Verniquet's survey, the metric grid, and the drawn projections all met on the map at the observatory's point; the orthogonal mode was what allowed for these different layers to cohere onto a single surface and communicate with each other through a grid.

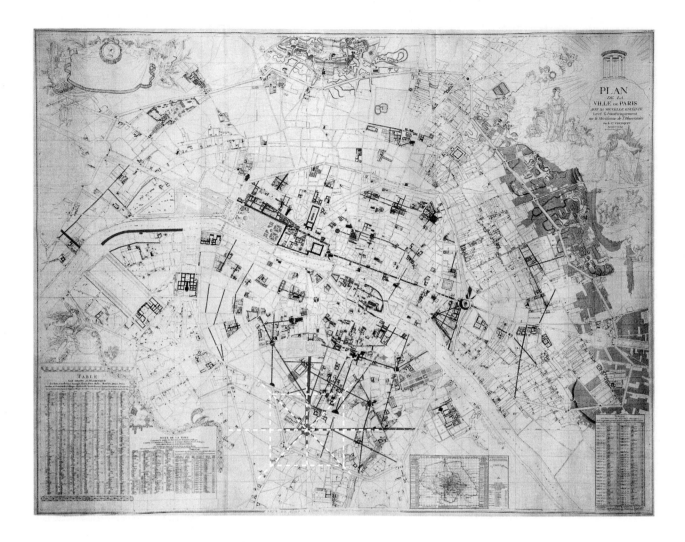

The observatory carried symbolic and instrumental significance as the anchor of a monumental reorganization of the city. Both astronomic and land surveys had begun at the observatory, a building that would symbolize the center of the scientific world and France's universalizing project. Its location came to denote the central point of the new metric system both in the world and on the map, and it would anchor the proportions and units of the grid. Accordingly, when Verniquet conducted his survey of the grand plan, he fixated on the meridian line and its perpendicular, using it as a reference for all of his triangulations.[52] His survey of the terrain, the metric grid produced by the Académie royale, and the drawn projections for Paris by the Commission des artistes all met on the map at the building's point. The orthogonal mode functioned as a common denominator between all these different layers, permitting the cartographic image to serve as an urban projection for engineers, architects, and administrators.

FIGURES 3.21 AND 3.22 From Adolphe Alphand, *Atlas des travaux de Paris, 1789–1889* (Atlas of the works of Paris, 1789–1889) (Paris, 1889). Bibliothèque historique de la Ville de Paris.

FIGURE 3.21 (opposite) Reconstructed map indicating the projected roads in red on a plan of Paris by Verniquet. White lines added by the author to emphasize the intersections between the Bastille and the Louvre.

FIGURE 3.22 (above) Reconstructed map indicating the projected roads in red on a plan of Paris by Verniquet. White lines added by the author to emphasize the intersections between the Observatoire de Paris and the cartographic grid.

FOLLOWING SPREAD

FIGURE 3.23 Part 3, plate 3, from Jean-Nicolas-Louis Durand, *Précis des leçons d'architecture données à l'École polytechnique* (Summary of the architectural lessons delivered at the École polytechnique) (Paris, 1802–5). ETH-Bibliothek Zürich, Rar 6766.

A . Grande Avenue.
B . Lac artificiel.
C . Pont triomphal.
D . Rochers d'ou sortiroient des torrens.
E . Parterres ornés de Fontaines.
F . Bosquets bas.
G . Terrasses.
H . Palais

I. *Iles*.

K. *Palais des Ministres*.

L. *Tête de la grande Cascade*.

M. *Hippodromes*.

N. *Canaux*.

O. *Menagerie et Haras*.

P. *Cirque et Naumachie*.

Q. *Bosquets hauts*.

Because there was no new survey of Paris when Napoléon III claimed the short-lived Second Republic in 1848, or when the Haussmann took office in 1853 under the Second Empire, administrators continued to use extant maps, of which there were many. The Commission des embellissements, headed by Henri Siméon and convened by Napoléon III before Haussmann took office, used Jacoubet's plates to draw their plans for Paris's reorganization. In fact, while the revolution and coup of 1848 and 1851 signaled dramatic political changes, there was remarkable continuity in the administrative practices of these official maps during this period. Jacoubet's plates were used until the annexation of the suburbs required Eugène Deschamps to draw up a new plan, and Jacoubet's were the sheets on which Haussmann's engineers drew their urban interventions. More than simply an issue of funding, the lack of a new survey also indicated the government's faith in the materials of the administration. With graduates of the École polytechnique and École des ponts et chaussées installed in most of the positions related to urban governance and administration throughout France, there was an entrenched institutional culture around these plans that persisted in the background of the major political events of the time.[53]

An important but not unconventional example of a French functionary is Chabrol de Volvic, who moved up the ranks in various outposts until being named prefect of the Seine in 1812. He remained in that position until 1830, employed through the Bourbon Restoration until the July Monarchy. As a graduate of the École polytechnique, he had attended Durand's lectures and synthesized Durand's architectural procedures into his own conceptions of urban planning. As sub-prefect of Pontivy, Chabrol de Volvic had drawn, for example, a plan for the new city of Napoléonville with a projected population of six thousand.[54] As Nicholas Papayanis has written, Chabrol de Volvic took the opportunity to organize the hospital, court, prison, school, church, and civic buildings on a grid composed of twenty rectangular units.[55] While earlier plans by Jean-Baptiste Pichot, chief engineer of the Morbihan region, sought to regulate both public and private buildings and spaces, Chabrol de Volvic's approved plan provided only a gridded outline, a framework of how different urban functions could communicate and relate to each other, with large parts left empty for private development. His plans signal how Durand's composition

system had been expanded to an urban scale. As he explained in a small text accompanying the plans, "We traced a light sketch of the proposals to show where they could be placed, but especially the way in which they could be established in each block."[56] Emphasizing emplacement and location, the grid became a structure to build in the functions of a city and, as Georges Teyssot and Paolo Morachiello argue, the plan demonstrated a "territorialization" of the nation by the central administration of the state. As a new military town, Napoléonville served as a gridded model for imperialist ambitions in Algeria and Indochina that sought to impose regularity—with little success—onto what was deemed open soil.[57]

If the grid had served to propose new French towns, by the time Jacoubet's *Atlas* was printed, a cartographic grid was a standard feature on urban plans, and functioned as more than a means to maintain an image's proportions and measures. As Chabrol de Volvic's commentary signals, grids became a base to situate, anchor, and generate new spaces in Paris. For the Commission des artistes, it served to mold spaces around monuments in the existing built environment. Under Haussmann's direction, the cartographic grid became an active agent in its own right, a sharp instrument to make the percements projecting new urban spaces through the city, not by law but in the engineering culture of the administration. Like any tool, the grid conditioned a practice and the results of a process. Long before urban planning was even considered a distinct discipline, a cartographic grid was outlining its norms by determining the relations between existing and proposed spaces, between street and block, between quartier and city. The grid ultimately pushed the needle of design away from an emphasis on embellissement and toward circulation.[58] Two examples—Les Halles Centrales and the Opéra—demonstrate how that instrument was wielded, and how the grid became integral to generating what would become understood as modern urban spaces.

LES HALLES CENTRALES

Contrary to the myth of Haussmann, there were several major urban interventions already in progress before his appointment as prefect that were precipitated by increasing urban congestion and the uneven displacement of the urban population.[59] One of the areas continually targeted for reorganization was the district of Les Halles Centrales, whose reorganization had been proposed as early as 1811

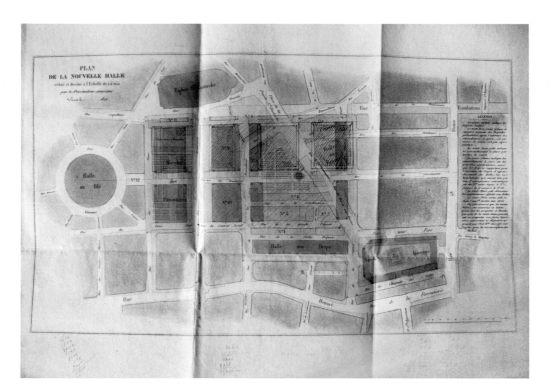

FIGURE 3.24 Préfecture du Département de la Seine, Commission des Halles, *Document à étudier, no. 2. Plan de la Nouvelle Halle*, 1842. Archives nationales de France, papiers Baltard.

under Napoléon I.[60] By the time a royal ordinance was declared on January 17, 1847, the call for a new market building had become more than just an architectural scheme: it presupposed the restructuring of the entire quartier, whose planning reflected an increasing differentiation of functions within the whole city.[61]

By the 1840s, the major question posed by Les Halles Centrales concerned the definition of the center of Paris. For the municipal councilors, the location of the center was framed cartographically: "When one takes a plan of Paris, one sees that the center of the city is close to the Pont-Neuf, and the center of the population at the Place des Victoires. Les Halles now approximately occupies the middle point between these two centers: [Les Halles] are therefore in the right position where they should be if they were to be created anew."[62] This geographical—and cartographic—assessment presumed that to locate a center was to see the city as a coherent object, distinct from its hinterlands and beyond. However, the municipal administrators also understood that this definition was at odds with another determined by the number of inhabitants, which located the population to the west of the Pont-Neuf. How to reconcile these two "centers," then?

The aim of the reorganization of Les Halles was to resolve these issues by using architecture to address the increasing displacement of the population to the northwest of the city, and to valorize the geographic center of Paris. An 1845 report from the Conseil municipal indicates that a municipal architect by the name of Lahure was asked to determine the spatial perimeters of this project.[63] He drew up plans and was directed to coordinate with other officials on the issues of property, distribution, and surveillance. The prefect of the police was responsible for indicating in plan the provisioning of the market, and the architect, Victor Baltard, was responsible for delineating the preliminary plans for the market building as well as the costs of its construction.

Lahure compiled all of the information into a *plan d'ensemble* for the session of the Conseil municipal on July 19, 1844. Edme-Jean-Louis Grillon served as a member of the council and argued with Jacoubet for the necessity of a total view of Paris. The boundaries of the immediate area precisely followed the boundaries represented in an earlier report from June 4, 1842, produced by the special Commission des Halles under the Préfecture de la Seine (fig. 3.24). The perimeter of the area was defined to the east by an extension of the rue de la Lingerie; to the west, the rue du Four-Saint-Honoré; to the south, a line formed by the rues des Deux-Écus, du Contrat-Social, and de la Petite-Friperie; and to the north, an extension of the rue Coquillière and the rue de Rambuteau toward Saint-Eustache.[64] The space between the two centers was to be

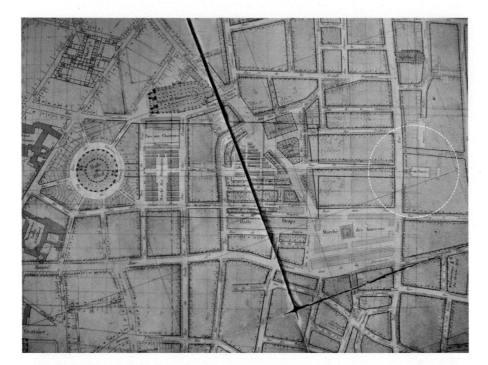

FIGURES 3.25, 3.26, AND 3.27 Composite of plates 23 and 32 from Théodore Jacoubet, *Atlas général de la ville, des faubourgs et des monuments de Paris* (General atlas of the city, suburbs, and monuments of Paris) (Paris, 1827–39). Archives nationales de France, papiers Baltard.

FIGURE 3.25 Detail showing the Cour Batave circled in a white dashed line.

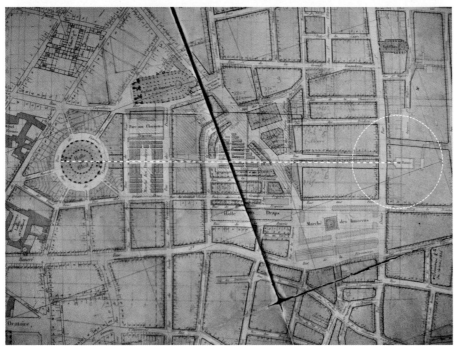

FIGURE 3.26 Detail with a white dashed line demonstrating the alignment between the Cour Batave and the Halle au Blé.

bridged by a straight line from the center point of the circular Halle au Blé to a point on the rue de la Cossonnerie, indicated on the plan.[65] While not shown on the Conseil municipal plan, the endpoint on the rue de la Cossonnerie corresponded to arches that led into the Cour Batave (fig. 3.25). It is unclear why the Cour Batave was not specifically named, but the arches that led from the rue Saint-Denis into the small court would have been a clear commercial anchor, indicating a link between private and public commerce. Batave was a privately developed new *cité* situated close to Les Halle aux Draps, constructed in 1795 during the revolutionary years to replace the church of Saint-Sépulcre with textile merchants from Holland. It was eventually demolished in 1858 as part of the construction of the boulevard de Sébastopol and the execution of the rue de la Cossonnerie's extension.

This court is, however, indicated on Jacoubet's plates used for the proposals for the new market by Victor

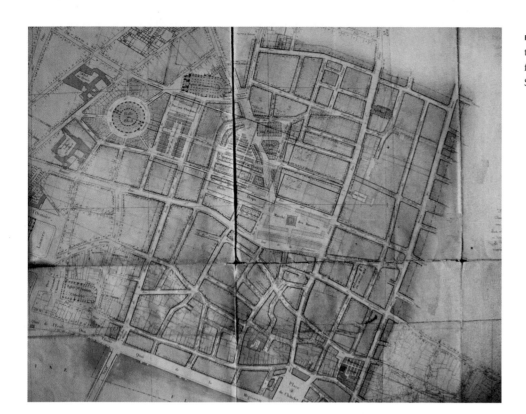

FIGURE 3.27 The two plates pasted together to show the entire area from Les Halles Centrales to the Seine.

Baltard, and it serves as the main axis for the quartier's development (fig. 3.26).[66] For the redevelopment of the quartier of Les Halles, Baltard's plans date to 1848 and are taken from Jacoubet's plates 23 and 32, visible through the maps' misnumbered graticules (fig. 3.27). The plates were pasted together to form one composite image of the immediate street network surrounding the market, because the *Atlas*'s plates had divided the area immediately around the Place du Châtelet. Baltard's project in fact only needed plate 23, but the previous years of debate and earlier plans by Hector Horeau and other architects had proposed locating the new market at the Seine, obligating Baltard to account for a larger area in his plan. Thus, the composite sheet defined the neighborhood by coloring its surface area pink, with the Seine to the south, the Halle au Blé to the west, the rue Saint-Martin to the east, and Saint-Eustache to the north. Baltard's projections for street alignments are outlined in red, and property expropriations are shaded in yellow.

The 1848 plans by Baltard indicate two grids: one printed and one drawn in graphite. The penciled grid is bound by the rue Saint-Denis to the east and the Halle au Blé to the west (and beyond to the rue du Louvre, not yet indicated), Saint-Eustache to the north, and the Fontaine des Innocents to the south. The Cour Batave

line functions as its east–west axis; the western limit of the drawn grid is aligned to the interior wall of Claude Perrault's east façade of the Louvre on the far left side and aligned to the Place du Châtelet on the far right side of the plate (fig. 3.28).[67] The graticules of the penciled checkerboard establish a new spatial network upon the preexisting medieval fabric. This drawn grid became the basic geometry for Baltard's designs for Les Halles from 1844, 1847, and 1851—designs that were the result of a new urban gridded space anchored by preexisting monuments.[68]

In these plans, a drawn grid is employed to create this new terrain; however, in other instances, the cartographic grid itself anchors the design of the monument (fig. 3.29). Also using a Jacoubet plate as its base, the plan uses the graticules of the map to anchor the main axes of Les Halles. A detail of this plate (fig. 3.30) shows the intersections of the east–west axis of the market; another detail (fig. 3.31) displays the intersections of the north–south parallels; and another (fig. 3.32) indicates the central interior "street" intersecting with the map's grid. The cartographic grid thus anchors and orients the market that is inscribed within an overall spatial logic of the percement defined by four diagonals (fig. 3.33). The four straight lines enter the market at an angle connecting the district to neighboring public spaces: the Louvre on the southwest corner;

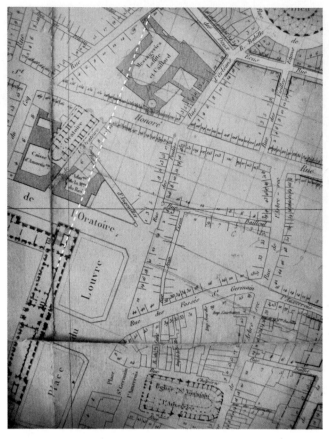

FIGURE 3.28 Detail of Plate 23 (fig. 3.29) showing a white dashed line that articulates the alignment between the drawn grid and Claude Perrault's east façade of the Louvre.

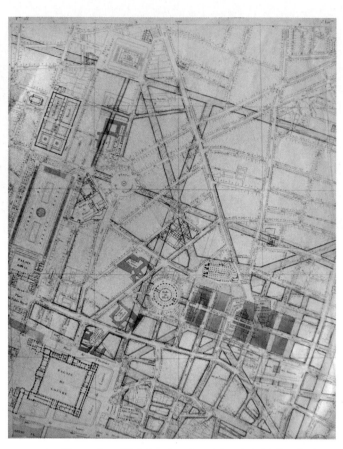

FIGURE 3.29 Plate 23 from Théodore Jacoubet, *Atlas général de la ville, des faubourgs et des monuments de Paris* (General atlas of the city, suburbs and monuments of Paris) (Paris, 1836). Bibliothèque nationale de France.

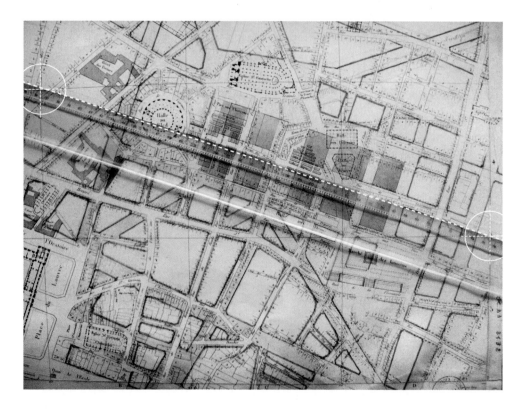

FIGURE 3.30 Detail of Plate 23 (fig. 3.29) with a ruler and white dashed line displaying the east–west axis of the market and its intersection with the cartographic grid.

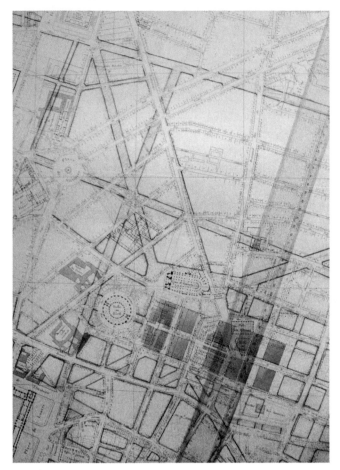

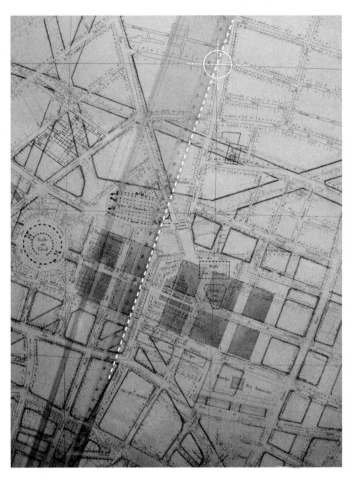

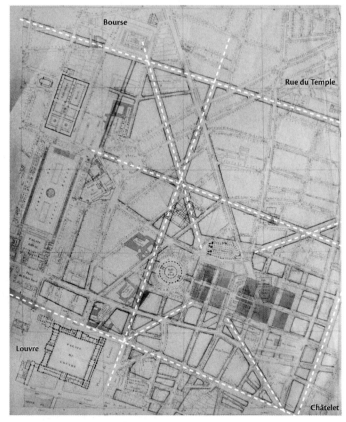

FIGURE 3.31 (above, left) Detail of Plate 23 (fig. 3.29) with a ruler displaying the north–south axis and its intersection with the cartographic grid.

FIGURE 3.32 (above, right) Detail of Plate 23 (fig. 3.29) with a ruler and white dashed line displaying the intersection of the market's internal "street" with the cartographic grid.

FIGURE 3.33 (right) Detail of Plate 23 (fig. 3.29) with white dashed lines indicating the rectangular bounded area and the diagonals that feed into the market quarter.

the Place du Châtelet on the southeast; the Bourse in the northwest corner with the outlines of the rue Réaumur; and a diagonal that extends onto another Jacoubet plate, eventually passing the Conservatoire des Arts et Métiers to merge with the rue du Temple and feed into the Place de la République (fig. 3.34). Here, the penciled drawings demonstrate the ways in which the new market district was conceived through fixing a spatial system oriented to the map's grid and organized around a principle of circulation. These diagonal percements would define the city's monuments as much as the cartographic grid, and both would determine the destruction of Paris's historic blocks.

While scholars have focused on the change in architectural design and material from stone to iron and glass, regardless of the building's design, across all iterations, the gridded structure remained the same (figs. 3.35, 3.36). These early drawings show a cartographic grid that conditions the building's schema. The market's spatial organization is composed of interior "streets" that conform to the paper grid of a cartographic terrain.[69] Over the course of

the second half of the nineteenth century, the district of Les Halles became a physically gridded space that benefited its commercial character. This Paris privileged the new openings through which commodities would move and be exchanged easily—market and street became fused through these architectural passages (fig. 3.37). The building's grid of wrought and cast iron framed forty thousand square meters of open space under a single protective roof. This market integrated within a new urban fabric represented, as Christopher Mead describes, a conception of space that was an open-ended system of repeating, interchangeable units, which could be extended indefinitely.[70] It was a conception framed by and anchored on the cartographic grid.

PLACE DE L'OPÉRA

Even prior to the official announcement of the Opéra competition, there had been a century-long history of debates, proposals, and competitions concerning theater architecture, all mediated through plans. Beginning in

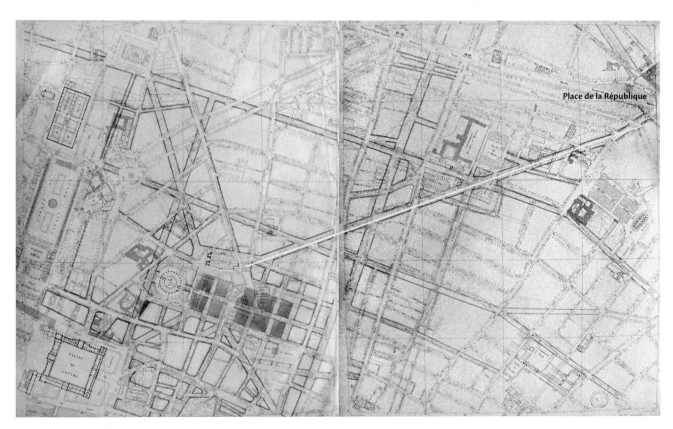

Place de la République

FIGURE 3.34 Detail of figure 3.29 showing a composite image with a white line indicating the connection between the market district with the Place de la République.

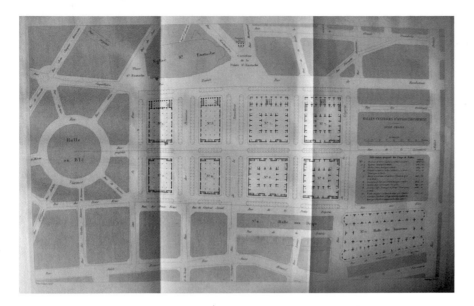

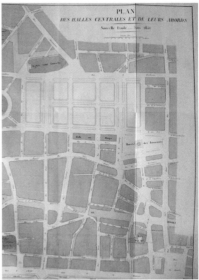

FIGURE 3.35 (top, left) *Agrandissement et construction des Halles Centrales d'approvisionnement, Rapport fait au Conseil municipal* (Enlargement and construction of the central markets for provisioning, Report for the Municipal Council), 1845. Archives nationales de France.

FIGURE 3.36 (top, right) Préfecture du Département de la Seine, *Rapport sur l'emplacement des Halles Centrales de Paris, Juin 1851. Plan des Halles Centrales et de leurs abords* (Report on the enlargement of the central markets of Paris, June 1851. Plan of the central markets and their surroundings). Archives nationales de France, papiers Baltard.

FIGURE 3.37 (above) Bird's-eye perspective. Plate 1 from Victor Baltard and Félix-Emmanuel Callet, *Monographie des Halles Centrales de Paris* (Paris, 1863). Bibliothèque nationale de France.

the 1840s, the two principal issues that the proposals consistently had to address concerned the location and shape of the site. Most of the proposals were situated in the established theater district where the Salle Le Peletier was located, and that served to temporarily house the Opéra after the Théâtre des Arts was demolished in 1820.[71] In the end, with the construction of the new rues Auber and Halévy and the funding for the Place de l'Opéra, the area next to the boulevard des Capucines was chosen as the site, and by April 14, 1860, Haussmann submitted his plan for public comment. The commission that reviewed the proposal and oversaw the comments mandated some provisions to his plan: his proposal of a rectangular site was rejected and led to the approval of an earlier plan by Charles Rohault de Fleury, the official architect of the Opéra, that outlined an irregular polygonal shape. Eight months after the funds were received, on September 29, 1860, Haussmann declared the site for "public utility," or eminent domain.[72] The projected location of the site and the form of the block was publicized in *Le monde illustré* on October 13, 1860, which showed the symmetrical framing of the front façade on the boulevard des Capucines, but a still unresolved rear situation (fig. 3.38). In the illustration, Rohault de Fleury's design, with its dramatically proportioned lateral wings, stands in for a proposed building occupying the entire lot.[73] In fact, however, the announcement for the architectural competition had not been made yet. The relationship between the building, the lot, and the district was still very much undetermined when the competition was finally announced on December 31, 1860, in

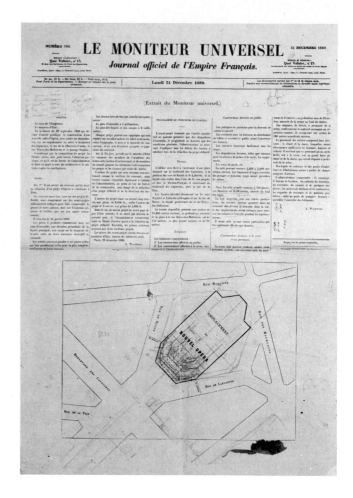

FIGURE 3.38 (above, left) "Emplacement et Abords de l'Opéra projeté" (Location and surroundings of the projected Opera). From *Le monde illustré,* October 13, 1860. Bibliothèque nationale de France.

FIGURE 3.39 (above, right) *Le moniteur universel; Journal officiel de l'Empire français,* December 31, 1860. Ryerson and Burnham Art and Architecture Archive, Art Institute of Chicago.

FIGURE 3.40 (right) Plate 13 from Théodore Jacoubet, *Atlas général de la ville, des faubourgs et des monuments de Paris* (General atlas of the city, suburbs and monuments of Paris) (Paris, 1836). Bibliothèque nationale de France.

Le moniteur universel (fig. 3.39). What was clear in the various versions of the proposed site was not only the matter of the new Opéra building's architecture and its symbolic value for the Second Empire, but also the issue of how this building would relate to its surroundings and anchor this newly developed district of the city.

On the plans from Jacoubet's plates are colored markings that date between 1857, when Rohault de Fleury submitted a proposal on the boulevard des Capucines with an irregular plan, and 1860, when the site plan was officially published in *Le moniteur universel.*[74] The building's undetermined form does not include the lateral

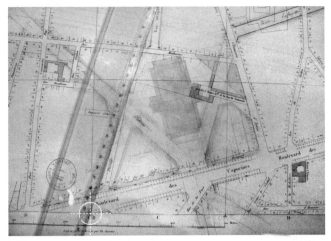

3.41A

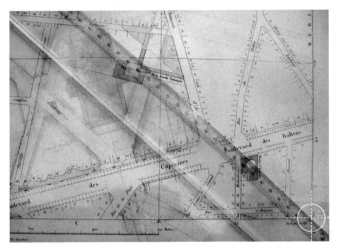

3.41B

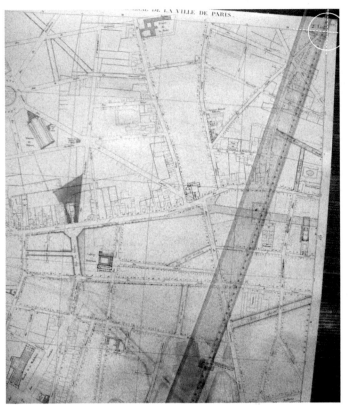

3.41C

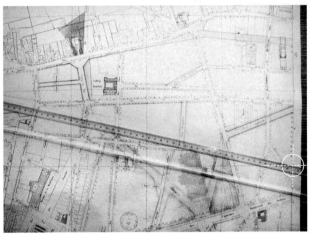

3.41D

inner courtyards that Rohault de Fleury's various versions include (fig. 3.40). Instead, the red-shaded area conforms closer to what would eventually be the final architectural plan by Charles Garnier. The drawn outline of the lot describes the approved proposal for the rues Auber and Halévy as well as the open plaza in front. While the approximately diamond-shaped lot was ultimately preferred, its orientation and angles were not necessarily evident if the building had not yet been designed, as is visible with the light sketches to the rear of the proposed Opéra. The shaping of this block needed a reference other than the Opéra building itself.

That reference was the printed orthographic grid of the map (fig. 3.41a–d). The drawn angles of the road network and the shape of the block for the Opéra correspond to the cartographic grid of plates from Jacoubet's *Atlas*.[75] The straight lines intersect with the grid and show how a ruler would have been placed to discover possible boundaries. The yellow represents the expropriated parts for

FIGURES 3.41A–3.41D Ruler indicating the intersections of the proposed lot's boundaries with the cartographic grid. Plate 13 from Théodore Jacoubet, *Atlas général de la ville, des faubourgs et des monuments de Paris* (General atlas of the city, suburbs and monuments of Paris) (Paris, 1836). Bibliothèque nationale de France.
3.41A Ruler showing the alignment of the proposed western diagonal to the cartographic grid on the lower-left side.
3.41B Ruler indicating the intersection of the eastern diagonal to the cartographic grid on the lower-right side.
3.41C Ruler indicating the intersection of the southeastern diagonal to the cartographic grid on the upper-right side.
3.41D Ruler indicating the alignment of the proposed diagonal at the rear of the building to the cartographic grid on the right side of the map sheet.

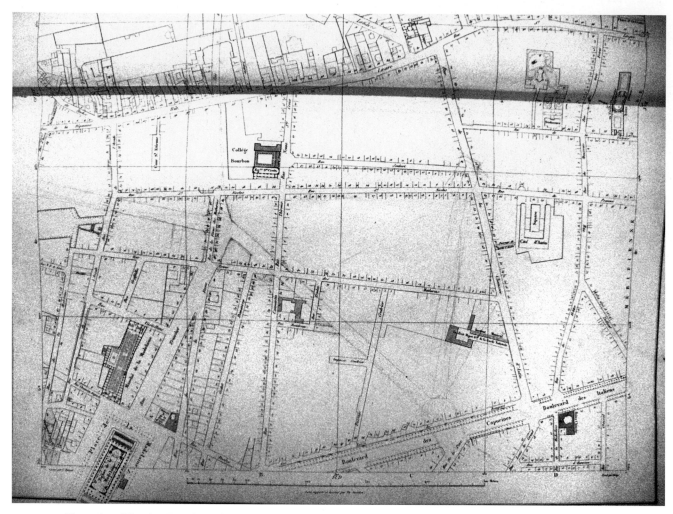

FIGURE 3.42 Plate 13 from Théodore Jacoubet, *Atlas général de la ville, faubourgs et monuments de Paris* (General atlas of the city, suburbs and monuments of Paris) (Paris, 1836). Bibliothèque administrative de la Ville de Paris.

the Service de la voirie (Roads department); the red, those that are to be built; and the graphite, potential proposals. In other sketches from a separate print of Jacoubet plates, there are parallel attempts to determine the orientation of the building and its lot (fig. 3.42). The penciled sketches show the tentative lines that were drawn to determine the Opéra's location by avenues that extend outward and connect to the lot. These plates were marked before the exact site was announced in 1860, and show a proposed plaza incorporated into the boulevard des Capucines. The alignments correspond to various intersections of the map's grid that are shown through a ruler laid on top of the sheet (fig. 3.43a–d). In both instances, the points of reference for an urban situation emerged from paper, and the composition of an urban plaza from a cartographic grid.[76]

In his study of the development of the quartier de l'Opéra, Jean Castex remarks that plans conditioned the edifices and the structure of the blocks.[77] He explains that while the *plans des quartiers* by Alexis-Hubert Jaillot and the *cadastre par îlots* (cadastre by blocks) of Vasserot and Bellanger were available, with no plan d'ensemble, no plans that conformed in scale, and no directed interventions by the state, there was no conception of a monument to anchor the district's development. Moreover, this urban fragmentation led to the city's lack of regulation and allowed for disjointed growth into odd property lots determined by individual preferences. During the Second Empire, the use of plans that offered a synoptic view and the specific planning methods of the percement and the *immeuble-îlot* (building block) were developed and deployed to provide some visual cohesion to a densely packed district. In his reading, Castex explains that the

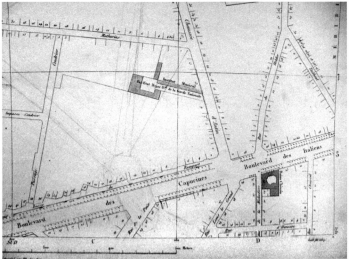

3.43A

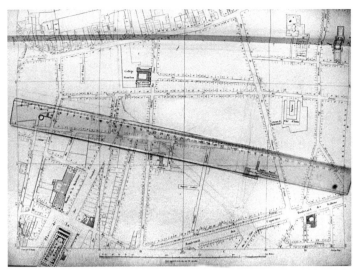

3.43C

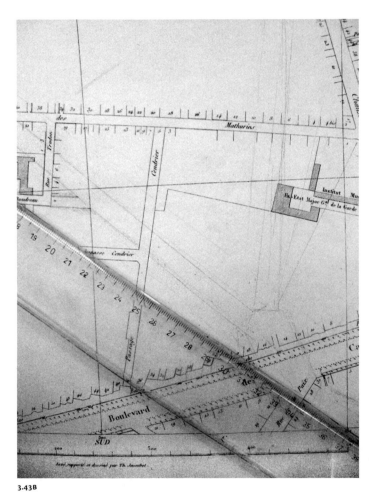

3.43B

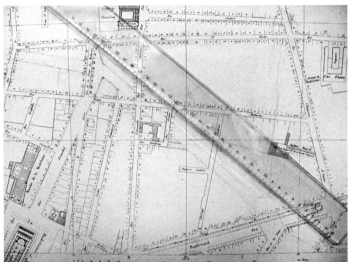

3.43D

FIGURES 3.43A–3.43D Ruler indicating the intersections of the proposed lot's boundaries with the cartographic grid. Plate 13 from Théodore Jacoubet, *Atlas général de la ville, des faubourgs et des monuments de Paris* (General atlas of the city, suburbs and monuments of Paris) (Paris, 1836). Bibliothèque administrative de la Ville de Paris.

3.43A Faint graphite markings indicating proposed lines for the streets around the Opéra.

3.43B Ruler showing the alignment of the proposed southeastern diagonal line intersecting with the printed cartographic grid on the lower right.

3.43C Ruler indicating the proposed street north of the boulevard des Capucines intersecting with the printed cartographic grid on the right.

3.43D Ruler showing the proposed northwestern diagonal line intersecting with the printed cartographic grid on the upper left.

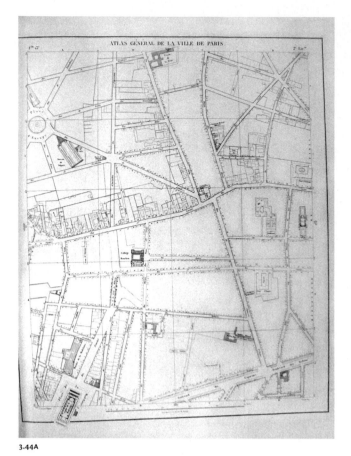

3.44A

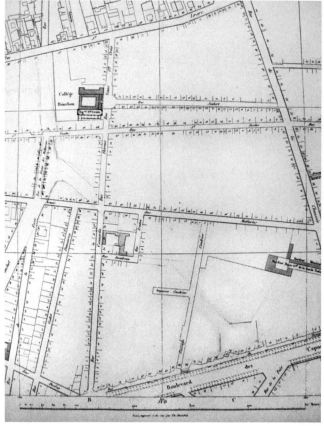

3.44B

FIGURES 3.44A–3.44B Plate 13 from Théodore Jacoubet, *Atlas général de la ville, des faubourgs et des monuments de Paris* (General atlas of the city, suburbs and monuments of Paris) (Paris, 1836). Ryerson and Burnham Art and Architecture Archive, Art Institute of Chicago.

3.44A Map sheet indicating the faint and tentative drawings that mark a proposed site for the Opéra.

3.44B Detail of the area around the proposed site indicating the drawn diagonal line intersecting with the cartographic grid at the center of the image.

building and the block became a single unit, whereupon the street network established "an immediate relation, between building complex and the cutting [*découpage*] of the street that surrounds it."[78] Accordingly, François Loyer discusses the tension between the status of the monument and the function of circulation in nineteenth-century planning.[79] One concerned the location of structures to fill and anchor a given space; the other referred to the urban voids that had to be created. These two concerns were not mutually exclusive, but they posed the relationships between the primary urban forms—building, block, and street—in ways that viewed the shaping of the city from the building outward or from the street inward. David Van Zanten sums up the theoretical situation: "The important point in the history of the shaping of the quartier de l'Opéra is that it was not conceived fully formed and imposed on the site, but instead that it evolved, inflected

by both practical and monumental concerns."[80] Still, this urban shaping by different hands under different administrations was mediated through plans, and in the case of the quartier de l'Opéra, through the specific plans of the Jacoubet plates. The interrelationship between street, block, and building was ultimately determined by and on a cartographic grid through an iterative process indicated by the several plans with drawn proposals (fig. 3.44a–b). The drawings show varying degrees of certainty of the site and its lot; however, all use the grid to orient possibilities for its emplacement. The oddly angled avenue de l'Opéra that was realized connected to the section of the Louvre completed under Napoléon III by Ludovico Visconti and Hector-Martin Lefuel. As the monumental symbol of the Second Empire, ultimately, it was fixed by the straight lines of Jacoubet's printed plans, drawn from Verniquet's ground survey.

FIGURE 3.45 Charles Marville, *Boulevard Saint-Michel, vers le sud* (Boulevard Saint-Michel, toward the south [from Place Saint-Michel]), 1870–77. Photograph on albumen paper from a glass negative, 8¾ × 14⅝ in. (22.3 × 37 cm). Bibliothèque historique de la Ville de Paris, Fonds Charles Marville.

THE GRIDDED PHOTOGRAPH

In the cases of both Les Halles and the Opéra, the grid, the plan of the monument, and the percements were aligned through drawing on a cartographic terrain to generate a new urban fabric. The printed grid determined projected axes that composed Les Halles and oriented the Opéra, and set the spatial logic of urban space that would condition the experience of the modernizing city. If Jacoubet's plans were the ground for Paris's planning and the foundation that made building possible, then photography became the means by which that building program and objective progress were conveyed to the public. The municipal government under the Second Empire and the Third Republic undertook comprehensive urban building projects using photographs for their projection and their historicization, officially commissioning Charles Marville to document the city under construction and putatively completed.

Through maps and photographs, urban development relied on a gridded flatness. Yet if the actual planning of Paris rested on an orthogonal representation of space, its visual reception and experience relied on its perspectival insistence. The construction of perspective is based on the optical illusion that parallel lines extending into absolute space are perceived to meet at the horizon. New spaces were drawn by architects, engineers, and administrators from a gridded surface of parallel lines on an orthogonal plan and, once built, were represented as a receding volume in the photograph. The perspectival views of the transformed city by Marville were conditioned by the administration's orthographic bias of their maps and plans for urban development.

Marville's photographs from 1870 to 1877 of the new boulevard Saint-Michel, whose construction was executed by 1860, present spaces of volumetric extension through perspectival projection (fig. 3.45). The boulevard unfolds infinitely into the distance, all the way to the sloping horizon line. As with most of Marville's compositions, while the street is centered, it is not symmetrical, creating a dynamic effect of movement into the pictorial space.[81] Two columns of the Fontaine Saint-Michel designed by Gabriel Davioud are clearly visible at the right margin. But the photograph only hints at this new monument, because the real subject is the opening in the urban fabric, a percement that became the emblem of the new city.

In another photo of the same boulevard, taken from farther south and looking north toward the Seine, there is a similar composition (fig. 3.46). The boulevard is again not centered, and the converging point is obscured by the offset point of view of the camera and the tops of the recently planted trees. The new alignment of the boulevard would have framed a view of the spire of Sainte-Chapelle, recently erected by Eugène Viollet-le-Duc and designated an official historic monument in 1862. Yet Marville hides it by taking the photograph from the side of the street rather than the middle. The effect of this off-centered position places primary focus on the opening that is given shape and contained by the buildings and guided by the lines of the curbs swinging deep into the pictorial space. This spatial void is the central object of the photograph; it empties out

FIGURE 3.46 Charles Marville, *Place Edmond-Rostand du boulevard Saint-Michel*, ca. 1877. Collodion glass negative, 11¼ × 15 in. (28.8 × 38 cm). Bibliothèque historique de la Ville de Paris, Fonds Charles Marville.

FIGURE 3.47 Charles Marville, *Rue de la Tonnellerie (de la rue de la Poterie)*, 1865. Photograph on albumen paper, 8⅞ × 14¹¹/₁₆ in. (22.5 × 37.4 cm). Musée Carnavalet.

the foreground and middle ground of the image into a vast plane, pushing the built forms into the background, transforming the surrounding buildings into a scenography that supports and envelops the volume.

Marville's photograph of the rue de la Tonnellerie from 1865 displays the city as a gridded form (fig. 3.47). On the right side is a façade of the new markets, and on the left are the old buildings, which would eventually be cleared. The image presents a point of view that is compositionally and temporally in the middle of the old fabric of

insalubrious and chaotic buildings, while suggesting the promise of a newly encroaching, regularizing system of gridded lines. It depicts the built environment as masses with partial spatial extension. There is some perspectival recession, but it is limited by the cramped space. This improves the quality of the photograph's description: individual buildings are distinguishable in their particular heights and widths; corners are articulated; and, generally, more details are legible because of the compactness of the area. However, as seen in another photograph of

FIGURE 3.48 Charles Marville, *Les piliers des Halles, rue de la Tonnellerie*, 1855. Photograph on albumen paper, 9⅝ × 14⅝ in. (24.5 × 37 cm). Brown University Library.

FIGURE 3.49 Charles Marville, *Rue de Rivoli*, 1877. Negative on collodion glass, 11⅜ × 15 in. (29 × 38 cm). Bibliothèque historique de la Ville de Paris, Fonds Charles Marville.

the same street, there is only enclosure (fig. 3.48). The buildings are not yet aligned by continuous walls and the enclosed spaces do not allow enough distance to create full perspectival illusions. Unlike the photographs of the transformed Paris, such as those of the boulevard Saint-Michel or rue de Rivoli, images of an earlier Paris depict the built environment as masses with no spatial extension through perspective.

With the development of the district of Les Halles, the individual shops, with their smaller sizes and distinct façades, gave way to the gridded scenography. Marville's photographs of the new Parisian buildings were taken from a distance and a position that was only possible with the construction of the broad and straight percements. Marville's 1877 photograph of the rue de Rivoli presents the completed street with the uniform system of façades that came to be associated with architectural development under Haussmann, even though the construction of Rivoli spanned many decades (fig. 3.49). Its repetitive order of streetlamps, windows, rectangular lots, and uniform lines

FIGURE 3.50 Charles Marville, *Rue de la Lingerie*, 1866. Negative on collodion glass, 10⅝ × 15 in. (27 × 38 cm). Bibliothèque historique de la Ville de Paris, Fonds Charles Marville.

marking each story emphasizes the perspectival composition of the image. Shadowy traces of horse carriages racing down the street only draw more attention to the ordered and straight lines extending into the horizon that could allow for such speed. While the lines are more insistent than in his photograph of the boulevard Saint-Michel (see fig. 3.45), the two are similar in composition: two columns are placed at the margins of both images, and a void is created in the foreground and middle ground of the photograph. In the case of the rue de Rivoli, two pilasters of Claude Perrault's colonnade frame the left side, with the curb bending around the Louvre's corner and drawing the viewer's eye deeper into the photographic space. By picturing not one column but the pair on the right corner of the colonnade, Marville continues Perrault's motif of repetition on the Louvre's eastern façade. With the right part of the colonnade exposed, there is a suggestion that the repetition is to be picked up—reading left to right—by

the building façades on the opposite side of Rivoli, stretching into and beyond the horizon so that the extension is not lateral but recessive. The viewer could imagine walking directly onto the street's cobbled stones into the open space, following this linear urban rhythm into the horizon.

Compare, for example, the image of the rue de la Lingerie, where there is no spatial extension beyond the frame of the picture (fig. 3.50). Instead there are built masses pushing into the space. The buildings are not yet aligned screens, and the tight spaces do not allow enough distance to create full perspectival illusions. In a photograph of a nearby intersection, carrefour Sainte-Opportune, there is a small opening between two buildings that would just barely fit the width of a wagon (fig. 3.51). Marville composed the picture with part of a building corner pushing in the photograph's frame. The enclosed space held by these buildings is relieved by this bottleneck that leads to yet another tight enclosure, walled

FIGURE 3.51 Charles Marville, *Carrefour Sainte-Opportune*, 1865–68. Photograph on albumen paper, 9⅝ × 14⅝ in. (24.3 × 37 cm). Musée Carnavalet.

FIGURE 3.52 Charles Marville, *Carrefour Sainte-Opportune (de la rue des Halles)*, 1865–68. Photograph on albumen paper, 8¾ × 14⅜ in. (22.3 × 36.7 cm). Musée Carnavalet.

in by another mass of buildings visible through the small, lit passage. Even the wares and merchandise seem to push out of the buildings, begging for an opening to circulate. Thus on the other side of the street is an egress, through the arches toward the Louvre palace onto the newly extended rue de Rivoli (fig. 3.52).

AN IMAGE OF PARIS

In his description of Paris published in 1843, urban commentator Hippolyte Meynadier wrote, "To better appreciate the organization and the direction of the new streets, one can place, between the marks at their ends, a narrow strip of paper, where one can draw, from one point to another, a line in pencil."[82] He understood the plan as a material means to see and draw through a problem. As an official in the Beaux-Arts administration during the Restoration and a contributor to the *Revue générale de l'architecture et des travaux publics* during the July Monarchy, Meynadier was experienced in the use of images to produce architecture. His ideas responded to a concern for both *embellissement* and infrastructural changes that would allow increased circulation through the city, a tension that could be resolved on paper. In this vein, Meynadier insisted on the value of a graphic overview of the city as opposed to textual description. Having noted that numerous descriptions had already been written about future plans for the city, he asked in the opening of his text, "But, where is the general plan, drawn by the municipal authority and the government, to form the organization of all these great works of art and public utility in the city of Paris?"[83]

The necessity of a total view was widely shared in the press of the time. When the *Gazette municipale* announced the publication of a plan by Jacoubet, it wrote: "We had long believed that the administration was on the wrong track in persisting with deplorable stubbornness in a system that lacks unity."[84] The lack of an overall and comprehensive plan and the piecemeal interventions were precisely what Jacoubet, Grillon, and G. Callou cited in *Études d'un nouveau système d'alignemens et de percemens de voies publiques faites en 1840 et 1841* (Studies for a new system of alignments and openings for public roadways built from 1840 to 1841) (1848) as the reason for the city's major problem concerning the displacement of the population from the center of Paris. They wrote, "We have had to seek the cause for these deplorable changes, which are causing disruptions to private fortunes. It's in the lack of an overall study of the [street] alignments that we found it."[85] For them, the fragmentary logic of interventions through alignments was hazardous in terms of commerce, hygiene, and security, and had only exacerbated the problems of urban congestion.[86] A total view was essential: "In our studies, we have embraced Paris as a whole and in its relations with roads and railways, which enter or reach there."[87] This position was no doubt informed by Jacoubet's career as an architect in the administration of Paris since the early 1820s, and particularly his activities dedicated to producing a standardized image of the city. Accordingly, the publication of this report set Jacoubet apart from previous mapmakers, in that he had overseen the production of a standard map of Paris *and* provided his ideas for the city's improvement based on those plans. Verniquet had combined surveying and mapmaking tasks, but Jacoubet went one step further. For Jacoubet, drawing the maps gave him the comprehensive knowledge to offer his planning ideas.[88]

In many ways, this opinion reiterated an earlier call for the necessity of a total view by Marc-Antoine Laugier in *Essai sur l'architecture* (1753). Remarking on the disorder of the city, he argued that a general plan would help design a wide, long, and straight avenue as a monumental entryway into Paris, in turn creating a regular distribution of gates into the city and organizing the roads into a "convenient" system. He wrote:

> It is therefore no small matter to draw a plan for a town in such a way that the splendor of the whole is divided into an infinite number of beautiful, entirely different details so that one hardly ever meets the same objects again, and, wandering from one end to the other, comes in every quarter across something new, unique, startling, so that there is order and yet a sort of confusion, and everything is in alignment without being monotonous, and a multitude of regular parts brings about certain impression of irregularity and disorder which suits great cities so well. To do this one must master the art of combination and have a soul full of fire and imagination which apprehends vividly the fairest and happiest combinations.[89]

Published almost a century apart, these two planning texts promoted different guiding principles to design and order the city: the former based on the value of hygiene, the latter on the value of pleasure and sensibility. However, they shared an insistence on the need for a general plan of the city's total assemblage that could allow an architect, engineer, and administrator to see its parts and set about practicing the "art of combining"—words that were eventually repeated by Haussmann while gazing upon a map of Paris in his office, albeit motivated by functionalism rather than pleasure.[90]

Laugier's emphasis on the importance of a city providing delights and not succumbing to monotony was shared by Napoléon III more than a century later. In the 1840s, members of the government and Louis Lazare, director of the *Revue municipale,* argued for the Conseil général de la Seine to convene a commission to study and propose a new *Plan d'ensemble d'alignement des rues de Paris* in 1850. Administrators, including the prefect Claude-Philibert Barthelot, comte de Rambuteau, were nervous to take up big projects in the context of a shifting political terrain and large economic uncertainties, but nevertheless called for a new plan.[91] Napoléon III finally took up the matter of executing a new plan for Paris on August 2, 1853, in a letter to the minister of the interior, Jean-Gilbert-Victor Fialin, duc de Persigny, who convened the Commission des embellissements de Paris headed by Henri Siméon.[92] De Persigny wrote to Siméon with a list of demands from the emperor. One demand asked "that in the drawings of the main streets the architects make as many angles as necessary so as not to knock down either the monuments or the beautiful houses, while conserving the same width of the street, and so that one is not enslaved by the trace of the straight line."[93]

The commission was under immense pressure to prepare recommendations; he worked rapidly and submitted the final eight-plan report on December 27, 1853. Jacoubet's plan, which represented a reproduction of Verniquet's surveying calculations supplemented with plans parcellaires, became the basis for the commission's recommendations. Having produced the report in less than five months, the commission had no time for a new survey and had to rely on extant plans. Siméon explained, "No one is better able than Jacoubet to provide the materials for this study. He is the owner of the copper plates of three plans of Paris, superior to all others in existence because they are drawn exactly to scale."[94] Commenting on the final report submitted to Napoléon III, Siméon referred to the original directives, repeating similar sentiments: "Another useful rule is not to become a slave to straight lines. It is necessary to respect the monuments and even the small constructions that have their own grandeur and interest."[95] The directive was for monumental spaces to be respected, and for the city to not fall under the tyranny of the straight line. While the commission's plans may have avoided the large percements through the center of the city, the image of Paris, to which all administrators, engineers, and architects would refer, was fixed to a measured grid.

No one understood the desire to fix an image of Paris better than Haussmann, who saw the value of plans to both determine and address a building program to the public. In his address to the Conseil municipal de Paris about the opening of new public streets, he stated, "It was first of all important to determine with precision the direction, the extent and the number of new streets to open, which offer to the highest degree possible a character of national utility, and among all those which can be drawn, from this point of view, on the plan of Paris, to make wisely a limited choice."[96] He was conscious of the rhetorical power of the map to naturalize proposals and turn possibilities into inevitabilities through the map's synoptic quality. "At a quick glance," Haussmann continued, "you will be struck for sure, Gentlemen, by the overall harmony of the project, by the perfect arrangement of the various groups of public roads, and by the superior view that the details provide."[97]

There are two related points to be made concerning these statements. The first, which is not to be taken for granted, is that the shaping of Paris happened on and through plans; the second is that the development of urban planning and its administration in Paris were conditioned by maps and the specificities of the cartographic medium. The image from which Paris was built was not a spectacular one, but a rather mundane matrix of orthogonal lines, composed of a grid that allowed for projective and descriptive marks to communicate with each other. Altogether, they would form the basis of a thorough reorganization of the city, in terms not only of design but also of administration. These orthographic plans and the practice of drawing on them were no longer simply architectural in intention but, as is clear in Haussmann's discourse, they also defined the government's planning culture.

Yet, despite how fundamental this grid was as the support structure of projecting new spaces, it did not present itself in the physical environment, contributing to what has been referred to as modernity's uncanny experience.[98] There was an uncoupling of structure and image, of interiority and exteriority, of plan and elevation that was not resolved by a finished and completed object, however much Haussmann and his administration desired. Accordingly, commentators would remark on the loss of an old Paris, but it was difficult to fully articulate what was happening on the ground, leaving one only with sensations of estrangement and alienation.[99] The modern city retained—or, as Walter Benjamin would remark, "has given material existence to"—labyrinths in spite of the regularity and rationality that seemed to structure it.[100] Paris did not manifestly succumb to a grid, and seemed to resist it by producing new kinds of spaces. It was this tension that drove the activities of the urban administrators: the constant need to align streets, which in turn simultaneously created ever more anomalies.[101] These straight lines offered legibility and rationality, but they also introduced a whole new set of social variables whose relations could not find stability in the disposition of monuments. The modern built environment that emerged was no longer an expression of a fixed society, in which urban elements of building, block, and street were distinct and architectural façades secured a fixed cultural identity or history. Instead, urban modernity became defined by an ever-constant striving to conform to the grid and its promise of control, while at the same time exceeding or overwhelming it.

RUE St MARTIN

RIVOLI

VICTORIA

DE

AMEL

E Nas

RUE A.

RUE

AVENUE

Statue
de Pascal

TOUR St JACQUES

BOULEVARD DE SÉBASTOPOL

ÉCHELLE DE 0.0015 POUR 1m00.

4 The Bureaucracy of Plans

Architecture is nothing other than administration.
—Eugène Viollet-le-Duc (1872)

Throughout the nineteenth century, the proceedings of the Conseil des bâtiments civils, the governmental agency in charge of overseeing all public building projects in France, consistently referenced documents related to reports, registers, and plans. No one questioned the necessity of these documents for seeing, recording, and managing urban spaces, and their ubiquity made them simultaneously indispensable and easy to overlook. The ordinary procedure started with the council approving the production of a plan, often indicating not only the building but also its surroundings and transportation connections. Once plans were drawn and submitted to the Préfecture de la Seine, the committee members would convene to discuss proposals. Routine references to measures, orientation, and dimensions in the meeting minutes testify to the administrators' facility with this graphic medium and to the plan's complete integration within the bureaucratic process.

PLANS AND PAPERWORK

The emergence and growth of bureaucratic agencies and institutions in Europe was fundamental to the development of the modern nation-state in the 1500s. These administrative structures were essential to the consolidation of political power, whether in the form of a rising merchant class, an absolute monarchy, or a centralized governmental body.[1] In France, Louis XIV's reign is cited as the beginning of an "information state" in which the development of a centralized administration was driven by the need for effective methods of tax collection.[2] Formed by the king's finance minister, Jean-Baptiste Colbert, a new administrative apparatus included new ministries and new positions—such as *intendants* (bursars), *subdélégués* (subdelegates), *maires* (mayors), and *échevins* (aldermen)—organized into a hierarchical structure that was then linked to officials in the provinces and centralized around *le roi administrateur* (the king as administrator). What held this increasingly extensive and complex structure together was the singular demand to produce detailed textual reports about the provinces.

The creation of these reports played a constitutive role in the formation of bureaucracy. Administrative offices and positions materialized through the production and circulation of paper documents, creating an "information ecosystem" that connected spaces, things, and people: provincial administrators, Parisian offices, foreign diplomats, physical desks, even the king's own body.[3] Paperwork moving through this entire system allowed the system to justify and sustain itself. In his classic study of the structure of bureaucracies, *Wirtschaft und Gesellschaft* (Economy and society) (1921), Max Weber writes: "The management of the modern office is based upon written documents (the 'files'), which are preserved in their original or draft form, and upon a staff of subaltern officials and scribes of all sorts. The body of officials working in an agency along with the respective apparatus of material implements and the files makes up a bureau."[4] Weber outlined a basic theory arguing that documents not only define the bureaucracy itself but also control the processes within organizations and beyond—both through their links to the entities they document, and through the coordination of activities around and through those documents. Indeed, the plan, as one kind of document, was not the consequence of the bureaucratic process, but rather its basis.

For Lisa Gitelman, the genre of the document is composed of forms, memos, and reports—the paperwork of bureaucracy.[5] While documents were long produced by institutions such as churches or guilds in the premodern era, Gitelman draws on John Guillory's argument that "the dominion of the document is a feature of modernity."[6] This dominion is based on Guillory's definition of the document as a "carrier of information and so the object of knowledge rather than knowledge itself."[7] This distinction between knowledge and its paper medium is the basis for the authority of the document as neutral. Guillory's differentiation, however, omits the purchase of the document's materiality in the shaping of information carried on it: what Gitelman has termed "the affordances of paper."[8] From new methods of printing to mass-produced paper sheets, the ability to physically produce documentation coincided with the impulse to quantify and measure the terrain, creating volumes of files. If the paper document was the infrastructural material that allowed for a bureaucratic system to cohere, then the paper surface was the medium for communicating the methods and knowledge that defined bureaucratic activities.

In his analysis, Weber assumes the primacy of text ("written documents"), when, in fact, these files were diverse: tables, charts, diagrams, maps and plans, and photographs. As such, questions about the methods and means of representation mediated by the paper document, as well as its legibility to officials inside and outside a given agency, are significant.[9] If we admit to a dominion of the document, it is because of the greatly increased production of paper and the reproducibility of the document's forms, as well as the systematic methods that emerged in the cataloguing, organizing, and archiving of the document. These practices defined the modern bureaucracy. Drawing from Michel de Certeau and Mary Carruthers, Guillory understands written, as opposed to oral rhetorical, strategies to be constitutive of the modern bureaucracy, but he also expands on their work by suggesting the qualities of scale and means as specific to the nineteenth century. It was not simply that text no longer referred to oral traditions of address and style, but rather that, through the large-scale growth of an organization's reach and internal structure, text developed a technical style of its own. Yet "text" is a broad category, and the role of graphics and images in this emergent bureaucratic dominion is only cursorily addressed in the literature. Guillory refers to the "graphically organized page" as one that formats text and obligates writing to adhere to the technical values of brevity.[10] While scholars of material and media culture have critically addressed the function and history of paperwork, their work has largely overlooked images among the various types of documents.[11]

In one ordinary example concerning the Hôpital Saint-Louis, discussions about the surrounding area of the building could not even begin because no site plan had been submitted. The committee members revisited the discussion a week later, when plans were drawn and presented.[12] The plan begot meetings and reports that led to proceedings and summaries, spawning ever more paperwork. The plan was also an aggregator, bringing together social, legal, intellectual, spatial, and representational practices, as well as diverse people and objects. Commissioners met around plans and ordered studies and additional drawings; they responded to proprietors who were affected by street openings, took their testimonies, and drew up new responses; they produced records for the prefect and other administrators, experts, and assistants. The ubiquity of graphic paper documents was not distinctive to the council: it typified every building agency associated with the Paris municipality. Moreover, the very process of building modern Paris was contingent upon the production and circulation of a growing amount of graphic images. Especially in the postrevolutionary years, as Paris grew in population and area, there was a concomitant growth in the bureaucratic infrastructure to manage the numerous building projects and the diversity of actors involved in those practices.[13]

Starting in the revolutionary period, drawn plans became a means of consolidating and organizing various activities related to building in the capital. This "paper-driven bureaucracy" of Parisian architectural culture and the endless parade of construction drawings and paperwork played regulatory and legal roles in state building campaigns during the nineteenth century.[14] "Nothing was improvised," writes Barbara Shapiro Comte. "A hierarchy of functionaries imposed a four-phased system of checks and balances . . . driven by a regimented sequence of drawings, segregated into conceptual and functional divisions, according to chronology, responsibility, and expertise."[15] In recognition of the diversity of interests and parties involved, laws were implemented in France that not only required an image in order to build, but also regulated how those images were drawn.

By the end of the nineteenth century, the link between plan and building was firmly secured by a highly regulated process. Though Weber's theory of bureaucracy does not mention images, there is an important point to be made about the extent to which drawn plans supported the ideological intention of bureaucratic regimes to appear objective and scientific. They were the specific means of constructing the administration as a series of technical operations, outside of the social life that they simultaneously sought to shape. In France, the primary mode of these legally promulgated images was orthographic. This chapter examines the specificity of the urban plan within new forms of urban bureaucracy, as well as the intellectual values that informed the drawn plan's authority and use in nineteenth-century Paris. The central argument is that the emergence of the orthographic plan in building and architectural practices was a consequence of the specific exigencies of construction per se and was informed by the need to develop a centralized administrative structure for managing the city's development. In sum, not only the city but its urban administration were constructed by and out of the plan.

THE REGULATED AND REGULATING IMAGE

The legal requirement of a building plan was underpinned by precedents dating from the seventeenth century. A historical analysis of the laws demonstrates that the plan did not emerge from architectural or construction necessities, but instead from the need to outline and manage changing private and public obligations. In France, the first ruling regarding the management of buildings was registered in an edict of 1607, when all proposed buildings had to obtain an *alignement,* or boundary line demarcating public and private space.[16] The original language of this ruling simply required that the alignment "straighten the walls where there are folds and bends, and . . . ensure the streets are embellished and enlarged as well as possible."[17] The ruling assumed a system of demarcation, but did not textually specify the drawing of a plan. In the proceeding period, there was no standardized process when it came to drawing plans and maps. Mostly, such images were determined by compositional rules espoused in the military and technical schools. Depending on the preferences and priorities of rulers, as well as rules negotiated by local authorities, practices were not prescribed. For example, when Colbert, under the authority of Louis XIV, instructed all provincial officials to survey their territories, no specific method of surveying was prescribed.[18] Moreover, with no standardized measures until the adoption of the meter, images were neither relatable to each other nor scalable without great mathematical efforts.[19] While Colbert had begun creating a networked bureaucratic infrastructure, the unique status of each image of the province undermined the intention of a relational system. Thus, the adoption and use of the metric system in 1793 created a shared language to compare and scale all images and was an essential element in the stabilization of the state's bureaucratic infrastructure.

The plan of Paris surveyed and directed by Edme Verniquet began in 1775 and was presented to the authorities in 1783, when a royal declaration established for the first time a regulation connecting the width of the roads to building heights.[20] That same *lettre patente* of April 10, 1783—a royal decision in the form of a letter—specified an image.

> Article 1. The minister of the interior is authorized to rule on the plans of the streets of Paris the enlargements and the adjustments required by each of them.
> Article 2. Only one alignement shall be drawn on said

plans, which shall be definitive, and the resulting cuts shall be not more than 10 meters of the width of the roads that may not have reached that dimension and that do not form an extension to the major routes.[21]

This order's significance, beyond the stipulation to draw up a plan, lay in the way that the practice of drawing directly onto the plan acquired an administrative and legal value as an instrument used by the minister to control the form of streets.

On September 16, 1807, a law was passed that specifically addressed the need for a *plan général d'alignement des rues* (general plan for road alignments) for urban development in Paris.[22] This plan was intended to mediate between public and private interests when it came to the management and construction of space. According to Henri-Jean-Baptiste Davenne, administrative director of the Paris public assistance office and author of many volumes concerning the history of the Paris administration, "The plans, fixed upon discussions among at least two opposing parties, become a sort of contract that reciprocally binds the administration and individuals, and it is their rigorous and impartial execution that guarantees the interests of everyone."[23] The plan was given a statutory value as a contract between the different parties, based on defining an alignment as "the limit established between public ways and private properties."[24] In this sense, the 1807 law addressed a central question about the relationship between public and private interests that had arisen during the revolution, which required the plan to be the mediating document.[25]

Defining the relationship between public and private property in a plan was a critical element in a law governing expropriation for public use (*pour cause d'utilité publique*) that went into effect on March 8, 1810. The concept of eminent domain had a longer history and was the primary tool used to justify the major road openings that would displace thousands of residents and reshape the city's blocks.[26] In the second section of the law, under "Of Administrative Measures Relative to Expropriation," a requirement stipulated that engineers draw up "a land terrier or figure plan of buildings whose assignments are recognized."[27] Once a plan was drawn and distributed to all property owners (as well as posted in a public place), all communications, including any contestations, were to be made through the plan of the area to be expropriated. The significance of the plan in the expropriation law was that all official actions were mediated through these images,

communication was primarily conducted through the images' exchange, and, most importantly, the images were legible to administrators and to the public, defined mainly as property owners.

Once the exchange and markup of plans became part of the administrative process of shaping and ordering Paris (and also entered the public discourse about the latter), it was necessary for the plans and their graphic language to be standardized. The plans had to be legible and consistent across all parties and interests involved in the management of urban properties. The legal instruction of October 2, 1815, regulated the way all maps were to be drawn, constructing a standard cartographic vocabulary and practice for the administration.[28] The use of colors to differentiate between buildings to be built and demolished was already a common practice for plans submitted to the Bureau de la Ville (Office of the city); however, its specific legal codification reveals a new will to standardization. There are several layers to this law that relate to the regulation of plans' graphic language, as well as to their use, access, and distribution.

The law's text begins by stipulating the scale and detail in the *plans généraux* (general plans, which provide the overview) and the *plans de division* (division plans, which provide the property details).[29] With regard to scale, the requirement of the metric measure confirms the legal adoption of the metric system. The text also points to the graphic representation of time through color. In article 5, red and yellow are designated as the primary colors for spatial propositions, layered on top of the black lines demarcating lots and roads. The third article also shows how multiple temporal values occupy the same graphic space: gray is used for buildings actually built, distinguishing them from the earthen-colored buildings that are not. Through color, the concepts of percement and embellissement become operational within the general plan's drawing.[30] Color was also a primary means of communication among the administrators, engineers, architects, and surveyors who were accountable for the accuracy of the cartographic representation.[31] Thus, the plan conditioned more than these parties' interventions in the city; it also conditioned their relationships with each other. Each agency was given a set of official maps in the form of an atlas, marked with the official stamps and seals of the different offices, based on one official survey. Gilbert-Joseph-Gaspard, comte de Chabrol de Volvic, as prefect of the Seine, remarked that he communicated through plans with Jean-Baptiste Sylvère Gay, vicomte de Martignac, minister of the interior under King Charles X, and the plans served as a shared reference among many parts of the government.[32]

Collectively, the official instruction shows the establishment of a cartographic language no longer based on pre-eighteenth-century pictorial aims that emphasized the topographic landscape dotted with architectural monuments and seen from a particular perspective. Instead, the text assumes a planimetric view and focuses on the functional requirements to describe the terrain, delineating public buildings from private properties through the outline of roads, plazas, squares, and quays, as well as distinguishing the natural topography from the built environment, all with lines and color blocks.[33] The majority of the other articles concern markings for new roads or for alterations to existing ones.[34] Location was prioritized over architectural expression.

Once a map was created and established, it was then linked to on-site building operations in a law passed on July 18, 1837. This law essentially obligated the reverse: building must conform to plans. All street improvements were mandated to be drawn on a *plan général d'alignement* by the Ministère des travaux publics.[35] While plans had previously been used for urban improvements, the 1837 law required *all* urban interventions—not just expropriation cases—to use plans. This was a detailed extension to an 1833 effort overseen by the Conseil général des ponts et chaussées to outline a plan for the whole country. In his record of administrative practices, Maurice Bloch explains that earlier alignments to an existing street had been approved by textual descriptions that were often not accompanied by a plan. These alignments were done on the ground, based on the interests of the landowners along a given street.[36] The law was supported by the minister of the interior, who on October 25, 1837, required municipal budget offices to allocate funds to support the drawing and distribution of plans.[37]

On April 3, 1841, the procedure to acquire land by the government was articulated by the legislature, defined in detail by prefectorial decree, and ordered by the judiciary if controversial. The expropriation law, passed one month later, established the procedure to begin with a plan.[38] Drawn up by a small committee of engineers, surveyors, and administrators, these plans were required to describe both the interior spaces of blocks and throughways.[39] Five articles address the objectives, production, and

distribution of plans parcellaires, and, specifically in article 37, the plan was acknowledged as a primary means of understanding the city, as well as determining the price of the properties. As Léon Daffry de la Monnoye explained, "The law considered the delivery of these plans as the surest way to make clear to the jury the situation, the capacity, the nature of the parcels for which it was responsible for assessing their value."[40] Once the interiors of blocks were made visible, it was inevitable that the May 3, 1841, law would be extended to allow for the expropriation of entire blocks, and not just the affected portions.[41] Thus, on March 26, 1852, a law was passed to do exactly that, and, in addition, it required all plans to include levels, which had not been mandatory before.[42] In this form, the expropriation law, which gave a central role to the plan, became the primary instrument of Haussmann's administration.

As the various laws chronicle, the idea and form of the plan evolved in relation to its changing use. This had little to do with prescribing architectural expressions— the kind of prescriptions that would eventually be made by Haussmann in his capacity as prefect and by Adolphe Alphand as director of public works at the start of the Third Republic. Instead, the laws were more concerned with determining responsibilities, financing, and jurisdictions, and communicating within a state administration with a growing number of agencies and personnel. As more laws were enacted to articulate these nuanced relationships between private and public interests, more plans were drawn. However definitive and standardized these maps appeared, their production and proliferation— apparent in the archival records—signaled a bureaucratic system that was itself growing ever-more complex given the increasing number of projects taken up by the state. This administrative ecosystem of building thus not only included graphic images, but also required specialized knowledge and accumulated experience in order to read and interpret them. The use and circulation of these increasingly specialized documents, determined through graphic regulations, created a system of power that established the bureaucrat's expertise over urban space.

Yet the burgeoning details and technicalities of these cartographic requirements also contained an inherent admission that they were incapable of capturing the particularities involved in these urban projects. Sometimes these particularities were features of the terrain, especially elevations; sometimes they involved differing interests between the state and property owners regarding property values. There was no image that could describe the terrain accurately in all its multidimensional values, and its inability to completely capture the world, which was a central claim of the rational methods of triangulation adopted by the government, paradoxically warranted an ever-increasing number of regulations to administer them. Beyond its descriptive claims, then, the regulated plan, whose drawing and circulation organized the bureaucratic apparatus, ultimately involved establishing the norms of a building culture and controlling who participated in that culture.[43] As much as the images themselves, the laws about drawing plans regulated the spatial and social relations around those images.

GRAPHIC GOVERNMENTALITY

When Verniquet arrived in Paris, the political climate was tense. Debates over tax reforms, from a *taille personnelle* (personal tax) to a *taille réelle* (actual tax) system, had already caused Henri-Léonard-Jean-Baptiste Bertin, the general controller of finance, to be forced from office by provincial states and parliaments. Fears of losing tax privileges, and unclear explanations of the role of local governance bodies in relation to centralized collection, fostered strenuous opposition to any changes. Furthermore, if there was to be taxation linked to land, there were questions about which models could be applicable: the Spanish method of declarations, the Piedmont method of a cadastre of only cultivated areas, or the Milanese model of a geometric plot survey? The *généralité* of Paris faced the worst complications when it came to determining taxation because of its dense layers of inheritance rights, dues, and exemptions, and it was also a major source of decreasing revenue for the state. In 1763 the physiocrat Anne-Robert-Jacques Turgot organized a visit to Italy to meet with Pompeo Neri, a legal consultant in charge of the Milan cadastre, and he compiled his arguments for taxation based on land instead of title in *Réflexions sur la formation et la distribution des richesses* (1766).[44] For Turgot, changing to the Milanese method of surveying all land would simplify and increase the state's revenue, as the existing personal tax was yielding decreasing amounts. He assumed a centralized operation, outside the scope of provinces and communities, under a single authority that could impose overall consistency.

The reforms took several decades, and in the ensuing debates we can see how the correspondence between land and representations was made within administrative

and legal domains, as well as how those representations shifted from textual to graphic modes. This process makes apparent how plan and bureaucracy did not develop independently of each other: the structure of the Parisian urban administration was conditioned by the logic of the plan, and the plan's authority to represent the city was supported by administrative practices and legal regulations. Cadastres were often the first administrative images of land, and, in the case of Paris, were closely aligned with the maps and plans that were eventually used to shape the city by architects and engineers, whose work was framed by administrators.

Before the eighteenth century, the record of properties was largely overseen by local parishes, and their descriptions were most often text-based; in the feudal system of land division, seigneurial rights and obligations were not based on the terrain's physical characteristics or even acreage, but on ownership by a specific person.[45] Accordingly, these written registers of land ownership and the accompanying legislation did not require images, mainly because of the high costs of production.[46] Indeed, cadastres were legal designations not necessarily related to geography, let alone cartography. Maps were only drawn for litigation concerning tithe payments to the local parishes, particularly when the density of churches in each area demanded a clear delineation of boundaries. Even in these cases, however, they were often crude images with each parcel numbered, referencing a separate textual list of properties. Litigation between landowners may have compelled the drawing of maps, but these plans parcellaires and the *plans d'arpentage* (surveying plans) were always meant to accompany written *censiers* (registers) that recorded the feudal rent due to the seigneur. Consequently, the map's image was always tied to a legal text if it was produced at all.

By the mid-eighteenth century, the link between image and text and their respective descriptive values began to decouple because of the scientific values permeating the administrative culture.[47] Convinced of the superior value of graphic images, one notary from Marcigny, Claude Edme de la Poix de Fréminville, sarcastically asked, "How can one conduct a land survey without drawing plans? It's impossible."[48] In his comprehensive practical guide, which covered topics from budgeting to persuading tenants to open their front doors for inspection, he explained that there were two types of plans that could be used in determining a lot: a "*plan visuel*" and a "*plan geométrique*." The former required no set rules and required only that the notary verify a visual description. The latter, which was preferred, was based on geometric principles.[49] Notaries advocated the use of graphic images produced using scientific principles, allowing images to remain consistent regardless of the reader and user.

That there were published treatises by notaries on property assessment speaks to the variability and irregularity of the practice. Bertin's cadastre sought to remedy the inconsistencies produced by this decentralization. The aim was not only to produce a single and shared image of all French territories, but also to centralize its production by training personnel, using a single measure, and normalizing surveying methods. While this standardization comported with emergent scientific values, the political and economic consequences were very risky. First, the costs of a plot survey could only mean greater expenditures and thus new taxes. Second, a centrally controlled cadastre would allow direct links between individuals (now taxpayers) and the state.[50]

Reforms were eventually pushed through by the Assemblée constituante. The consequences were that surface area was made the determinant of taxation, and qualitative distinctions such as titles were negated. Moreover, the revolutionary government's 1790 law determined that the assessment of property taxes "will be distributed by proportional equality on all properties at the rate of their net income."[51] The goal was to have taxation defined by land but also levied proportionally on real wealth. By relativizing tax contributions, this law promulgated a concept of space that was conceptualized relationally, in which no one property was considered in isolation. Each was connected to others, and their proportionality assumed space as unified *and* interchangeable, based on a shared common denominator of surface area. In 1791 the Assemblée constituante stipulated a single land tax that was to be shared in equal proportions by all landowners according to their net income. The decree from September 23, 1791, outlined the procedure: "When surveying a community's territory, the engineer in charge of the operation will first make a master plan that will show the community's constituency and its division into sections. He will then draw detailed plans that will outline the plots of the community."[52] Although the surveying project was stalled due to a lack of financial resources, the perimeters were established. In the end, each department was charged with drawing up maps at its own expense; however, every department was ordered to follow prescribed methods and standards so that they could compose a whole picture at a uniform scale. Ultimately, these tax reforms placed descriptions of property under the purview of the state and linked taxation to maps.[53] In order to maintain the standards, the government called on October 5, 1791, for the creation of a cadastral office "to link the survey of these plans to more extensive operations and to direct them all towards the preparation of a general cadastre which will be based on the large triangles of the map from the academy of sciences."[54] The aim of this new administrative office was to maintain the link between the plan and its actions, and its guarantee was based on triangulation.

Gaspard François de Prony was appointed the Bureau de cadastre's first director. He would later become the director of the École des ponts et chaussées. As part of the Ministère de l'intérieur, the bureau's task was to draw up and supply all the maps and geographic information to the legislative and administrative institutions. For the various administrative necessities, it prescribed a centralized mapping system that stipulated three different types of maps at various scales: 1:20,000 scale maps in full-size folios; 1:2,500 scale maps for each commune, indicating only rivers, roads, and administrative boundaries; and 1:666 scale maps showing individual properties.[55] The demand for these maps necessitated the training of mapmakers, and ultimately the creation in 1794 of the École des géographes du cadastre, which subsequently handled all the geodetic work related to the establishing of the metric system.

The office, however, was unsuccessful at controlling the diverse practices of independent surveyors. This was due to financial problems, power issues among the different authorities, and major delays in establishing the metric system.[56] Moreover, property owners who found the individual tax distribution unfair frequently became embroiled in disputes with tax authorities. Thus, on January 27, 1808, the Ministère des finances outlined instructions in case of disputes: "If a portion of land is claimed by two or more people, the surveyor will seek to reconcile them; if it does not succeed, it will indicate on plan, by dotted lines, the apparent limits and will temporarily assign to each owner the portion of which that appears to be in possession at the time of the survey."[57] These legal regulations were significant because they made it clear that negotiations were to occur on and through a graphic image. No longer was text or even oral argumentation the means of resolving disputes: instead, lines became the arbiters, assumed to be authoritative and neutral.

In 1807 Napoléon decreed the creation of a plot survey cadastre. Martin-Michel-Charles Gaudin, duc de Gaëte, the minister of finance, was directed to oversee the operation, which included not only measuring the plots but also evaluating the taxable products and income from that production.[58] Graphic images were preferred, but there was still the question of how detailed these plans should be and which details were necessary to include. By the early nineteenth century, the focus was no longer on particular and localizable visual topographical detail, but rather on the system as a whole and the relationship between parts. Early cadastres had included depictions of each building's massing and exterior façades. However, these quickly gave way to the *cadastre napoléonien,* which outlined only property lines and the interior plans of buildings. Its representational value was in displaying invisible connections, such as the hidden relationships between wealth production and the internal organization of buildings behind the façades. The graphicization of the terrain through surveying thus refers to geometric representations and diagrammatic compositions themselves as well as to the way in which these lines proposed social and spatial relations between producer and consumer, public and private, state and subject.

While the aim was to create a more efficient and standardized means of land management, the production of the cadastre napoléonien created its own array of property disputes after it began in 1807. As the departments were charged with surveying all of the properties in their respective administrative territories, multiple discrepancies arose. A whole new survey had to begin again in 1826, and the last departmental cadastre was completed in 1850.[59] By law, only a map was stipulated. Moreover, only through the cadastre's survey and drawing were property owners able to see and know boundaries, a visuality defined and controlled by the state administration.[60]

This visuality was essential to the governance of urban space and represents the modality used for the "rationalization of governmental practice in the exercise of political sovereignty."[61] Michel Foucault argued that power is not an object to be possessed or exchanged, but instead is a series of diffused and diversified relational processes; governmental institutions and their technologies then become a significant site to investigate the means in which political power is formed and manifested.[62] Important to Foucault's analysis is how rational knowledge is both produced by and a product of the procedures and discourses that compose governance, but he offered few specifics on the materiality and modality of that knowledge.[63] During the eighteenth century, the graphic terms of this rationalization were negotiated through the debates over the representation of land and their new relation to property value. This included the question of who had the authority to draw these images and, eventually, the question of how the authority of specific individuals was displaced by a standardization that created an illusion of no specific authorship. The regulation of the plans and their purported advantage over text as descriptions of territory was thus a technique of political power.[64] Moreover, this power was no longer centered in the body of Louis XIV, but instead dispersed through a burgeoning bureaucratic apparatus. The bureaucracy's growing complexities and the anonymity of its administrators concealed the power that these images had over the economic and social lives of the people, who were precisely not pictured in these orthographic representations.

ORTHOGRAPHIC ADMINISTRATION

What these laws and regulations indicate is the increasing convergence of governance practices and mapping, and, significantly, their formulation as orthographic plans. Leon Battista Alberti first outlined the principles of orthography in his fifteenth-century *Descriptio urbis Romae,* whose ideas were implemented by Leonardo da Vinci in his 1502 plan of the city of Imola—often cited as being among the first orthographic urban plans.[65] Due to the territorial divisions, however, their commissions were discrete, and these images were largely independent from the development of a centralized bureaucracy, not comparable to the French administrative apparatus.[66] Conversely, as Hilary Ballon explains, stereoscopic and perspective views were often used to develop military strategy, as well as to begin creating a coherent image of France by demarcating its boundaries on paper for the first time.[67] Thus, while orthographic images of cities had existed earlier, and while urban maps using other representational modalities had been used for administrative purposes, the practice of governance and the making of maps largely remained distinct operations until the eighteenth century.

Significant early urban orthographic plans include the plan of Rome by Giambattista Nolli from 1748—initiated because the reorganization of administrative units prompted the boundaries to be accurately indicated—and the 1793 *Plan des artistes* based on Edme Verniquet's surveys and plans of Paris, which marked the revolutionary government's urban projects for the city. In both examples, the ground plan was prioritized over elevations. The city's orthographic representation flattened the surfaces of the city to ninety degrees and projected them as planes on a single surface. Unlike a bird's-eye view that privileges the singularity of buildings through the identification of their façades, an orthographic plan reduces all forms to a shared notation, while privileging location.[68] It serves as an instrument that produces abstract graphic knowledge of the city, based on the image's geometric correspondence to the actual terrain. The linear qualities of orthogonal images no longer reproduced a direct visual experience of a built environment; instead, they represented the structure of a city whose graphic validity lay precisely in its inability to be compared to direct experience. Thus, the shift from bird's-eye to orthogonal views marked a change from drawing what could be seen, to revealing that which could only be seen by drawing. Indeed, along with the development of this mode of representation were new subjects of mapping that were not visible, including geology, hydrology, and, eventually, social information and statistics.[69]

Orthographic plans had been used in the practice of architecture and military engineering since the Renaissance, but they were still often tied to textual description.[70] By the eighteenth century, however, the link between image and text began to decouple, and the map's utility for description shifted.[71] As Gilles Palsky argues, it signaled an important change from "*la carte spectacle*" to "*la carte instrument*": from the map as spectacle to the map as instrument.[72] The increasing dominance of the image (and specifically the orthographic image) over text in geographical description was part of a phenomenon referred to by historians as the "geometric spirit."[73] The adoption of a cartographic language that prioritized the voids, lines, and points that were produced through geometric concepts of space was related to a growing culture of diagrams and an appreciation of quantitative representations and methodologies. Theodore Porter and Ken Alder, among many historians of science, have demonstrated that this impulse to quantify was based on an administrative desire to acquire information about the environment, its people, and resources to make judgments based on comparable data.[74] As maps became diagrammatic in their expression, they became instruments that were simultaneously products and productive of urban knowledge for the administration. As Hélène Vérin argues, the map "is not a representation of a space but recognition of a space with a view towards action."[75] In its use by the administration, the propositional statement of the orthographic map compelled locative action by the administration. The map shifted from being a statement of "this is" (description) to an action of "this is *there*" (prescription) that was ultimately the aim of planning: to know, control, and distribute resources.[76]

The promotion of the map as its own autonomous instrument took place within the context of tax reformation efforts led by the physiocrats. The physiocrats argued for an *impôt unique* (exit tax) paid to and collected by the central state, and held that land was the source of wealth. In 1776 a *taille tarifée* (use tax) was introduced by the government official, Louis-Bénigne-François Bertier de Sauvigny.[77] A taille tarifée or *taille proportionnelle* was taxation based on real wealth determined by land. The taille in Paris was radically different, as it linked taxation to a measured surface area; it required a map that was known as a *plan d'intendance* (stewardship plan).[78] Eventually, these plans were combined into a system by an ingénieur-géographe, Jean-François Henry de Richeprey.

In de Richeprey's 1782 *Projet de règlement pour les ingénieurs-géomètres* (Draft regulation for surveying engineers), he outlined the methodology and elements for the French cadastre, a map defining property for the purposes of taxation.[79] The goal was to be able to see a given territory in its totality at once. In 1789 the concept of "*synoptique*" officially entered the *Dictionnaire de l'Académie française*, meaning a "general view, in a single glance"; it was cited frequently by administrators and technocrats as a necessary quality for making decisions based on visual comparisons.[80] On December 1, 1790, a law was passed abolishing all the old taxes, replacing them with a single property tax to be divided equally among all properties. The discussion surrounding the execution of this law made the use of orthographic maps indispensable for government.[81]

Claims of the maps' synoptic representation were belied by their incapacity for total representation, as more information had to be continually gathered and documents had to be constantly produced to explain and augment the images. Their proliferation during the nineteenth century contradicted the map's supposed brevity and efficiency. Graphic compositions were supposed to be universally and immediately legible, and it was because they were not texts that they were appreciated as objective. Yet their insufficient clarity only compelled the creation of further paperwork. Moreover, these papers, produced to visualize, manage, and control land and resources, came in such great quantities as to be often unmanageable. The prevalent administrative and subsequent archival category of "*divers*," which included a myriad of papers that did not fit into prescribed categories, points to this phenomenon.

This contradiction between the representational claims of the document and the document's inability to represent did more than determine the organization of the urban administration; it also conditioned practices of urban planning. However, the pervasiveness of this paperwork alone—and specifically, the graphic images that made up this paperwork—was not what made it significant. As Lisa Gitelman has argued, documents are important for their epistemic power: their capacity to frame facts and to make them operational through their "know-show function."[82] The link between knowing and showing was specific to the intellectual values of the nineteenth century and its increasing appreciation of objectivity, such that the document's showing became knowing.[83] This slippage occluded the epideictic strategies of persuasion through blank spaces, numbers, and lines, ultimately allowing for their mediating function to become invisible as the plans aimed to make the world more and more visible.

"IN A SINGLE GLANCE"

By the end of the eighteenth century, the geometric form and graphic mode of the map was not simply a cartographic and technological development, but was embedded in a changing meaning of land. Within the context of tax reformation efforts led by the physiocrats, who believed land was the source of wealth, the map and the diagram as related geometric figures held particular significance. Paul Alliès has observed that the measure of the terrain was both a technical and an economic endeavor.[84] It was ultimately the graphic image that mediated the land's value. Physiocratic philosophy relied on numbers and the composition of numbers in graphic diagrams. One fundamental formulation was the *Tableau économique*,

drawn by François Quesnay, who trained as a draftsman and sought to provide a synoptic view of their proposed economic system.[85]

At the top of the table were three categories of expenditures, and centered below them was revenue. From revenue, two dotted lines spanned outward and inward, connecting different annual sums that formed two columns. Writing to his most ardent supporter, Victor Riqueti, marquis de Mirabeau, Quesnay provided an explanation that emphasized the table's graphic composition:

> It lets us apprehend in a single glance the sources and product of earthly riches, the links and interactions among its various parts, and the principles behind the "economic governance" of agricultural nations; thus the zigzag, once properly grasped, condenses a great deal of details and paints for the eye complicated processes that are difficult to grasp by mere intellect, or decipher and make sense by means of discourse; such ideas would soon disappear, but by means of the Tableau these relations are verified [*apurées*] and fixed in the imagination and will not fade away, or at least they will be easy to conjure up again in their proper order and correspondence, to be meditated upon at leisure without losing sight of anything and without the mind having to organize this material.[86]

As Liana Vardi argues, Quesnay privileged visual comprehension and advocated for graphic over textual representations in his emphasis on the relations of parts. This diagram would better allow for individuals to repeat a mental process, and ultimately to reach the same conclusions—which is to say, it would model the scientific method for discovering truth through reproduction.[87] That truth would be so clear as to be unmediated and immediately graspable in a "single glance."

However, it was not simply that the diagram, in its reduced graphic syntax, could efficiently and effectively communicate complex economic relationships. The tautological and rhetorical advantage of the graphic form was its specificity, both in rendering complicated discourse synoptically and in offering evidence by means of following the zigzag. The diagram was on the one hand a mediating instrument and method, and on the other hand, unmediated in its reception. As Catherine Larrère argues, "The study and use by the Physiocrats of measurement and calculation is thus inseparable from that of their theoretical approach."[88] The same held for graphic images and why diagrams maintained such an important place in the body of their economic arguments. The calculations organized into numbers were understood and used as evidence, which presupposed a kind of "arithmetic image."[89] Diagrams and geometric maps had the same capacity to present their forms as given through their synoptic composition. Their reductive forms suggested that no interpretation on the part of a producer had been necessary for their production and precluded any need for interpretation by the reader. The graphic line was as natural as the terrain itself, especially when drawn to the metric system that claimed to be nature's own measure.

Quesnay's goal of creating an image that could reveal the relations of the whole economic production process was supported by Karl Marx's reading of his work as:

> an attempt to portray the whole production process of capital as a process of reproduction . . . to present the circulation between the two great divisions of productive labor—raw material production and manufacture—as phases of this reproduction process; and all this depicted in one *Tableau* which in fact consists of no more than five lines which link together six points of departure or return. This was an extremely brilliant conception, incontestably the most brilliant for which political economy had up to then been responsible.[90]

What Marx considered remarkable was this diagram's ability to reveal relations as the basis of a political economy, rather than simply isolated individual categories of producer and consumer. It was the line and its synthetic and synoptic reception as established fact that made this an effective image for comprehending the economy as a system of production.

These qualities of coherence and immediacy expressed as "one Tableau" were reiterated in an 1831 editorial justifying the establishment of a specific maps and plans department in the Bibliothèque du roi. In the editorial, Edme-François Jomard, the director of the *Description de l'Égypte*, argued:

> We read a book, we perceive—so to speak—a map. A book is read word by word, and page by page; a map allows one to comprehend the whole subject at once. The maps are also a description like a geography book, but a graphic description. Each of these products address a different intellectual faculty. The geometric

figures and the graphic sections of a map are nothing other than geometry, which has the attribute of putting before one's eyes, simultaneously, the totality of the elements of the study's subject, and the perception of the whole in an instant.[91]

In order to justify establishing this cartographic collection, Jomard felt obliged to define the cartographic medium. For him, its specificity lay in its ability to present a figure in its totality all at once, in contrast with the narrative form of a book.[92] During the 1820s and 1830s, Jomard was centrally involved in debates around the creation of a distinct collection dedicated to maps and plans. He outlined the key characteristic of maps that gave them operative value, defining maps as synoptic and repeating the phrases: *"tout à la fois"* (at one moment); *"simultanément"* (simultaneously); and *"l'ensemble à l'instant complète"* (together in a total instant). In advocating for a collection apart from the prints department and the books collections, he outlined and emphasized the maps' qualities of graphicism, simultaneity, and geometry.

The value of the map as a representation apart from textual (and narrative) descriptions was shared by many of Jomard's contemporaries. One of them, Chabrol de Volvic, was a classmate with Jomard at the École polytechnique, a collaborator in the *Description de l'Égypte*, and the prefect of the Seine. His work on the *Description* as an engineer had been to determine the extent of the population of Egypt, the country's amount of cultivatable land, and the number of villages, in order to establish and draw a cadastre.[93] Chabrol's introduction and development of quantitative methodologies during his service in Egypt informed two significant administrative activities: the first, a plan for the new city dedicated to Napoléon Bonaparte; and the second, the establishment of the Recherches statistiques, a census of Paris that appeared from 1821 to 1860.[94]

The Recherches statistiques represented a new vision and knowledge of the city through numbers, shifting from the material qualities of monuments and buildings to climate, infrastructure, topography, and population quantities.[95] It was meant to facilitate the understanding of relations between units, defined in this case by persons, within a grid that allowed for the whole population to be captured. The census, he explained, took "the table form as the most concise expression and the most suitable for making comparisons easily."[96] It reduced the city to numbers, organized by arrondissements and quartiers, whose spatial distribution allowed for comparisons between parts within a whole based on a common denominator. In this way, the organization of the statistical data assumed a gridded logic that assumed a total view.

Furthermore, neither the orthogonal maps nor the census produced by and for the Paris administration expressed any authorial perspective or any specified viewer. For example, the format of the census was a collection of data in multiple volumes with no single author: Chabrol de Volvic had overseen the work with Jean-Baptiste-Joseph Fourier the mathematician and Frédéric Villot the statistician. This quality of being a collective product is part of the basis for scientific objectivity, in which knowledge is not based on a specific individual.[97] The shift from bird's-eye to orthogonal plans and planar views compelled a change in how the city was viewed, as well as a change in the status of who viewed the city. In perspectival maps, the buildings' façades and urban topography were made available for visual observation in a specific direction and accessible to anyone; however, the planimetric mode completely removed the mediating presence and position of an assumed observer. The orthogonal grid referred to the flat material surface and thus functioned as a self-referential system, making the urban image independent from its being seen. It no longer needed a viewer to activate and validate its description. And while the direction of the text on the map may have linked the top of the page to the northern cardinal point, the gridded image of Paris itself could be moved and turned around on a table while remaining constant, regardless of where the viewer stood. Ultimately, the gridded image became autonomous from any viewer, supporting the image's status as objective.

Using both tables and maps, Chabrol de Volvic promulgated a new quantitative approach to administration as prefect of the Seine, linking numbers to space and graphics to cartography.[98] Like statistical graphics, orthogonal plans presented each function as a distinct layer using the same modality, allowing for comparisons not only within each layer but also across them in their assemblage: "The widening of the narrow streets in the shopping districts, the sidewalks, the sewers, a distribution of abundant water, on all the points of the surface of Paris are, so to speak, public needs which must be satisfied above all and we have just seen that all these operations were successively studied as a whole."[99] As Chabrol de Volvic explained in his memoirs,

Paris was to be understood in its multiple levels, and its reorganization necessitated a total view. Furthermore, he remarked on the overlapping fabric of urban functions, demonstrating an approach to the reorganization of Paris informed by the emergent concept of "*réseau*," or network. Significantly, this articulation of a network was simultaneously linked to the delineation of a boundary and the definition of a coherent whole.

Haussmann would also remark repeatedly on the value of synopticism in his memoirs, especially with regard to a large map in his city hall office that allowed him to survey the city, where he "could at any moment, but by turning around, look for a detail, check certain indications, and recognize the topographical connections between the neighborhoods and districts of Paris."[100] This was supposedly a copy of the map shown to him by Napoléon III with colored traces for the planned transformation of the city—the map from which the modern city was "apprehend at once" and from which the city was built.[101] Maps and plans would, under Haussmann, become the primary medium of urban administrative practice. And this was, significantly, his legacy for the transformation of Paris, rather than the construction of the roads (since the last network was finished during the Third Republic). The creation of a new administrative structure made the Service du plan the central node—which was what allowed the projects to be completed after Haussmann was dismissed. Located in the Division des travaux publics municipaux (Division of public municipal works), this mapmaking service was charged with producing all of the maps for every urban project and provided the language of communication among administrators.

These forms of practice and the orthographic mode of representation testified to the immediacy upon which the map's truth value depended. This capacity to "apprehend the whole subject at once" was defended in opposition to the text; the map was a result of scientific methods and was also a scientific instrument itself.[102] Accordingly, the great number of services established in order to produce maps, as well as the technical training of the engineers and bureaucrats ("officials and scribes," in Max Weber's terms) who were necessary for those maps to be understood, contrasted with the simplicity of their reception.[103] The paradox was that the map was indeed a product of scientific methodologies (that is, mediated through geometry and physical labor), and yet immediate in its visual reception (that is, transparent). This contradiction permeated the organization of the Paris administration and its actions. The plan was considered scientific, drawn based on quantitative methods, and as a consequence believed to stand apart from politics. And because the administration was centered around this plan, its bureaucratic practices were also understood to be technical and objective, not political. The effacement of mediated production by seemingly immediate reception is, as William Mazzarella theorizes, "a political practice that . . . occludes the potentialities and contingencies embedded in the mediation that comprise and enable social life."[104] These graphic images ultimately offer and settle on a single solution, and in the case of Paris, that solution was administrative.

THE SERVICE DU PLAN

After the French Revolution, the Conseil des bâtiments civils emerged as a state authority under the Ministère de l'intérieur to administer and oversee all civic building projects. Replacing the functions of the office of the Bâtiments du roi, albeit not in all aspects, the body proposed a standardized construction process that would ensure financial accountability and aesthetic control. As both Lauren O'Connell and David Van Zanten have argued, this body would enjoy increasing influence through the nineteenth century and "attain the status of a full-blown bureaucracy."[105] Importantly, the council initially included six auditors in addition to its six architects. Their equal numbers speak to the growing value of the bureaucrat in the process of construction.

Haussmann was no novice at administration. To gain command of the prefecture meant to gain control of the bureaucracy, and by 1854 Haussmann controlled the municipal administration, having placed his men in charge of newly organized services under the umbrella of the Direction des travaux de Paris.[106] The Service du plan was at the top of this new administrative structure.[107] As the prefect explained, "geometry and graphic drawing play a more important role in governance than architecture."[108] The Service du plan centralized the production of all official maps and plans, and was kept independent from the other services in order to preserve a screen of confidentiality between surveyors and builders and avoid speculation.[109]

Producing maps had previously been the responsibility of the Bureau de la voirie (Office of roadways)—other responsibilities included surveying, laying out, and building the streets. There was a separate Bureau de géomètre

du cadastre that was concerned with mapping property limits and the divisions of urban blocks for the purposes of taxation.[110] Thus, the old administrative organization was based on one office that handled the mapping and building of roads—that is, exterior spaces between the urban lots—and another office that dealt with the interior spaces within the lot. What was significant about the Service du plan under Haussmann was that all map production and distribution was consolidated into a single governmental office that produced maps of property and roads—both private and public spaces.

The urban plans made by this agency promoted a synoptic view of the city that was consonant with the office's position at the head of an administrative and governance structure. Moreover, all the plans referred to Eugène Deschamps's survey that was executed in 1858 with the annexation of the suburbs, and drawn and printed in 1867.[111] With the city undergoing massive construction projects in its central neighborhoods, Paris was not the stable reference: the map was. This map of the city became the reference for all other plans, as well as the reference for administrative practices related to the planning of Paris. For forthcoming construction, the map functioned projectively, outlining the city's form; for construction underway or completed, the map functioned descriptively, as a means of orienting the context of those actions. Accordingly, the map represented more than a particular terrain: it represented the aggregation of a social and political system, one that linked different temporalities as well as different processes. With this image, an urban administrative structure was organized.

The duties of this agency were not themselves new. The task of making maps had already existed, albeit in a fragmented way, in several areas of the city's municipal administration. The significance of the Service du plan was not in its approach to the organization of Paris but rather in the administrative actions that became linked to the plans—bureaucratic actions that assumed the visuality and legibility of orthographic representations. The architecture services, which had previously been organized into six positions, under Haussmann swelled to eight subdivided ranks in the new office of the Direction du service d'architecture. In the Service des bâtiments of the national palaces, the architect was at the service's head, but immediately below were the inspector and the auditor, in charge of keeping the records.[112] The location of an auditor and accountant just below the senior architects, and the specific naming of the filing clerks, speaks to the bureaucratic logic that sustained this hierarchy, and to how much paperwork was produced through this labor, as well as how it conditioned the building processes. What held this network together, along with the many different personnel, was the shared plan.

This abundance of paperwork was in part due to the increasing trust in the process and in the methods of documentation itself—trust, that is, that observations of the world could be reduced to something marked on paper and read by an increasing number of often-unnamed officials. Frequently those marks took the form of a graphic plan. Thus, the trust in this graphic paperwork also owed to a belief in the plan's capacity to accurately describe the terrain and the world. This belief was not only located in the image itself, however, but also in the institutions created to produce and maintain those plans and their authority. These institutions were established through paper, whose material and content were in turn maintained and validated by those very institutions, such that they mutually constituted one another.

The extensive traces of the many and oftentimes unknown actors involved in the administration and in the production of documents, however, is at odds with how histories of planning and architecture have often been written. Any visit to the French archives will reveal cartons of letters, registers, reports, plans, proceedings, scraps, even doodles that are unsigned and unattributed. A register will indicate various authors identifiable by the different inks and handwriting, but no specific name. Paradoxically, it is precisely this quality of anonymity based on collective production that gives the documents their authority as fact. Extending Robert Beauregard's argument about the role of things in planning practices, it conveys the agency of the document in the process of urban planning. Beauregard sees planning through ontographies that include processes, methods, objects, places, and people that in their aggregate constitute its practice.[113] His new materialist approach to the analysis of urban planning is methodologically significant in shifting the focus away from individuals as well as erasing false distinctions and hierarchies between buildings and representations as separate sites of inquiry. He describes how objects such as documents determine and compel certain human actions related to planning, and it is precisely the anonymous quality of these documents that confers their social responsibility in the shaping of the

urban environment. For Ola Söderström, the plan was "the visual order of civil servants," calling attention to the fundamental role of municipal administrators in developing the rules and regulations that ultimately defined urban planning by the late nineteenth and early twentieth centuries.[114] These incremental movements over time by unknown men had an immense influence on establishing normative practices of design that required the production of a massive amount of paperwork.

NEGOTIATING STANDARDIZATION

While examples of preprinted forms for routinized administrative tasks date to the ancien régime, by the nineteenth century, the shared document came in a standardized form. Earlier, general subject titles, department names, the lines of a given form, and all of its specific contents were individually handwritten and drawn on irregularly cut sheets of paper as manuscripts (fig. 4.1). Ledgers such as the one pictured here show that even before printing was adopted for these bureaucratic forms, the necessity of standardizing information was present.[115] Cornelia Vismann notes that this standardization extended to style and rhetoric, such that even elaborate ceremonial forms of address were reduced to a graphic vertical line drawn in the margin of the document whose length marked the difference in status between the sender and receiver of a form.[116]

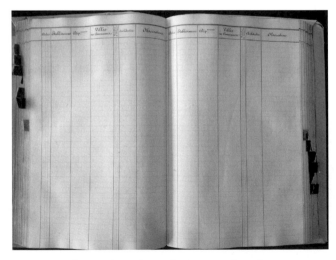

FIGURE 4.1 Before mechanical reproduction was adopted, administrative documents in manuscript form demonstrate that standardization was an important priority for the municipal bureaucracy. Directories of plans concerning construction projects submitted to the Conseil des bâtiments civils, 1800–1851. Archives nationales de France.

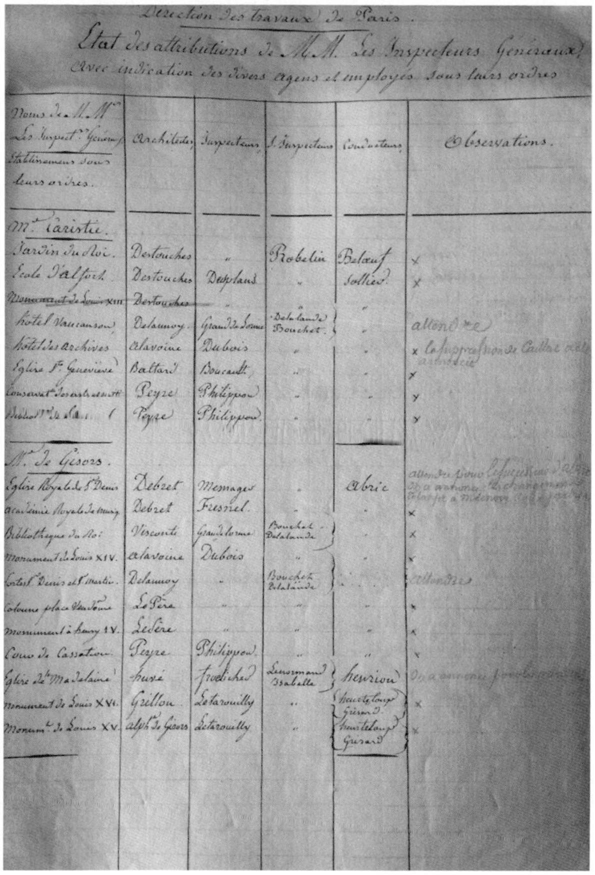

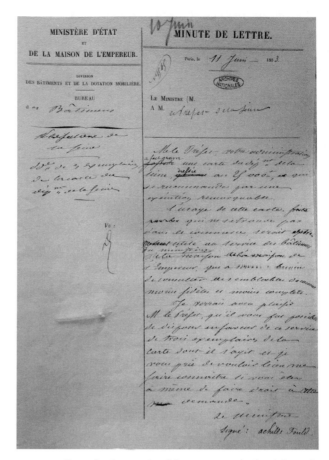

FIGURE 4.2 Example of a handwritten manuscript sheet on a customized sheet of paper. Organizational structure of the Direction des travaux de Paris (Department of public works of Paris), inspectors general, 1830. Archives nationales de France.

FIGURE 4.3 Example of a preprinted form on a standardized sheet of paper. Organizational structure of the Division des bâtiments et de la dotation mobilière (Division of buildings and of staffing), letters to the Préfecture du Département de la Seine, Paris, June 11, 1853. Archives nationales de France.

By the early 1800s, however, headers, titles, the words "*fait à ___*" (made on), and the outlines of text boxes often came printed on the sheets. Compared to a manuscript form (fig. 4.2), a preprinted form (fig. 4.3) was enabled by new technologies in job printing and facilitated much of the communication among the bureaucrats, who were tasked with completing them. An extension of blank spaces that appeared around diagrams during the eighteenth century, these blank forms –blank because they needed to be filled in—set up the parameters for the entries and their use.[117] However mundane and ephemeral, these paper objects taken together testify to the enduring rules and categories that controlled and regulated interactions within, and knowledge of, the government.

Standard forms also attest to the number of actors involved in creating a document. There were those completing the form and adding cumulative pieces of information, but there were also the readers of the forms, those who catalogued and filed those sheets, and the job printers who created the forms and distributed them. These actors' mutual anonymity, along with the eventual replacement of handwriting by type in the 1870s, contributed to the document's value as objective. As Gitelman explains, this genre of the blank form is an "expression of the modern, bureaucratic self," conveying information without inspiring the "readerly subjectivity" of identification.[118] The forms "didn't have readers, then; they had users."[119] Like orthographic maps or censuses produced by and for the Paris administration, they expressed no authorial point of view or specified viewer. While most atlases for sale named their publisher and director, the various men involved in the tasks of their production, from surveying to drawing and printing, were left unknown. It was via their capacity to routinize tasks and mental processes through standardization—as Weber argues, via their capacity to dehumanize—that they acquired their value as objective.[120]

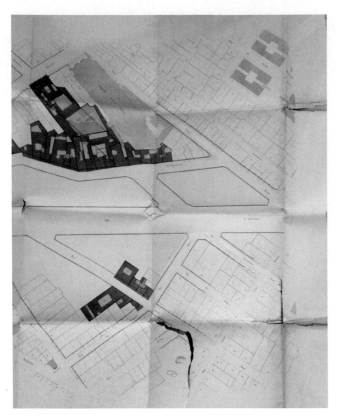

FIGURE 4.4 Plans related to the leveling of the Butte des Moulins, 1877. Archives de Paris.

They did not elicit any affective relationship but rather the contrary—insofar as the presence of an individual person was "refracted," as Ben Kafka writes, "through the medium of paperwork."[121]

While these textual forms were easily standardized through job printing and mass-produced paper sheets, maps were much more difficult documents to standardize. Each particular urban site had its own set of issues and tasks associated with any intervention, and each map offered a necessary and specific view, with the result that there were frequently multiple kinds of map drawn for the same area. Often the plans associated with a dossier were of various dimensions, and then folded in different ways to fit into the file folders. The archives contain a great diversity of maps and plans. Some are simple line drawings, while others are fully watercolored by hand, even if there is no maker credited.

In a dossier associated with the construction of the avenue de l'Opéra, the Butte des Moulins was slated to be leveled. Included in the dossier are several types of maps and elevations, often colored and annotated by hand. One plan consisted of a survey of the properties immediately

affected by the extension of the rue des Pyramides between the rue d'Argenteuil, the rue Saint-Honoré, and the alignment of rue Sainte-Anne. Specific lots were identified in red, orange, yellow, and green (fig. 4.4). This plan accompanied an expropriation list of addresses, associated owners, precise surface areas, and prices. In response to issues raised by property owners, an engineer was asked to provide further firsthand observations to justify the project.[122] The main point of contention was how to level the steep hill. A response was offered that included a simplified and hand-drawn site plan that added a second street to divide the block and separate it from the church of Saint-Roch (fig. 4.5). A comparison shows the difference in scale, measures, colors, and sizes, which was meant to highlight competing arguments about how the blocks should be planned. If maps resisted standardization, what was normalized were certain types of plans, as they appear on preprinted text forms, listing the requirement of a site plan of the given area ("*plan annexé*") and a plan of the lots ("*plan parcellaire des propriétés*") (fig. 4.6).

Nevertheless, there were attempts to create professional standards for mapmaking. Since the 1840s, there had been a determined effort to establish professional standards and outline regularized methods in surveying. The *Journal des géomètres* printed its first issue in May 1847 with the aim of bringing some measure of official recognition to the profession—particularly in light of the government having just announced the resurvey of the country's cadastre. Admittedly, the surveyors did not even know their own numbers, as the profession was wholly decentralized.[123] Outside of the engineering schools, there was no school dedicated to the training of surveyors, and consequently no uniformity to their education. Indeed, that first issue of the surveyors' journal published a petition to the minister of the interior to establish an official Corps des géomètres-arpenteurs, aimed at creating standards for their methods, gaining some protections, and limiting their numbers. No corps was established, but the definition of administrative standards brought recognition of their value in urban development, as well as imposed regularization on their training. To gain government employment, a surveyor had to sit for an exam, which required specialized knowledge in a number of subfields, including the use of surveying instruments, trigonometry, descriptive geometry, and orthographic drawing.[124]

These basic standards were supported by Haussmann's successor, Alphand. If, in earlier decades, maps were

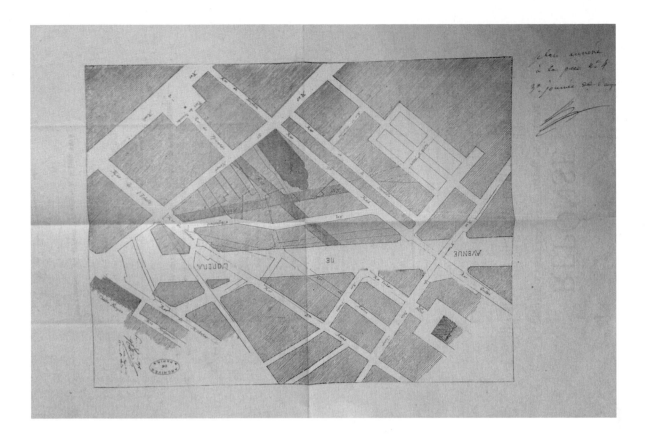

FIGURE 4.5 (above) Appended plans related to the sanitation and leveling of the Butte des Moulins and the planning of the avenue de l'Opéra, 1876. Archives de Paris.

FIGURE 4.6 (right) Dossier related to the sanitation and leveling of the Butte des Moulins by the Préfecture du Département de la Seine, March 3, 1877. Archives de Paris.

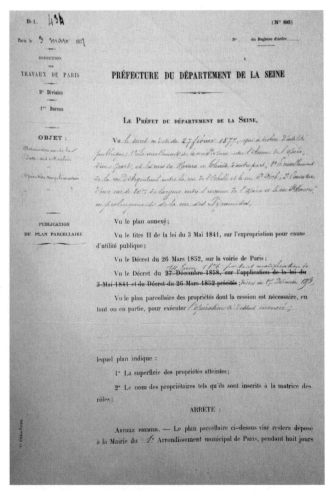

produced in light of the needs of the site, by the end of the nineteenth century, the sites were made to conform to the framework and medium of the map. One of Alphand's major publications as head of the service was *Les promenades de Paris* (1867–73). It catalogued all the green spaces and parks designed, developed, and built during his tenure, including the Bois de Boulogne and Bois de Vincennes, as well as the trees and street furniture lining many of the new boulevards. Each section's organization begins by describing the process of construction, and then the typology of the structures and objects that define the space (fig. 4.7). In the urban plans of the parks, Alphand represented the surrounding blocks orthographically, and emptied them of any architectural objects. This highlighted the picturesque depictions of the green spaces, in which the trees and planting were depicted in three dimensions and the buildings in plan (fig. 4.8). Through

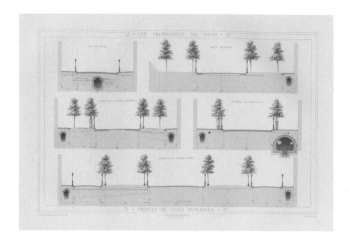

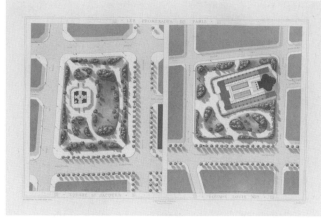

FIGURES 4.7, 4.8, AND 4.9 From Adolphe Alphand, *Les promenades de Paris: Histoire, description des embellissements, dépenses de création et d'entretien des Bois de Boulogne et de Vincennes, Champs-Élysées, parcs, squares, boulevards, places plantées, études sur l'art des jardins et arboretum* (The promenades of Paris: History, description of improvements, expenses for the creation and maintenance of the Bois de Boulogne and of Vincennes, Champs-Élysées, parks, squares, boulevards, green spaces, studies of the art in the gardens and arboretum) (Paris: Rothschild, 1867–73). ETH-Bibliothek Zürich, Rar 10062.

FIGURE 4.7 (top left) "Profils de Voies Publiques" (Elevations of public roads).

FIGURE 4.8 (top right) "Square St. Jacques and Square Louis XVI."

FIGURE 4.9 (above) Examples of several green spaces.

detailed views, plans, and topographical studies, the text sought to provide a record of design solutions arrived at through rational and standardized processes. The trees and greenery were depicted individually, but their shadows were always angled at forty-five degrees to the lower-right corner of each sheet, regardless of the actual location of the site and its relationship to the sun's arc (fig. 4.9). The regularization of the shadow angles froze the seemingly unique objects in a timeless and placeless state.[125] As Alphand explained in the introduction, "It is necessary to improve the methods and to introduce exactitude into the empirical processes."[126] The use of fixed scales allowed for comparisons across the projects, and their regularized representations presented a standardized rationale in garden design.

Alphand's *Atlas des travaux de Paris*, published for the Exposition universelle of 1889, while Alphand was the director of public works, used many of the same graphic techniques as the previous publications. Additionally, improvements in print technologies—especially lithography—allowed for the production and exact reproduction of plans, eliminating any trace of the human hand's mediation. Composed of sixteen plates, organized around five themes into four periods, and printed on the regularly dimensioned paper sheets, these maps sought to present the current state of Paris's built environment as a consequence of technical endeavors that began with the French Revolution (fig. 4.10a–c). All printed at the same scale and sharing the same base map, the comparable orthographic images advanced the argument that the modernization of Paris was a cumulative effort with no specified author, and indeed, no particular user. Not only did the *Atlas des travaux* and *Promenades de Paris* demonstrate a standardization of the representation of urban space; they also illustrated, under Alphand's tenure, the routinization of the administrative procedures that drew and employed these images. Even if discrepancies persisted among the various maps and plans within the bureaucratic process, these publications displayed a strong desire to construct a normative image of the modern city, both for the public and for the administration itself.

To focus on the materiality, composition, and organization of these documents is to address how they were themselves objects of knowledge within a system that was shaped by and through them. Features of this paperwork, namely their standardization in format and style, were tied to an administrative structure based on increasing distances between people and things, and the subsequent anonymity of authors and users. Moreover, as the French administration specified more objects and practices, such as drawings that were to be managed and controlled by the state, its internal organization became correspondingly more complex and impersonal. Relations were codified through regulations, laws, and rhetorical rules, based on a concept of centralization and hierarchy. But this process was not without complications, and centralization was always ongoing and never achieved—especially when it came to the production and circulation of maps and plans.

Graphic representations, including diagrams, numbered tables, and geometric plans, were exemplary products of scientific methods advanced by the French administration. As Theodore Porter has duly pointed out, "Quantification is well suited for communication that goes beyond the boundaries of locality and community. A highly disciplined discourse helps to produce knowledge independent of the particular people who make it."[127] Yet as the regulations surrounding map production show, the drawing of urban plans was not truly about mathematics, but instead about creating a standardized and shared compositional code. The rules for drawing a measured line, for example, were not so much based on mathematical principles as they were on expressing a rhetoric of science. This "grammaticality," as Gilles Deleuze and Félix Guattari explain, "is a power marker before it is a syntactical marker."[128] Indeed, the claims of the map's objectivity obscured the ways in which economic and social value were ascribed to space.

As plans gained legal status as the means through which different and multiple agents communicated, they were limited in their capacity to resolve the urban issues they set out to address and shape. In the many reclamations by private landowners placed against the Préfecture de la Seine, the Conseil des bâtiments civils often recommended that "a site visit shall be made."[129] The recommendation implied that plans could misrepresent or falsify the site conditions; even if they could be generally trusted, plans could never fully represent a given site and supplant the physical experience. Nonetheless, additional plans and documents were frequently requested upon these visits, further perpetuating their determinative use and role in the relationship between site and administration.

The paper material of the plan that made these documents so easily disseminated was, in the end, the plan's greatest vulnerability. When the renovations of the Hôtel de Ville were completed in 1842, the prefect and his offices were consolidated and located at this new building. The second floor was reserved for the bureaucratic offices related to all the building services. When Haussmann took power, his offices were placed along the building's north side.[130] On the opposite end, the architectural offices occupied the length of the quai de la Grève. The offices of maps were placed between them and the offices of the Service de la voie publique (Service for public roads) and the grand Salle de Fêtes, located along the rue du Tourniquet Saint-Jean. The area was composed of three major rooms that included storage and archival spaces, as well as the office for the conservateur du plan.[131] This space was occupied by the Service du plan until the fire of 1871 during the Paris Commune, which would consume the building and the paper documents contained within (fig. 4.11).

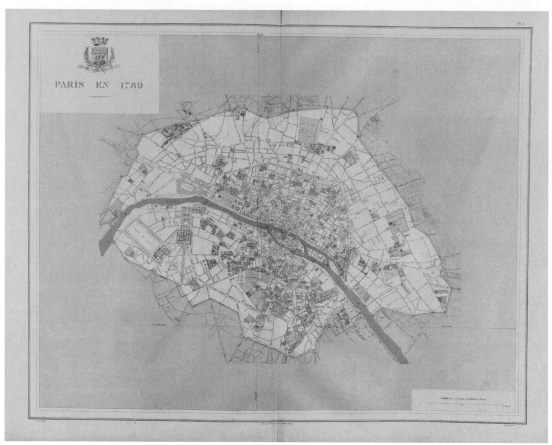

4.10A

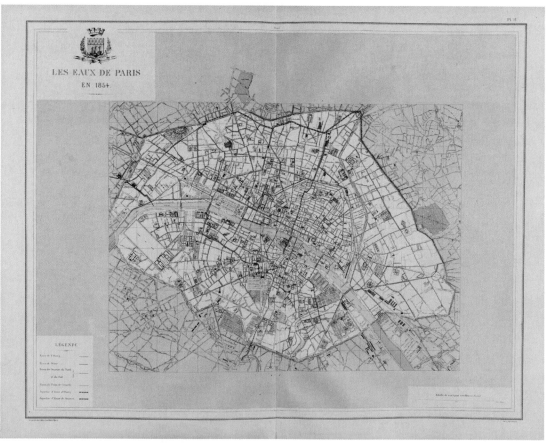

4.10B

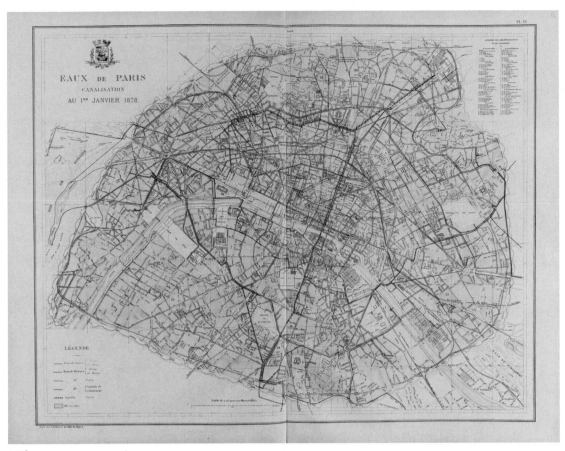

4.10c

4.11

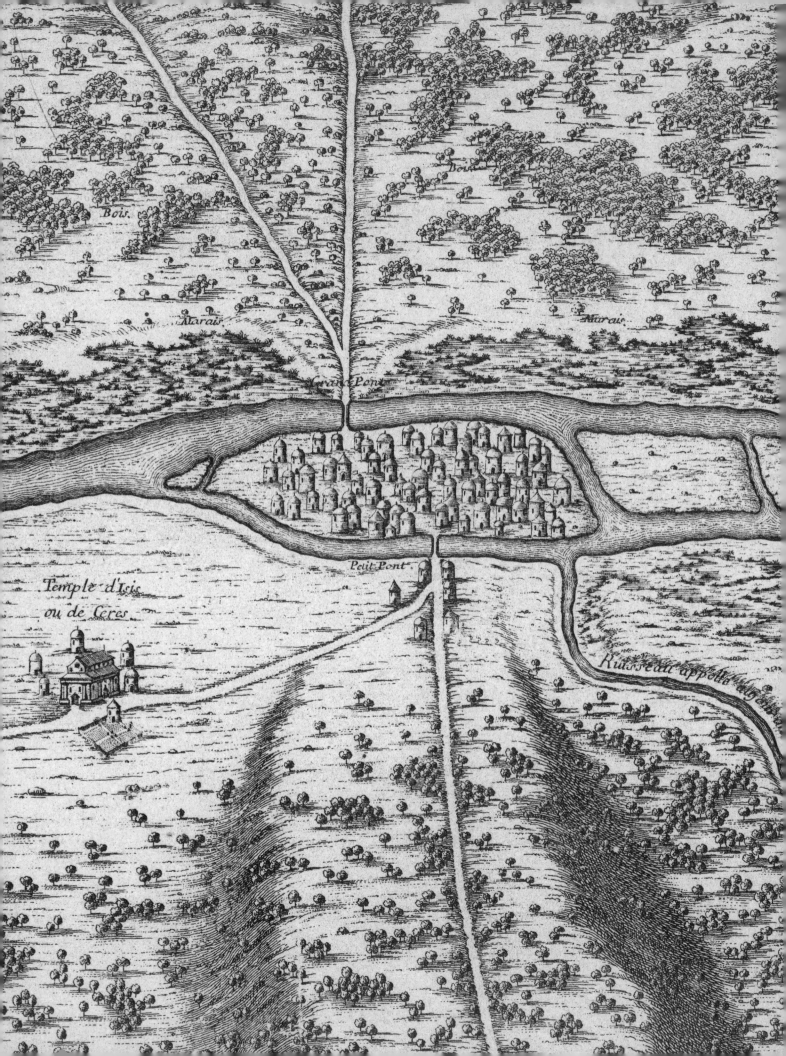

Bois.

Bois.

Marais.

Marais.

Grand Pont.

Temple d'Isis
ou de Ceres.

Petit Pont.

Ruisseau appellé ...

5 Cartographic Presentations

A map of the city represents the production of discourse on a city.
—Louis Marin, *De la représentation* (1994)

In the nineteenth century, the proliferation of cartographic materials converged with emergent technologies of quantification, and together these materials and technologies were incorporated into the political apparatus of the administration of Paris. Importantly, the impulse for quantification did not emerge from mathematicians or scientists but from administrators needing to acquire reliable information about their environment and resources in order to make judgments based on comparable data.[1] Quantifying, counting, and calculating became a modality. These activities generated a whole new genre of maps as a powerful tool for the government, which saw its obligations expand beyond territorial matters into biopolitical spheres.[2] This relationship between information-visualization techniques and administrative and regulatory practices, however, was one of not causal but contingent development. That is, the capacity to visualize and spatialize concepts of population and education, for example, were developed in concert with notions of management and control. Governmental agendas were formed and organized through the media that they used, and maps held a unique position as a means of visualization and an integrated tool of governance. Maps represented the object of governance, the means of governing that object, and the governmental body itself—all in spatial and geometric terms.

MAPPING UNITY

For the Third Republic, cartography offered an important technique to shape and drive its political and social priorities. From its inception, when the Third Republic was proclaimed on the balcony of the Hôtel de Ville on September 4, 1870, its legitimacy had been contested from both external and internal forces. During the Franco-Prussian War, the French army had no choice but to accept defeat with its miserable losses at the Battle of Sedan and the capture of Napoléon III. Paris, however, refused to surrender, and a terrible siege ensued. After negotiating a humiliating peace treaty, in March 1871 the Orléanist Adolphe Thiers, as head of state, marched the French military into Paris against its own citizens to tamp down what had begun as resistance to the Prussians but had transformed into political insurrection. The conquest led to the *semaine sanglante* (bloody week) that crushed the Communards, who eventually left Paris and its Hôtel de Ville burning. Not surprisingly, even after the fires were extinguished, the early years of the republic were immensely turbulent, with discord in many cities and constant challenges by Bonapartists who favored French imperial interests, and monarchists who sought to reinstate Bourbon or Orléanist royal lines.

Against the background of defeat and strife within French society, republicans organized around the goal of universal manhood suffrage. It was a powerful idea around which to unite, yet the actual meaning and practice of republican politics proved to be difficult to define.[3] Third republicanism assumed the accommodation of mass participation in politics, yet there was no guidance as to who these new participants would be and what their citizenship would mean more broadly; republicanism offered no consistent ideology.[4] More than simply offering the right to vote, this political project involved the creation and support of institutions that could allow for participation as well as defining the terms of that participation. Thus, the invention of symbols, ceremonies, rituals, and sites also became critical in the effort to secure and represent a coherent republic, symbolically and geographically.

As a consequence, more so than at any other moment during the nineteenth century, geography, specifically cartography, was used to shape and drive a specific political and social agenda. Unlike earlier times, when maps and plans were incorporated into administrative practices based on positivist terms that sought an expedient means to achieve territorial goals, the republican government produced an increasing number of maps representing its capital city with political intentions. Exemplified in two large-format atlases—the *Atlas des anciens plans de Paris* (Atlas of old plans of Paris) (1878) and the *Atlas des travaux de Paris, 1789–1889* (Atlas of the works of Paris, 1789–1889) (1889)—the urgent and explicit motivation was to preserve valuable documents whose archival value and fragility were exposed by the fire of the Hôtel de Ville in 1871.[5] But by gathering original cartographic documents from the various governmental agencies and assembling them into historical atlases, the administration sought a cartographic narrative of effective administrative control over the city and, hence, France. Through historical maps, the Third Republic sought to establish its historic legitimacy. Via the production of statistical maps, social and spatial concepts were linked, and the cartographic images aimed to define the republic, establishing territorial and social coherence. Citizenship and civic participation were then explicitly tied to map literacy, such that geography and cartography

became integrated into school curricula, and a profusion of atlases both official and popular were produced for national events.

The most significant official events were the Expositions universelles, which—when reinstated—were done so with propitiatory fervor. They served to galvanize resources and activities on many different fronts, both private and public, to a broad audience within and beyond France. Encouraged by rail travel in France and colonial conquests beyond, the geographical reach of the expositions' displays and reception reinforced a cartographic culture in everyday life.[6] The railroad network, in the midst of expanding, brought people from distant places to the capital, while providing a sense of territorial connection between cities and regions. The colonial campaigns in newly acquired African and Southeast Asian territories were brought into the public imagination through the exhibitions, such that the French Empire was neither distant nor trivial, but rather essential to the republican identity as a civilizing project.[7] In all of these arenas, from tourist maps of Paris and illustrations of the colonies to the official atlases of the capital, maps proliferated, such that their power of persuasion moved beyond professionals and technicians.[8] A new cartographically literate public was aligned to citizenship, and the argument pushed by the government—made through maps—was that France had recovered triumphantly into a unified republic.

MAPPING HISTORY

The *Atlas des anciens plans de Paris* and the *Atlas des travaux de Paris* were both published in the early decades of the Third Republic to coincide with two Expositions universelles. Initially, events celebrating the republic's establishment were avoided.[9] The defeat by the Prussians and the bloodshed of the Commune left both the country and Parisians in a somber mood. Moreover, during its early years, the newly formed government tried to distance itself from the excesses of the Second Empire, including the expositions. However, when the republicans definitively secured control after 1876 and 1877, some self-declaration seemed to be in order, and celebrations were planned in the revolutionary traditions of the 1790s and the Second Republic in 1848.[10] It was in the context of these national festivities that two atlases of the French capital were produced for the Exposition universelle of 1878 (the *Atlas des anciens plans de Paris*) and the centennial Exposition universelle in 1889 (the *Atlas des travaux de Paris*). They represented public occasions for the French republic to present its modernization efforts to the international community and, importantly, its own citizens.[11]

The intentions of the organizers and the government for the 1878 Exposition universelle—the first of the Third Republic—were moral and pedagogical in tone. The exposition was meant to carry symbolic weight and to recall a republican identity in its eighteenth-century purity. Historian Jules Michelet, an ardent supporter of these national celebrations, employed historiographic and emotive arguments, claiming that they would provide a civic education and that the gatherings themselves would form an image of patrimony: "Man's education will always take place through festivals. Sociability is an eternal sense that will awaken. . . . Until his last day, he will carry an image of this beautiful living homeland."[12] These sentiments were echoed by government officials and in texts such as Léon Gambetta's journal *La république française* (founded in 1871) that called for celebrations on the centenaries of the deaths of republican heroes such as Voltaire and Rousseau.[13] While these specific public commemorations never occurred, the Exposition universelle was open from May 1 to June 30, 1878, with an intentionally uncontroversial and apolitical theme of "Peace and Work." Popular writer Frédéric Bernard assured in *Fêtes célèbres* (1883) that at the exposition "inequalities are effaced, or at least attenuated for several hours; a common goal, a common idea brings close together the different classes of citizens and develops in them the sentiment of unity which makes the strength of nations."[14] There was a shared belief that such festivals would cultivate republican values, coalesce the people, and offer common political symbols.

Unity was the message, in political and historical terms, and the map was an effective way to broadcast it. The advantage of the map as an argument for French solidarity was precisely its synoptic view. Making a similar point about North American cartographic practices, Margaret Wickens Pearce compares text to map: "The unique power of the map, over the word, is its ability to depict simultaneity."[15] Any specific temporality is erased in the putative coherence of the cartographic image—a stark contrast to the reality of major parts of Paris still under construction in 1878. Thus, cartographically, Paris became a pictorial and temporal unit in which all spatial elements were described in a uniform graphic language. Precisely because the city was visually ungraspable under the repairs from war and the continued infrastructural

works, recourse was to the mapped city. It projected the completed city as contained and organized geometrically, and representable as a unified totality. Significantly, this desire for visual coherence extended into a need for historiographic continuity.

The combined goal of linking the Third Republic to the revolutionary values of 1789 and creating a unified sense of the city for the reading public was evident in the historical scope of the *Atlas des anciens plans de Paris.* Directed by the Conseil municipal de Paris, this atlas was one item in a larger program to preserve cartographic materials and to delineate a narrative of Paris's technological development. The project was managed by the Commission municipale des travaux historiques (Municipal commission for historical works), formed by prefectorial decrees on January 22, 1862, and March 26, 1863, and executed by the Service des travaux historiques (Service for historical works) in coordination with the Commission des recherches sur l'histoire de Paris (Research commission on the history of Paris).[16] The importance of the project merited the formation of a commission with top administrators: Adolphe Alphand, newly named director of public works from 1871 to 1891; L. Michaux, division head at the Préfecture de la Seine; Émile Hochereau, curator of the *Plan de Paris;* L. Fauve, chief surveyor of the *Plan de Paris;* Jules Cousin, librarian of the city of Paris; Alfred de Champeaux, deputy of the Bureau des Beaux-Arts; and Lazare Maurice Tisserand, chief inspector of the city's historical service. They were supported by Eugène Viollet-le-Duc, who had been invested in preservation concerns throughout his career, and by Charles Marville, who was charged with photographing the documents.

The decree of March 10, 1879, specifically describes their mission as being concerned with "all the studies and research aimed at fixing and preserving the memory of the events and men whose history is so closely tied to that of the city of Paris."[17] The writing of history here was understood as a significant administrative function, and as reported earlier in 1867, the conservation and reproduction of rare and important plans was one of the most important tasks in establishing this historical record. Bureaucratic culture embraced historiography: it was aligned with record-keeping and organization, and thus was within the ideal purview of the various clerks, secretaries, and administrators who constituted the expansive bureaucracy. This chronicling impulse was specifically tied to the preservation of documents, including texts, images, and plans. The document preserved a record of the past, and it was assumed they would remain legible as transparent communicative acts regardless of how much time had passed between the document's creation and being read. In a sense, they held a double status as archived objects: historical as an authentic record of the past, and atemporal as an ever-static document, regardless of vantage point. Thus, the preservation of documents and their declaration by the Assemblée nationale as public property to be placed into an archive, initiated by the French Revolution, became under the Third Republic the basis for several historiographic projects.[18]

Work on the *Atlas des anciens plans de Paris* began in 1877, and the Exposition universelle of 1878 gave the project a publication goal. Related works produced by the municipality since 1865 had also demonstrated an enthusiastic concern for urban historiography. Research was conducted under the title *L'histoire de Paris,* which was published as one series dedicated to the reproduction and dissemination of rare and original documents (supported by the resources of the department in charge of maps and plans at the Bibliothèque nationale de France, established by Edme-François Jomard), and another series devoted to historical monographs of Paris. As a pamphlet from 1867 sentimentally interprets: "In recent years, the prefect of the Seine, moved by perhaps the disappearance of so many memories of the Paris of yesteryear, thought of undertaking, supported by the city, a series of publications retracing the history and physiognomy of old Paris."[19] The series included titles by administrators who had also contributed to the historical atlas, such as *Introduction à l'histoire générale de Paris* (one volume) by Lazare Maurice Tisserand; *Topographie historique du vieux Paris* (eight volumes) by Adolphe Berty, H. Legrand, Tisserand, and Theodore Vacquer; *Paris et ses historiens aux XIVe et XVe siècles* (one volume) by A. Le Roux de Liney and Tisserand; and *Les anciennes bibliothèques de Paris* (three volumes) by Alfred Franklin.[20] The close relation between these maps and the growing historiography lay in how the maps, although derived from historical interpretations, were treated in a positivist manner as facts once drawn and printed, and thereupon used to make historical claims. And while many administrators in the early nineteenth century—most loudly Jomard—had argued that maps were distinct from text and narrative, now these cartographic representations became effective in constructive historical accounts.

Through these publications, the history of Paris itself was shaped into a visual object, one contained within a graphic form and the borders of the paper sheet. In total, the *Atlas* compiled thirty-one different maps, several of them composed on multiple plates. All were organized into two parts—*plans rétrospectives* and *plans contemporains*—with the second part subdivided into the chronologically ordered sections of *plans cavaliers* (perspectival) and *plans géométraux* (orthographic). The plans contemporains reproduced images of Paris intended to describe the city at the moment when it was drawn. The plans rétrospectives were designated as historical maps—that is, cartographic reconstructions of an earlier Paris. With the exception of the first image, the following six maps that comprise the plans rétrospectives were all drawn in the nineteenth century as reconstructions of how Paris—and, significantly, how a map of Paris—would appear in a given historical period.

The opening image of the first section and of the *Atlas des anciens plans de Paris* was drawn in the eighteenth century and offered a graphic conjecture of Paris during the Gallic era, which carried historically symbolic weight (fig. 5.1). This *Plan de la cité gauloise* (Plan of the Gallic city) was originally produced for the *Traité de la police* (Treatise of the police) by Nicolas Delamare, published in three volumes from 1705 to 1738. The *Traité* was a significant document that delineated the role of the police as an institution of urban order and administration.[21] The original image was one of eight historical maps drawn and engraved by Nicolas de Fer that were used by Delamare to argue for the necessity of street regulations.[22] The administrative treatise is not a historical account, but its images are, and these historical plans describe the growth of the built environment and its development into an organized grid, as well as the regularization of the territory not only through an accumulation of forms in the image but also through the cartographic modes of its representation. This first plate of the *Traité* indicates only one main road, running north and south with a fork on the Left Bank, while the Seine is drawn on an east–west axis—an anachronistic orientation that dates the production of the map to the eighteenth century (fig. 5.2). The buildings grouped on the Île de la Cité are drawn pictorially in perspective, and already by the second plate in the *Traité* the city is described orthogonally, the urban pattern shaded in outlined blocks surrounded by a thick black line that represents the fortifications. The city of Paris grows outward from this central island in

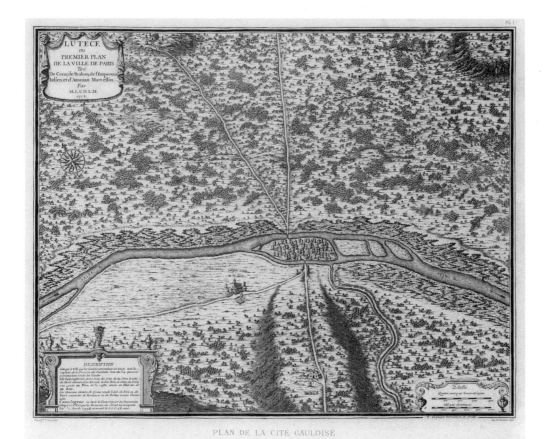

FIGURE 5.1 *Plan de la cité gauloise. Fac-similé du plan imaginé par la Commissaire Delamare et gravé par Nicolas de Fer* (Plan of the Gallic city: Facsimile of the plan imagined by the commissioner Delamare and engraved by Nicolas de Fer). Heliography, 14⅝ × 16½ in. (37 × 42 cm). Plate 1 from Adolphe Alphand, *Atlas des anciens plans de Paris* (Atlas of old plans of Paris) (Paris, 1880). David Rumsey Historical Map Collection.

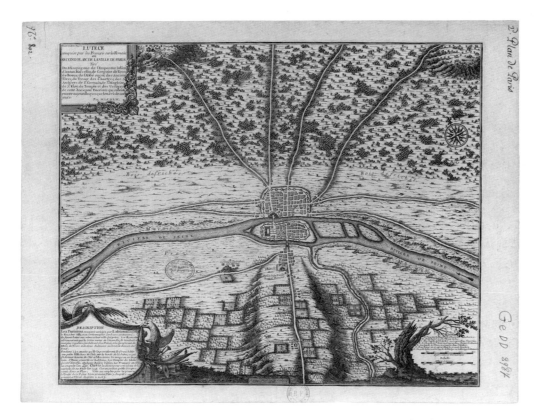

FIGURE 5.2 *Lutèce conquise par les françois sur les Romains ou second plan de la Ville de Paris* (Lutèce conquered by the French over the Romans or second plan of the city of Paris). Heliography, 17½ × 21⅝in. (44.5 × 55 cm). From Nicolas Delamare, *Traité de la police* (Treatise of the police) (Paris, 1705). Bibliothèque nationale de France.

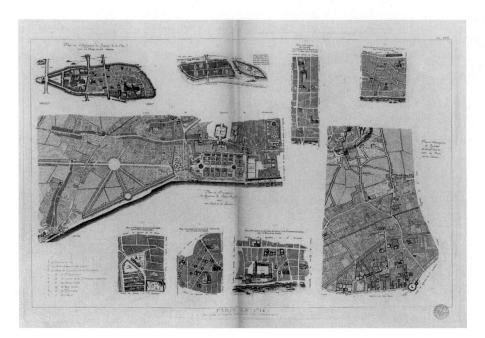

FIGURE 5.3 Jean de La Caille, *Plan de La Caille* or *Description de la ville et des fauxbourgs de Paris en vingt planches, dont chacune représente un des vingt quartiers suivant la division qui en a esté faite par la declaration du Roy du 12. décembre 1702* (Description of the city and the surroundings of Paris in twenty plates, each one representing one of the twenty neighborhoods following the division, which was made by the declaration of the king on December 12, 1702), 1714. Plate 24 from Adolphe Alphand, *Atlas des anciens plans de Paris* (Atlas of old plans of Paris) (Paris, 1880). Heliography, 19 × 28¼ in. (49 × 72 cm). Bibliothèque historique de la Ville de Paris.

concentric circles, keeping the main north–south road as its diameter through all eight plates. Settlements appear outside of the walls, but they are soon enveloped within the enclosures; and the country, drawn in perspective, is gradually conquered by an orthogonal point of view. By the last plate, a representation of Paris contemporaneous with the *Traité* itself, street names appear, making the city literally legible through the map.

The Delamare image is significant as the first plate in the *Atlas,* and the volume's editor, Adolphe Berty, made specific mention of the issue of accuracy: "The plans by Delamare escape criticism by the very excess of their imperfection."[23] As the first plate, it introduces the making of the image of the city as a governmental and republican project, with the Third Republic specifically referencing Gauls as part of a past that avoided mentions of the church and monarchy.[24] However, it had far-reaching implications, as the editors of the *Atlas* noted: the plate was "to give an idea of the efforts of attempted restorations in the last century."[25] There is a slippage in the reference to "restorations" that conflates the project of building Paris with the project of representing the city through time. Thus, this plate, more than a description of the capital, records a historiographic practice understood as being within the state's administrative domain.

The *Atlas des anciens plans de Paris*'s organization argues for constant progress in both historiographic and cartographic technologies, providing a catalogue of increasingly accurate images of the city and its past. The individual plans, because of their discrete origins in the archives of different governmental institutions, do not share a scale, iconography, style, or form. With the exception of the section on geometric plans, the historical maps all feature different perspectival modes, reproduced during a time when this mode was no longer conventional, making not only the reconstruction of the image but also the methods of reproducing its visual modality part of its historiographic process. Overall, with no consistency in visual modalities, including *vues cavaliers* (ground views) and *vues à vol d'oiseau* (bird's-eye views) as well as planar views, the focus alternates between a shifting Paris and the methods of the maps' reconstruction and production. Ultimately, the *Atlas* offers a history of its cartographic representations.

The reproductions in the *Atlas* do not follow a straight progression from pictorial to geometric images, from perspective to orthogonal visual modes. Graphic diversity is present, as each map was commissioned for a different purpose by divergent interests, such as the *Plan de La Caille* (1714) (fig. 5.3), made for Marc-René d'Argenson, police lieutenant in Paris, or the *Plan de Turgot* (1734–39), made for Michel-Étienne Turgot, *prévôt des marchands* (merchant provost) (fig. 5.4). Because of the differences between the people involved, the methods applied, and the instruments of surveying, drawing, engraving, and printing, not to mention the transfer of information that

PARIS DE 1734 A 1739

FIGURE 5.4 Michel-Étienne Turgot, *Plan de Turgot* or *Paris au XVIIIe siècle.*
Plan de Paris: En 20 planches (Paris in the eighteenth century. Plan of Paris: In
20 plates), 1734–39. Plate 27 from Adolphe Alphand, *Atlas des anciens plans de
Paris* (Atlas of old plans of Paris) (Paris, 1880). Reproduction of the plan by
Louis Bretez, called the *Plan de Turgot.* Scaled reduction at ²⁄₅. Heliography,
20½ × 25³⁄₁₆ in. (52 × 63 cm). Bibliothèque historique de la Ville de Paris.

would have also affected differences, each image was distinct. There were also different intended audiences for each of these maps. For example, a map of Paris by Robert de Vaugondy from 1771 was originally called the *Tablettes parisiennes,* and was intended to be a small, portable atlas (fig. 5.5). It is the only plan that kept its original measures in its nineteenth-century reproduction. In contrast, the *Plan de l'Abbé Jean Delagrive* from 1728 measured two meters by one-and-a-half meters and was reduced in scale on four plates for the *Atlas.*

Thus, while each individual image tells a nuanced story of cartographic production, the nineteenth-century

commentary and the images' compilation into an atlas form superimposed a linear narrative onto them. With each discrete description that accompanies the plates, the same phrases are repeatedly used to cite the precision and accuracy of the methods used to produce the images. The *Grand Plan de Jouvin de Rochefort* (1672), whose "geometric value is incontestable"; the *Plan de l'Abbé Jean Delagrive* (1728), which follows "new and conscientious geometric operations, also interesting in its details as accurate in its whole"; the *Plan de Roussel* (1730), which is "accurate in its details, regularly oriented and engraved with care"; and the final example in the section, the *Plan de Verniquet*

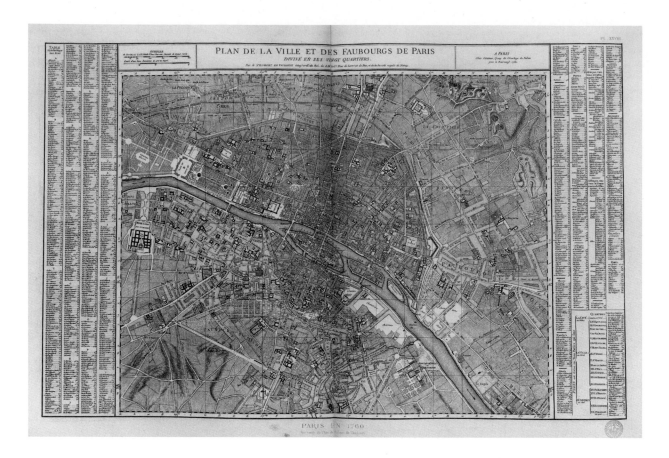

FIGURE 5.5 Robert de Vaugondy, *Plan de la ville et des faubourgs de Paris divisé en ses vingt quartiers* (Plan of the city and the surroundings of Paris divided in its twenty neighborhoods), 1771. Plate 28 from Adolphe Alphand, *Atlas des anciens plans de Paris* (Atlas of old plans of Paris) (Paris, 1880). Heliography, 22 × 32¼ in. (57 × 82 cm). Bibliothèque historique de la Ville de Paris.

(1789–98), of which "nothing was neglected to produce a masterpiece of geometric accuracy."[26] The claims of each map's exactitude highlight the elevated value of geometric correspondence through mensuration. Yet in their progression is a tension: on the one hand, each discrete image is supposed to accurately describe how Paris appeared at the time of the map's making. On the other hand, the cumulative narrative, that of reducing the representational gap between image and city, undercuts the individual images' claim to be accurate depictions. Cartographic accuracy and historiographic accuracy are thus suspended against each other.

The tension between each discrete map as an accurate image of Paris and the collection of maps as a historical narrative follows the basic logic of scientific discourse. For Jean-François Lyotard, the legitimacy of scientific knowledge depends on its denial of narrative, and yet narrative is also what provides for its legibility as scientific. Scientific knowledge, he wrote, "has always existed in addition to, and in competition and conflict with, another kind of knowledge, which I will call narrative."[27] He argues that scientific discourse encloses narrative within a rubric of fact, which is defined as a denotation. "Scientific knowledge requires that one language game, denotation, be retained and all others excluded," meaning that factual referents are emphasized at the expense of their narrative or social meaning.[28] However, "the game of science thus implies a diachronic temporality, that is, a memory and a project."[29] The production of a new scientific statement depends upon being acquainted with preceding statements that share its referent—what is meant by "memory"—as well as upon being different from those previous statements, in order to "project" new information. This oscillation between discrete units and

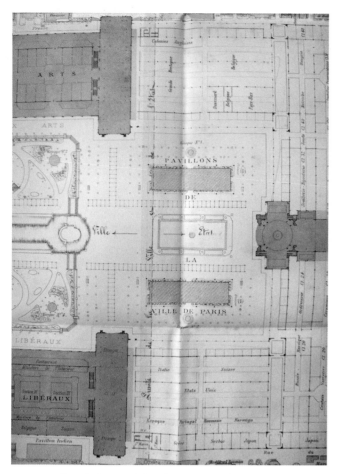

FIGURE 5.6 Detail of *General Plan of the Grounds of the 1889 Universal Exposition* (fig. 5.8).

the connections between them that define those very units, between past and future, between events and narrative, all constituted the scientific discourse of the *Atlas.* "Paris" became a given object that was simultaneously unique in each map but changed and developed across all the maps.[30]

The final image in the *Atlas des anciens plans de Paris* was the *Plan de Verniquet,* providing a triumphant conclusion that progress had been achieved. As Joseph-Jérôme Lalande, director of the Observatoire de Paris, wrote in the final accompanying text: "The plan by citizen Verniquet appears to me as a work the most perfect of its kind that has ever been executed."[31] Considered the first geometrically exact plan of the city, within the *Atlas's* narrative, it is the culmination of the work of those who had labored to produce a map with accurate fidelity to the territory of Paris. It is displayed as an opening onto the future; the blank spaces within the outlines of the blocks are waiting to be filled. It also implied a future with

allusions to its republican status, for as Alfred Bonnardot emphasizes to the reader, "The *plan de Verniquet* captures the state of the capital at the very moment when the Revolution broke out."[32] Far from the truth, the orthographic plan is translated into a singular event.

AN ADMINISTRATIVE HISTORY

Much has been written about the "invention of tradition," and how activities organized and promulgated by the Third Republic were essential in creating the very idea of a republic for a society that could cohere around its symbols.[33] Among the invented traditions, these activities took the form of new didactic artifacts, numerous public monuments, and public ceremonies, including the Expositions universelles, that ultimately concerned the institutionalization of republican values. Yet while the production of these rituals and symbols was explicit, historiography also became a more subtle but nonetheless negotiated site of politics.[34] Numerous historical publications were commissioned by the Conseil municipal de Paris, and the institutionalization of historiography took place through administrative agencies such as the Commission municipale des travaux historiques, the Service des travaux historiques, and the Commission des recherches sur l'histoire de Paris. In the early years of the Third Republic, these projects sought to establish a clear link between the revolution and the current government.

As Maurice Agulhon has emphasized, the republic was made by the city in the city, as a republican island in an ocean of French conservatism, making the representation of Paris, as opposed to the nation of France, a way to demonstrate the republic's authority.[35] For the Exposition universelle of 1889, to celebrate the centenary of the French Revolution, the location of the pavilions of the Conseil municipal de Paris, colored gray on the plan, was precisely on the territorial border of the city and the state on the Champ de Mars and exposition grounds (fig. 5.6). Its location emphasized Paris as a municipality and a capital. The grounds of the exposition occupied the Champ de Mars and traversed the Seine up the bank to the Musée d'Ethnographie (now the Trocadéro), hugging the Left Bank of the Seine eastward to the Esplanade des Invalides. The city's two pavilions were placed along the main axis of the Eiffel Tower and the grand Palais des Machines, close to the halls dedicated to the liberal and fine arts. Their location at the center of the exhibition grounds suggested that the representation of the capital city served

FIGURE 5.7 Hippolyte Blancard, *Universal Exposition of 1889: The Pavilion for the Préfecture of the Seine, City of Paris. View Taken from the Terrace of the Central Dome*, 1889. Photograph, 6½ × 8¼ in. (16.5 × 21.1 cm). Musée Carnavalet.

to enshrine the idea of a coherent republic, of both city and state.

Standing in the shadow of the newly erected Eiffel Tower, two rectangular pavilions of one hundred meters by twenty meters were modestly designed by the architect Joseph-Antoine Bouvard (fig. 5.7). They held displays from each of the services under the direction of the urban administration (fig. 5.8).[36] The east pavilion, next to the Palais des Beaux-Arts, was dedicated to displays of urban infrastructure: public services including "Travaux historiques" (Historical work), architecture, fine arts, roads and promenades, water, sanitation, the Observatoire municipal de Montsouris, quarries at the Seine and local services, and maps. The west pavilion, beside the Palais des Arts Libéraux, was dedicated to displays of urban information, including the services of municipal and statistical libraries and the departments of education, finances, municipal affairs, departmental affairs and administrative annexes, public assistance, and the police. Recalling Foucault's formulation of governmentality, the number of services indicated the expanded responsibilities of the municipality, as well as the extensive oversight the government had gained over a city that was defined not only by its territory but its population.[37]

Altogether, the pavilions outlined what had been consolidated as the administrative, social, and infrastructural responsibilities of the municipal government under the Third Republic.[38] As cited in an 1887 report to the

Conseil municipal de Paris, every agency included plans and maps in their pavilion display, such that these cartographic images were how each agency met its responsibilities. These plans and maps conditioned bureaucratic activities as well as represented their results. The display of "Travaux historiques" included historic plans, photographs, and albums; "Architecture" included plans in frames and albums; "Voie publique" (Public roadways) displayed plans, photographs, models, sample materials, and even example building tools that were used; "Promenades et plantations" (Promenades and green spaces) had an exposition of plans and perspectival views of the new cemeteries in Pantin and Bagneaux and plans of the crematorium at Père-Lachaise, among other new public spaces; "Service vicinal" (Local services) included relief plans and cartographic murals; and the sections "Canaux et dérivations" (Canals and diversions), "Eaux" (Water), and "Assainissement, égouts et salubrité" (Sanitation, sewers and hygiene) all displayed plans to show their interventions in the city.[39] The various displays illustrated the graphic visuals that composed the administrative language and the spatial thinking that undergirded their logic. These were images that were thoroughly normalized with governance practices, and understood as objective tools that allowed these agencies to function technocratically.

Significantly, what was exhibited was not merely the process of the city's construction, but how the municipality defined the city to and for the public. The displays were concerned with infrastructural matters, but the exhibition also spoke to the ways in which the government was concerned with managing the population within the city and, moreover, defining the practices of the urban citizen. The exhibits of parks and cemeteries related to norms of leisure and death; sewage and sanitation presentations clarified hygienic standards; and even the public roads division sought to demonstrate the paths of circulating through the city. Thus, the pavilions called attention to the brick and mortar aspects of the agencies' work, but simultaneously, each section also established its target as the urban population, and its principal form of state knowledge as calculation.

As one administrator summarized, "The Centenary festival gave all the works a statistical look," demonstrating the pervasive use of maps and information graphics in the urban administration, and the way in which it was integrated into all of the different services.[40] The displays

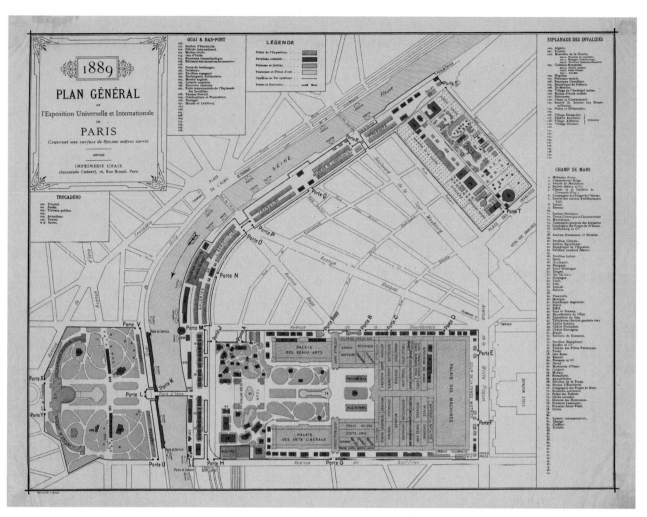

FIGURE 5.8 Paris's pavilion is located in the inner courtyard between the Palais des Beaux-Arts and the Palais des Arts Libéraux at the entrance of the Palais des Machines. *Plan général de l'Exposition universelle et internationale de Paris couvrant une surface de 850,000 mètres carrés* (General plan of the Universal and international exposition of Paris covering an area of 850,000 square meters), 1889. Map, 17³⁄₈ × 22 in. (44.2 × 56.2 cm). Musée Carnavalet.

of the Service de la statistique, of course, included graphics, but so did the display by the "Bibliothèques municipales," which included a map of all the municipal libraries, as well as an excess of statistic charts noting the number of loans, collection numbers, acquisitions, and so on. Another example included the rather unknown agency of municipal affairs that displayed all manner of statistical charts for the number of burials, consumption of different goods and resources, and donations and bequests to charities, as well as maps that showed the locations of food markets and cemeteries.[41] This abundance of statistical and graphic representations signaled the enthusiasm for and convergence of developments in statistics, data collection, and technology in graphics.[42] Significantly, this

fervor for statistics was tied to both the expanding role of administrators in governance and to the democratic reforms of the Third Republic, in which quantification was understood as equalizing.[43] Just as the French language was promoted as a common language for all citizens and regional dialects were discouraged, statistics also sought to create a flat representation of the population under the French state. Each citizen had a quantitative place that was locatable on these maps.

Plates of the *Atlas des travaux de Paris* were found in the east pavilion under the "Plan de Paris" section (fig. 5.9). Their display was located at the pavilion's central intersection between the sections "Travaux historiques" and "Voie publique," functioning as "before and after" urban

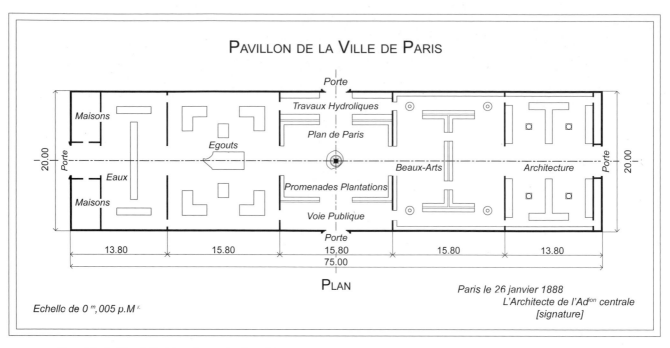

FIGURE 5.9 *Plan of the Pavillon de Paris* based on a plan found in the Archives de Paris. Drawn by Sony Devabhaktuni.

Theme	Date	Plate
Plan of Paris	1789	plate I
Water and sewage	1789	plates II, III
	1855	plates IV, V
	1878	plates VI, VII
	1889	plates VIII, IX
Public roads	Reconstitution of the *Plan de la Commission des artistes* from 1793	plate X
	Plan of Paris in 1854	plate XI
	Plan of Paris in 1871	plate XII
	Plan of Paris in 1889	plate XIII
	Method of paving in 1889	plate XIV
Public transportation	1889	plate XV
Public buildings, university and school buildings constructed	1871 to 1889	plate XVI

TABLE 1 Analysis of the plates in the *Atlas des travaux de Paris, 1789–1889*. Colors differentiate the three different base maps used for the various plates.

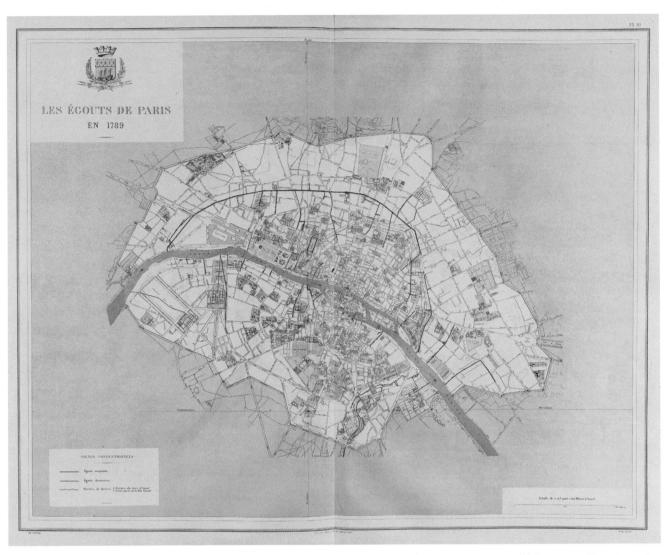

FIGURE 5.10 "Les égouts de Paris en 1789" (The sewers of Paris in 1789). Plate III from Adolphe Alphand, *Atlas des travaux de Paris, 1789–1889* (Atlas of the works of Paris, 1789–1889) (Paris, 1889). Bibliothèque historique de la Ville de Paris.

representations.[44] As outlined in the notes of one administrator in the Service du plan, "The atlas thus composed provided the graphic history of the city."[45]

The *Atlas des travaux de Paris* was composed of sixteen plates and organized around five themes into four periods drawn on three different base maps (see Table 1). They sought to link the current state of Paris's built environment to the ideals of the French Revolution, maintaining the idea that the modernization of Paris over the course of the century was a technical rather than political endeavor.

The *Plan de Verniquet* was reproduced as the first map in the *Atlas des travaux de Paris,* and the plan was further reduced to a single sheet that served as a base map for four different plates (figs. 5.10–5.13). The first base map—using

the example of the development of the sewage system— shows the city bounded by fortification walls. Streets extending from the walls are drawn in, but with a different ground color from the area within the walls (plates I, II, III, X) (see fig. 5.10). The second base map delineates more roads outside of the fortification walls, with the image of the city bounded by a gridded artifice (plates IV, V, XI) (see fig. 5.11). The city falls off the page in the third base map, with only portions of the Bois de Boulogne and Bois de Vincennes visible (plates VI, VII, VIII, IX, XII, XIII, XIV, XV, XVI) (see fig. 5.12).

Sharing the same base image, all of the plans in the *Atlas* were conceived of and produced as a set and organized into a continuous narrative of Paris's urban

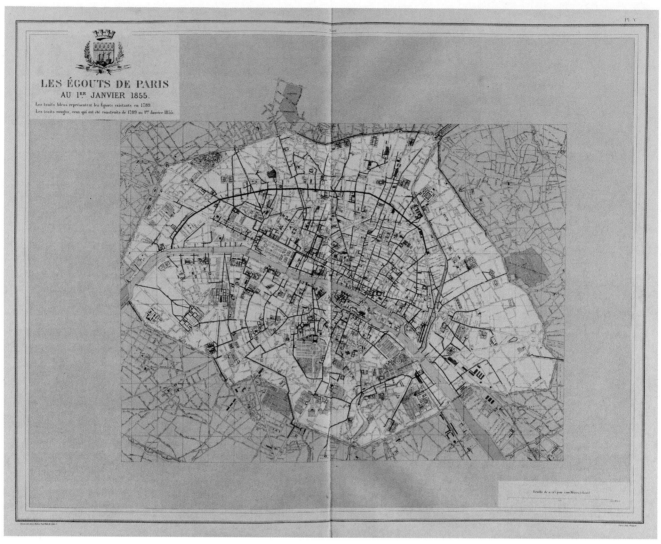

FIGURE 5.11 "Les égouts de Paris au 1ᵉʳ janvier 1855" (The sewers of Paris on January 1, 1855). Plate V from Adolphe Alphand, *Atlas des travaux de Paris, 1789–1889* (Atlas of the works of Paris, 1789–1889) (Paris, 1889). Bibliothèque historique de la Ville de Paris.

morphology.[46] However, unlike a textual source, this narrative is built on a series of discrete graphic events.[47] Notes from the production of this atlas repeatedly highlight its graphic quality, which refers to the "geometric" methods that were used to survey the terrain and that were drawn orthogonally into a linear composition. Descriptive texts associated with the image were minimized, and cartouches were replaced with legends that laid out the schema of the image's color and hatched patterns. The aim was to produce a diagram of the spatial relationships of the city that could be visually captured at once.[48] The discontinuities and gaps between the synoptic images within the narrative of the atlas permitted the administration to avoid and ignore political events that had shaped the city as much as the scientific and administrative techniques.

This composition of the images demonstrates the manufacturing of Paris as a technical object. In the first plate of *Atlas des travaux de Paris*, unlike the Verniquet plate in the *Atlas des anciens plans*, all cartouches are removed and replaced with rectangles at the extreme corners that feature the seal of the city, the title of the plate, and a scale (fig. 5.14). The image of the city, centered on the plate, is composed of a web of black lines, describing the road patterns and the outlines of the public buildings, and a network of green spaces, including the parks and Seine,

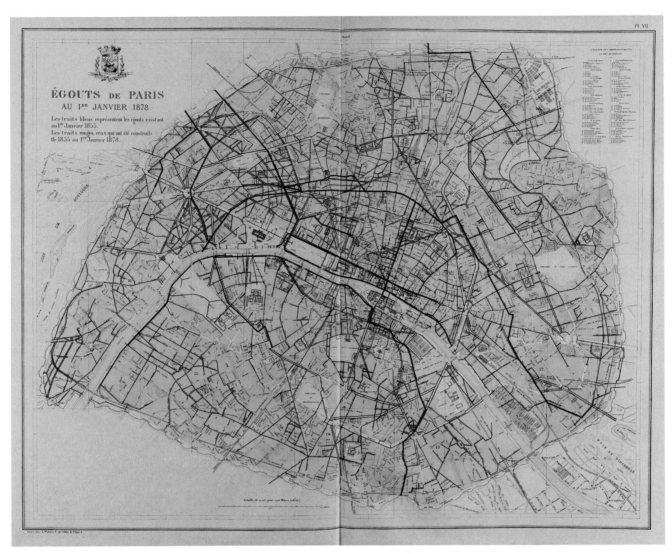

Pl. VII

FIGURE 5.12 "Égouts de Paris au 1er janvier 1878" (Sewers of Paris on January 1, 1878). Plate VII from Adolphe Alphand, *Atlas des travaux de Paris, 1789–1889* (Atlas of the works of Paris, 1789–1889) (Paris, 1889). Bibliothèque historique de la Ville de Paris.

all on a colored white area. Most of the sheet is left blank, revealing the paper support of the map itself. This blank material space functions to make Paris into a contained and constructed object in its representability and simultaneously serves to foster this floating object's potentiality to expand.

From the vantage point of 1889, Paris was to be read as a work in progress. The plates in the *Atlas des travaux* from 1854 and 1855 still show an expanded surface area of the city in white, now overlaid with a grid. However, in the plates from 1878 and 1889, the urban object finally conquers the sheet of paper, and the image of the city grows beyond the borders of the page, with the Bois de

Vincennes and Bois de Boulogne running off both ends of the sheet. Combined, the plates gradually create a layered image of Paris with surface and underground networks, like the plans of a multistoried building, insofar as the flatness of the respective plates was given depth by their accumulation and succession of colored lines. The city is in many respects an architectural object; just as ground plans of a single building show its spatial relationships, the maps conform to the representational strategies of architecture.

The orthogonal mode secured the map's synoptic quality and its consistency. The flatness, the completeness, the linearity were all elements that signaled its presentness

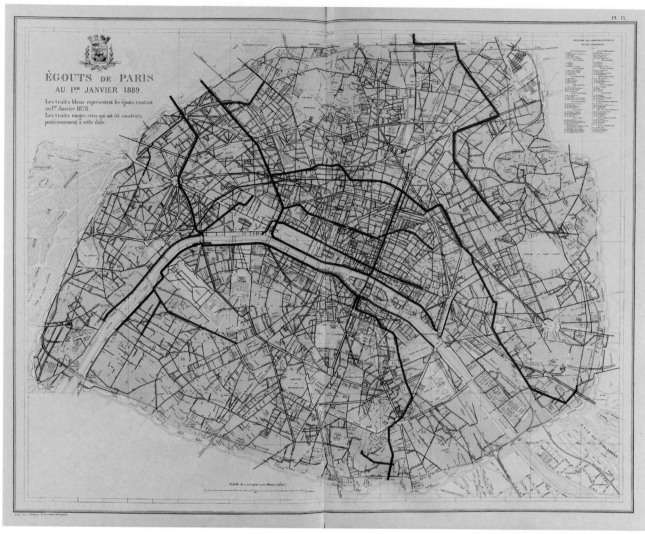

ÉGOUTS DE PARIS
AU 1ᴱᴿ JANVIER 1889.

Les traits bleus représentent les égouts existant
au 1ᵉʳ Janvier 1878.
Les traits rouges, ceux qui ont été construits
postérieurement à cette date.

FIGURE 5.13 The maps from 1878 and 1889 use the same base plan.
"Égouts de Paris au 1ᵉʳ janvier 1889" (Sewers of Paris on January 1, 1889).
Plate IX from Adolphe Alphand, *Atlas des travaux de Paris, 1789–1889*
(Atlas of the works of Paris, 1789–1889) (Paris, 1889). Bibliothèque
historique de la Ville de Paris.

and absoluteness—that is, its occupation of a moment out-side the contamination of subjective interpretation and time. The immediacy of the image's visual effect made it beyond the reach of human corruption. This deprecia-tion of the human hand in scientific practices was aligned with new technologies in image production during the mid- to late nineteenth century. As Lorraine Daston and Peter Galison have examined, the value and definition of objectivity came to center on the image reproduced by mechanical means.[49] The camera or the printed line transformed subjective experience into a fixed and stable composition that could in theory be reproduced indefinitely. Thus, it was the loss of aura around a sin-gular event through its reproducibility that elevated the

representation into an objective image, erased of its temporal specificity.

The *Atlas des travaux* was based on individual images that were designed to exclude time, and yet, as a whole it was sequenced and ordered to narrate the passage of time. The elements of event and continuity, of synoptic image and diachronic series, were balanced by a cartographic grid fixed at one particular building. In the first base map, two perpendicular lines that intersect at the Observatoire de Paris are drawn in black (see fig. 5.10). This expands to a full grid in the second base map, in which each graticule unit represents one thousand square meters, still anchored on the observatory (see fig. 5.11). While in this iteration the city is enclosed by the grid, on the third base map,

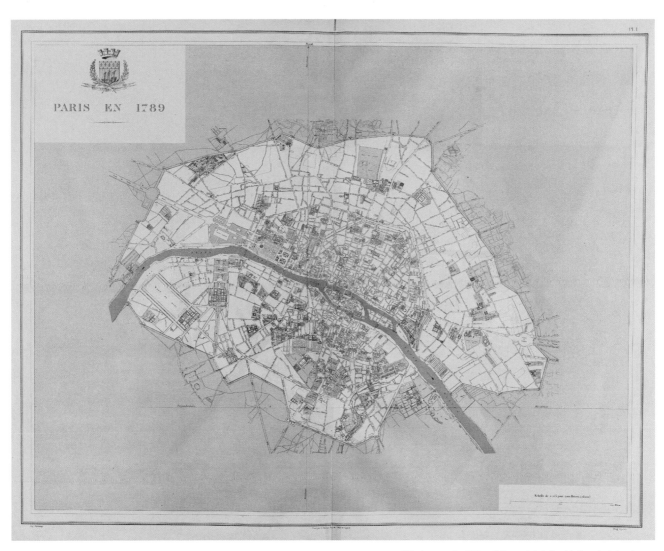

FIGURE 5.14 "Paris en 1789." Plate I from Adolphe Alphand, *Atlas des travaux de Paris, 1789–1889* (Atlas of the works of Paris, 1789–1889) (Paris, 1889). Bibliothèque historique de la Ville de Paris.

the city overtakes the cartographic grid (see fig. 5.12). The growth and development of Paris is measured in relation to the observatory, an institution that was the center of mapmaking and of scientific practice since its foundation under Jean-Baptiste Colbert. The longitudinal baseline established at the observatory served as the datum for surveys by the Cassini family, and it was where Jean-Baptiste-Joseph Delambre and Pierre-François-André Méchain set off to conduct the land survey of the meridian, as well as where the meridian was measured astronomically. Through its form, the grid bridges space and time, enabling spatial measurements within each image and temporal measurements across all of them. Paris, shown in the same scale in every map, serially expands across

the page and eventually surpasses its borders, as if to signify the capital of the republic as constantly growing through the accumulation of infrastructural networks in the city.

In his study of the state's control of temporality, David Gross writes that because of the imperceptibility of large-scale time in everyday and personal experience, events on this scale "depend upon external accounts of their factual validation"; it falls on institutions to decide their value as significant or insignificant.[50] The nineteenth century, Gross maintains, witnessed an active and ideological interest in the temporal orientation of the total population. It was important "to formulate the categories of 'pastness' most advantageous to the state, to push for state forms of periodization, or advance political conceptions of dura-tion."[51] As each synoptic image of the city follows another at the same scale, what emerges is a linear narrative of technical progress, whose categories of the past were pro-duced through images. It begins with the *Plan de Verniquet,* retrospectively dated 1789, and proceeds to Alphand's plans of 1889—from the storming of the Bastille to the Third Republic—fostering a stable and singular idea of the repub-lic that was realized. The smooth graphic expansion of the city across the sheet of paper, from the white figure in 1789 to the one in 1889 that overwhelms the plate, constructs time and space as linear and flat.

Accordingly, there is no indication of revolutions, restorations, or damages from war, including no distinc-tions in the final plate of the *Atlas des travaux* between new buildings and restored buildings such as the Hôtel de Ville after the Paris Commune. Significantly, the peri-odizations of the *Atlas* correspond to the dates of 1855 and 1878, the completion of the Bois de Boulogne and the previous Exposition universelle, respectively. They foreground infrastructural achievements as opposed to political regimes. By submitting these cartographic images to a narrative structure, the state constructed not only a history of Paris's development, but also its own myth. The graphic quality of the maps, their acceptance as technical and scientific, and their organization into the *Atlas* all function to transform the political, in terms of the actual city and its representation, into the natural. In this case, the origin was the *Plan de Verniquet,* whose date was changed to coincide with the French Revolution, and the myth of urban modernity was cultivated on historio-graphic and cartographic terms.

COLONIAL ABSENCES

The men who did the most to construct the institutions and procedures that would form the cartographic founda-tions of the city's bureaucratic practices had been trained a century earlier, under Napoléon Bonaparte's *mission civilisatrice* in Egypt. Mapping, surveying, and quantify-ing had been first tested in that imperial project, and the lessons were then brought back to Paris when men such as Chabrol de Volvic and Jomard headed the Préfecture de la Seine and the Description de l'Égypte, respectively. Jomard would go on to head the maps and plans division of the Bibliothèque nationale de France, controlling the cartographic archives of the state, and Chabrol de Volvic would instate the first comprehensive census of the city

and produce a series of prints by Théodore Jacoubet that would be used by Haussmann's administration to draw new circulation networks. Colonial campaigns were foun-dational to the cartographic culture of the administration, and yet there was no acknowledgment of their influence. Even a century later, when the Third Republic again took up the civilizing mission, it remained absent in the displays of the municipal pavilions.

Even before the great 1931 Exposition coloniale inter-nationale, the Exposition universelle of 1889 included colonial subjects imported to the exhibition as living examples of the civilizing mission underway.[52] The idea was that the republican message could be and should be applied to all, even these "savages."[53] They were essential to the message of progress, providing a useful foil to the examples of technological innovation and social advance-ment also displayed. Under the mantel of French national-ism, the Third Republic promoted its colonial ambitions, symbolically offsetting the forfeiture of Alsace-Lorraine and economically compensating for commercial losses from the war with Prussia.[54] Thus, while the Ministère des colonies was established in 1894, colonial projects were already incorporated into the French imagination well before.[55]

The 1889 Exposition universelle featured several dis-plays that sought to replicate indigenous habitations in the metropole.[56] Creating its own colonial geography, the Palais des Colonies housed replicas of villages from Senegal, Madagascar, Tahiti, and Gabon, palaces from Tunisia and Cambodia, and settlements from Guadeloupe, Cayenne, New Caledonia, and Indochina, as well as spe-cific colonial displays from other European colonies such as Java and Congo. The contrast between these exhibitions and the municipal pavilions was stark. Here, there were no cartographic displays to orient and contextualize the presentations, architectures, goods, and people on exhibit. Instead, the goal was immersive: "bazaars" offering objects made by locals, indigenous people performing tasks beside "primitive huts," and agricultural products such as sugar-cane, coffee, and spices arranged to show their transfor-mation into commodities for European audiences. Here the visitor was "lost" in an agrarian world, among these natives, who could not see beyond their "savage" situa-tion. They did not make maps; they were subject to them. The argument was that, left alone, these unnamed humans would not be able to improve their status, and it behooved France on moral grounds to help their lot.

The most immersive of the exhibitions was the "Rue du Caire" (Street of Cairo), which had been privately funded by Baron Alphonse Delort de Gléon and included a scenographic construction of a fictional street from Egypt's capital city.[57] The street included twenty-five buildings, all representing different styles and typologies, creating a collage of architectural elements taken from demolished buildings sutured together for this set, inhabited with musicians, artisans, entertainers, and even donkeys that would take tourists along the street.[58] As in the national exhibits of the colonial sections, the curation offered subjective experience, not distant and disembodied analysis.

Compared to the municipal pavilions, these colonial pavilions poignantly highlight the remove at which Paris was held in the arrangement and choice of exhibition objects: maps and images versus living humans. While the actual city offered visitors the immersive experience of the infrastructural works in their promenades beyond the exposition's gates, the display offered in the pavilions made Paris into a different kind of object than that of the colonies. Both objectified place, but in presenting Paris in these technical and historical maps, it showed a city that had made progress. It had been transformed and become modern, not relegated and fixed into a primitive past. No doubt, this claim was situated against a backdrop that represented Paris as a contest between order and disorder: the year 1789 had sealed Paris as a place of revolution, again in the June Days uprising of 1848, and then finally in the Commune of 1871.[59] Absent in the maps were the class divides that seemed to pose the greatest threats to the bourgeois control of the central parts of the city under the Third Republic. Instead, the municipal exhibits flattened the urban territory and reinforced the larger national aim that the metropole had achieved civilization and could extend its expertise beyond.

CARTOGRAPHIC LITERACY

The final plate in the *Atlas des travaux*, titled "Édifices de Paris construits de 1871 à 1889" (Edifices of Paris constructed from 1871 to 1889), is markedly different in its content from the other fifteen plates, whose focus pertains to road infrastructure such as street openings, paving, water, sewers, and public transportation systems (fig. 5.15). Here, the map records the vast number of public services under the responsibility of the state, demonstrating a reach beyond territorial concerns into the management of society and regulating its populations. In red, this last plate identifies public buildings: hospitals, markets, museums, and ministries. In blue, it identifies new educational establishments throughout the city, with a total of fifty-four new buildings on the Left Bank and around a hundred new buildings on the Right Bank, mainly clustered around the peripheral arrondissements.[60] The significance of this plate is to make these new agencies and the extension of the government into issues of social welfare part of the actual infrastructure of the city. The map makes legible these new roles and new spaces to a public who, in turn, was expected to participate in political life as part of the republican project. The number of buildings on the edges of the city related to the ways republicans sought to gain social control over the working-class periphery, created by displacements during the Second Empire. This map was a way to tame visually those arrondissements and establish order in all public spaces throughout the city.

The map's subject of educational institutions was especially significant as a reaction to the Franco-Prussian War. In editorials and in government assemblies, the reason pinpointed for the French defeat was that the Prussian army was perceived to be better prepared and better educated, especially in geography.[61] Ultimately, there was large support for educational reforms under the Third Republic, if mostly to assuage the urban bourgeoisie that democratic efforts could expand while still maintaining social stability.[62] The educational reforms were headed by Jules Ferry, who served as the minister of education for two periods between 1879 and 1883. These changes included mandatory primary school—free of charge—as well as curricular adjustments that included more emphasis on modern languages, physical training, and, above all, geography. The reformation plan was announced in September 1872, and it established a program for geography that moved from the concrete to the abstract, from the local to the global. Geographic study took a "bottom-up" approach, beginning in the immediate vicinity of the student's home, school, and neighborhood or "commune," gradually enlarging in scope to include France, continents, and ultimately planetary systems.[63] The program also specified wall maps and particular textbooks, with cartographic illustrations, for classroom use.[64] For example, nineteenth-century geography textbooks such as Joseph Gibrat's *Traité de la géographie moderne* (1813) did not contain any maps or cartographic representations of cities, only a list of longitudes and latitudes of major sites. As described by Paul Claval and Numa

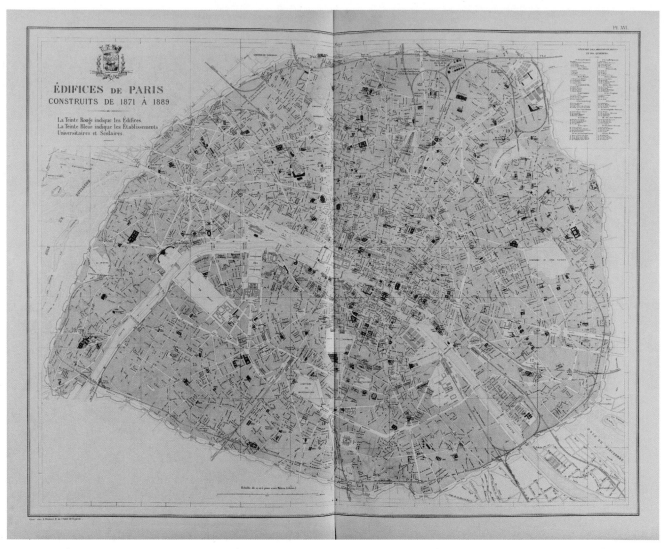

FIGURE 5.15 "Édifices de Paris construits de 1871 à 1889" (Edifices of Paris constructed from 1871 to 1889). Plate XVI from Adolphe Alphand, *Atlas des travaux de Paris, 1789–1889* (Atlas of the works of Paris, 1789–1889) (Paris, 1889). Bibliothèque historique de la Ville de Paris.

Broc, teaching geography during the Second Empire was limited to memorizing lists and was not linked to contextualizing place.[65] As one school inspector explained: "One does not teach geography; one teaches only the way to learn it. Composed only of facts, isolated from each other, or rather, not linked to each other by a chain as in other sciences, geography is largely the exclusive domain of memory: one learns it by reading history and traveling."[66] Here, the statement draws from the relatively recent conceptions of networks and the possibilities of travel. The notion that the meaning of places was interconnected and not discrete had long existed in religious pilgrimages. However, for the secular Third Republic, there was

a particular valence to the ways in which nationhood was imagined within a France whose regions and cities were tied together through the extension of the railways, the mandate of a shared language, and colonial occupations.

The new direction of geographical study in education was promoted at the Exposition universelle in 1889 using various displays of classroom material. The section of "Enseignement primaire élémentaire" (Primary elementary instruction) included an example of a classroom with chairs, desks, blackboards, and wall maps of France, prominently displayed. "Enseignement primaire supérieurs" (Primary superior instruction) presented a collection of scientific instruments as well as an installation of geometric drawings and models by students.[67] The display and use of orthographic maps were no longer used within exclusive administrative or engineering circles, but were integrated into general education, creating a populace sensitized to cartographic imagery. Moreover, the young population was also inculcated with the idea that the loss of Alsace-Lorraine—as a cartographic form—made the hexagon incomplete. The lost regions were often shown in differently shaded colors, and the bolded boundaries of "the Hexagon" were interrupted by missing lines to the north. That these large-scale maps were displayed on the wall recalled the ways in which powerful men in government had also displayed theirs. It trained children in a modality of presentation that instilled aesthetic expectations of power and control.[68]

Geography and, consequently, cartography expanded into society from many different directions. In addition to the educational reforms, there were also several commercial vectors, including the emergent tourism industry that was aligned to the popularity and accessibility of train travel, the reduced cost of printed maps, and the publication and sale to the public of maps by government agencies. Maps were related to a larger print culture that permeated all urban space.[69] Examples included the immensely influential and widely read *Le tour de France par deux enfants* (The tour of France by two children) by Augustine Fouillée, which offered a new conceptualization of the nation through the journey of two orphaned boys traveling through the different regions of France.[70] Many of these conditions existed in earlier moments during the nineteenth century, but the political will of republicanism aggregated them to turn the citizen into what Kory Olson has called "the map reader."[71]

Thus, when the plates from the *Atlas des travaux* were presented at the Exposition universelle, they confronted a bourgeois audience that was cartographically literate and understood the medium and imagery as a means of accessing Paris.[72] The detailed street names and dates of completion on the plate "Opérations de voirie exécutées de 1871 à 1889" (Road operations executed from 1871 to 1889) reassured its readers that Paris was ordered and controlled.[73] Moreover, its representation also fostered a culture of democratic citizenship, in which the individual understood his place and responsibility within the community, city, and nation. And after having gazed upon these maps, that reader could then be confident as a bourgeois *promeneur* (pedestrian) in the new modern capital.

The maps in the *Atlas des travaux* were thus an outcome of a concerted political effort that informed new pedagogies, papered the surfaces of public spaces, and consolidated administrative communication and practices.

In addition to the publication of the *Atlas*, Alphand oversaw another major publication by the municipality completed in time for the 1878 Exposition universelle: *Les promenades de Paris*, an album dedicated to his career within the urban administration under Haussmann. Distributed to administrators and governmental officials and sold to the public, the two volumes included 487 wood engravings, eighty steel engravings, and twenty-three chromolithographs, dedicated to detailed images and maps of all the parks, gardens, squares, and promenades of the city (fig. 5.16). The wood engravings were used to represent picturesque scenes of people utilizing the green spaces. The steel engravings and chromolithographs depicted the components of their construction and details of their design.

The format and organization were didactic. First, a series of prints from wood engravings showed how the space was intended to be used. This was followed by a site plan that included the various networks of paths and water. Then, plates focused on street furniture such as kiosks and benches in plan, elevation, and profile (fig. 5.17). This order was followed within each urban project, organized hierarchically from the two major works, the Bois de Boulogne and the Bois de Vincennes, all the way down to tree plantings along streets.[74] Each of these parks was slowly, through each plate, deconstructed into its component parts—recalling Jean-Nicolas-Louis Durand's compositional methods—demonstrating that Paris was a technical outcome of combining and connecting different green forms, independent of context. Parallel to the *Atlas des travaux*, the organization of *Les promenades de Paris* was based on the project's scale and function in the urban landscape.

If both publications chronicled a standardization of the production of urban space, then they also illustrated the conformity of the administration that employed these images. The major municipal services under Haussmann's tenure were the Service des promenades et plantations, Service des eaux et égouts, Service de la voie publique, and the Service du plan. Among all the directors of these services, Alphand exemplified the model administrator, not only serving as Haussmann's right-hand man, but continuing the urban building program as head of the

LES PROMENADES DE PARIS

BOIS DE VINCENNES. — VUE DE L'AVENUE DAUMESNIL.

Ministère des travaux publics during the Third Republic.[75] In addition to being left unscathed by the controversies of the Second Empire, Alphand was even honored and admired for his service, representing the ideal bureaucrat. His personnel files show no divergence from the career path of an officer in the Ponts et chaussées, recording his steady rise through a fixed bureaucratic hierarchy.[76] Paul Strauss, municipal councilor, shared and repeated this sentiment in his printed eulogy of Alphand from 1891: "He was above all a man of action, and he had nothing of an ideologue; all of his intellectual capacities had a practical destination."[77] His characterizations reinforce Alphand's lack of ideology, his practicality, and his utility, and they support the notion that the practice of urban planning during this period was not a political but a purely technical one, based on the figure of a dispassionate administrator. There were stakes to the portrayal of Alphand as apolitical. He had been intimately involved in Paris's modern development from 1854, when Haussmann nominated him to direct the Service des promenades et plantations, to his death when he was director of Ministère des travaux publics, and he managed the ways in which the development projects were packaged

and displayed to the public. If Alphand was exclusively a man of technical skill, it meant that the modernization of Paris was the consequence of technical achievement.

It is to Alphand's credit that Haussmann's modernization program was realized. Notably, the term "Haussmannization" was coined in 1892 and took on a connotation of effective modernization, shedding its association with displacements that had accompanied the large-scale demolitions.[78] Historians who have addressed the urban planning of Paris after the Second Empire generally argue that the continuation was inevitable, because the works were already underway, and construction could not be stopped abruptly.[79] On a historiographical level, the plan—the one Napoléon III handed to Haussmann when he received his appointment in 1853—seems to overdetermine the history of Paris's development. However, on a historical level, the plan for Paris was never a clear instrument, and only provided the conditions for and limits on how Paris could be imagined and constructed.

In one sense, then, the plans were political theater to make Paris graspable, all managed under Alphand's direction. Called to Paris on December 12, 1854, to improve

LES PROMENADES DE PARIS

PROFIL SUIVANT L'AXE DE LA VOIE.

PROFIL TRANSVERSAL.

PLAN AU DESSUS DU TROTTOIR

PLAN INDIQUANT LA DISPOSITION DES DRAINS
SERVANT À L'AÉRAGE ET A L'ARROSAGE.

Grille garantissant le pied
des Arbres.

Banc

Candélabre

BANC DOUBLE
Vu de côté.

CHASSE ROUE
Élévation.

GRILLE EN FONTE SERVANT À GARANTIR
LE PIED DES ARBRES.

Vu de face.

Plan

ÉCHELLE DES PROFILS ET DES PLANS DE 0,01 POUR 1.00.

ÉCHELLE DU BANC ET DE LA GRILLE DE 0.04 POUR 1.00.

VOIE PUBLIQUE DÉTAILS

E. HOCHEREAU & LABEYRIE, DEL.

J. ROTHSCHILD, ÉDITEUR.
Tous droits réservés.

E. LEBEL, SC.

the public spaces of the city, Alphand had worked for Haussmann earlier in Gironde. He had made an impression on Haussmann, whose recollections during the time in Bordeaux center on various celebrations or receptions: one in October 1852 for a visit by Napoléon III, and another for his talents in the organization and decoration of charity events in the spring.[80] Alphand was charged with engineering and construction projects as well, but the specific and repeated mention of his talents in arranging these events speaks to the admiration for his skills of framing and presenting the public works. These early experiences proved valuable for Alphand's work in the Expositions universelles during his tenure in Paris: director of the grounds of the Exposition universelle of 1867, member of the select commission of the Exposition universelle of 1878, and director general of the Exposition universelle of 1889. As affirmed in the organizing committee meetings in which Alphand participated, "The Universal Exposition of 1889 must have the dual purpose of showing the civilized world the ceaseless progress of our national industry, and at the same time celebrating in a peaceful way the centenary of our immortal revolution."[81] What greater opportunity to present the urban improvements of Paris, the city's shaping and modernization, to the national and international public than these urban spectacles? Moreover, the map was the medium of the city's public and global presentation.

To display Paris to the world—and, importantly, to its own citizens—Alphand oversaw major publications highlighting the city in conjunction with all the Expositions universelles in the second half of the nineteenth century. *Les promenades de Paris* was published to coincide with the Exposition universelle of 1867; the *Atlas des anciens plans de Paris* coincided with the Exposition universelle of 1878; and the *Atlas des travaux de Paris* coincided with the Exposition universelle of 1889. The arguments supporting the Expositions universelles paralleled the technical rhetoric of the publications. M. Berger, managing director of the 1889 exposition, rebutted claims from many contemporary monarchies that the chosen date of 1889 was political: "We have been accused of having chosen the anniversary of a revolutionary date. I absolutely refuse this position. I state that the date of 1889 is an essentially scientific and industrial date."[82] The introduction to the *Monographie de l'Exposition universelle de 1889* noted that the tradition of holding expositions in eleven-year intervals determined that the date of the next exposition would be

1889, citing as the first exhibition of its kind not London's "Great Exhibition" of 1851, but a display of manufactured goods organized by François de Neufchâteau, the French minister of the interior, in 1789. Its author, Alfred Picard, pointed out that Napoléon Bonaparte had continued this practice by organizing expositions in 1801, 1802, and 1806, with others following in 1819, 1823, 1827, 1834, 1839, 1844, 1849, 1855, 1867, and 1878.[83] Thus, the claim was that the Expositions universelles were demonstrations of technical and industrial progress, arguably not aligned with political regimes that served as an appropriate platform for Alphand's publications of Paris.[84] Moreover, the associated publications of the atlases provided a complementary scientific image and history of the city, and taken as a whole, showed a consistent message of administrative and technical control in the urban development of Paris.

Reproductions of both the *Atlas des travaux* and the *Atlas des anciens plans de Paris* were available for sale to the public. Collectable individual prints of the latter were sold at three francs per sheet or as a complete set for the bargain of one hundred francs. The first print run included two hundred editions, which was only made possible by the advent of new printing technologies. For the former, sheets were offered in limited editions. The first hundred prints were given freely to various parts of the administration, and four hundred additional sets were printed for the public to recoup the costs of its production. Although expensive and accessible only to a professional elite, the offer of these official maps of Paris for public sale created the impression that the cartographic view was no longer the privilege of rulers. The public sales gave the sense of a government that was transparent with its citizens, and suggested that the modernization program that was materially manifest in the city was part of an organized and directed plan.

The medium of these historical representations was significant because the map's synoptic qualities made the argument that the modernization of Paris was the result of "inevitable facts" rather than political choices.[85] The map, with its visual aesthetic resulting from accumulated layers of homogenizing categories, communicates urban space as objective, universal, and reproducible, and its history as linear. Accordingly, the publications offered an urban planning model that presented modern Paris as it was imagined first, then drawn, then built, fixing the narrative encapsulated in Haussmann's memoirs when he recounted Napoléon III handing him a colored map with his intentions for modernization.

Significantly, the publication and sale of various atlases of the city and its history, its new parks, and its development, along with the display of the large cartographic plates at the Exposition universelle, gave people access to an image of Paris that had previously been restricted. Verniquet's surveyed map, measuring a little over five meters in length and four meters in height, had been fixed to the surface of a large oak table behind closed doors at the convent at Cordeliers.[86] Napoléon III's map of Paris was stretched out behind his desk in his private office, and Haussmann attached his to a folding screen in his office at the Hôtel de Ville. These displays conferred power, and viewing maintained entitlement. Consequently, their public display extended that privilege to an enlarged audience. In doing so, both the visual and physical spaces of Paris became accessible and legible, teleologically framing the city for a bourgeois public.

This history of cartographic representation was enabled by the invention of photoengraving as a technology of reproduction. A copper plate was coated in a light-sensitive material exposed to the image to be reproduced and then etched, producing an intaglio print that could capture continuous tones.[87] The process essentially consisted of transferring an image across various media: from an etching to a copper plate to a chemically treated paper to a press and finally to paper. Photoengraving was particularly useful for cartography because it could retain the clarity of the line from the etching as well as the difference in tonalities; moreover, colors could be easily added, either by hand or by print.[88] The fidelity of the reproduction was of the highest importance, for it guaranteed the authenticity not of the original document, but of the copy.

The virtue of authenticity was imparted by mechanical reproduction techniques that eliminated the human hand, which had often drawn the original. As Alphand explained, "Whatever the degree of reduction, the photographic process we used, excluding all personal interpretation, guarantees in an absolute manner the perfect authenticity of the reproduced documents."[89] As Walter Benjamin has theorized, the technologies of mechanical reproduction allowed art to become a practice of politics and opened participation in art's auratic experience to a larger public.[90] For Theodor Adorno, mass reproductions produced not critical citizens but passive viewers or listeners, and subjected them to cultural production "stamped with sameness."[91] Mass media was not concerned

with the masses "of the spirit which sufflates them, their master's voice."[92] The argument was that through the collapse of distinctions within art, the development of mass media conditioned by industrial reproductions did not offer greater possibilities for diversity and critique, but rather, having emerged under authoritarian and totalizing regimes, forced members of the public into conformity and made them into "an object of calculation."[93]

What kind of politics did these cartographic presentations and reproductions then afford? What possibility of subjectivity was offered by these orthographic maps, teleologically arranged? The Third Republic saw in these objects a means to impart an ideological message of nationhood and republicanism, and they were deployed in the country's educational reforms for Paris's public presentations, and for private use in the map's sale. Accordingly, their claims to objectivity—argued through their orthographic composition—interpolated the readers as new types of citizens who would submit themselves to a city defined and navigated in these graphic terms. The reduction of Paris to geometry resolved the need to come to terms with France's recent historical failures, as well as with the colonial violence that undergirded its image as a peaceful republic.[94] The suppression of local and regional languages and indigenous customs in France, as well as the fact of its colonial ambitions, was obscured to the reader in the performance of democratic accessibility through the Expositions universelles.

In this context, the orthographic presentation of the French capital must be linked to the parallel formation of the country into another geometric form—the hexagon. While several references occur centuries earlier, in the mid-nineteenth century, this geometric shape is mentioned in at least seven different works, which were then adopted by 1870 in the Third Republic as natural.[95] The moral idea was that this form emerged from France's innate geographical symmetry and balance.[96] France's identity as "the Hexagon" was expedient in making the argument that the loss of Alsace-Lorraine was to break the nation's "natural" form, disrupting its geometric regularity. Simultaneously, in arguing for the nation's coherence, the form elided France's "supranational history."[97] It gave no indication that imperialism and colonialism had been essential in shaping Paris at its geometric and symbolic center.

Conclusion

The Total View of the City

Between the idea
And the reality
Between the motion
And the act
Falls the Shadow
—T. S. Eliot, "The Hollow Men" (1925)

The study of Paris's mapping and maps presented here is a history of the powerful: those who had the means to define the terms of the maps' use and to effect change and control over others through these representations. The men who have populated these pages participated in and maintained—even when critical of it—the cartographic discourse that came to define the urban administration and its command over the city. The defeat by the Prussians did little to change the trajectory of how maps were used. Indeed, in light of the humiliating loss, there was a sense that the government had to double down on its institutional and epistemological investment in geography more broadly. The installment of the Third Republic came with a cartographic fervor that sought to develop the graphic potential of maps through the incorporation of social statistics and all manner of quantitative measures. Ever linked to military efforts, even World War I did not offer a reconsideration of the ways in which mapping and its graphic language were complicit in violence. Instead, the newly formed conception and profession of urbanism saw in war the possibilities for building new cities; it thoroughly allied itself to representational conceptions that had originated in orthographic modes developed almost a

century earlier: immediacy, overviews, totality, functionalism, and precision.

At the start of World War I, the architects Donat Alfred Agache, Jean Aubertin, and Édouard Redont published a book, *Comment reconstruire nos cités détruites* (How to rebuild our destroyed cities) (1915), that introduced the term *"urbanisme,"* or urbanism. The text is careful to outline a definition for this new word:

> Urbanism touches on all questions which concern towns: circulation, hygiene, aesthetics. Urbanism is a science and an art. It requires in its practitioners precise knowledge and a special competence, but these qualities would not be sufficient unless they were vitalized by natural talent and intuition. The engineer may provide the logical solution, the architect may know how to endow the city with structures, noble or picturesque; but it is left to the Urbanist to coordinate all these values into a conception of totality [*une conception d'ensemble*]—in a word, to produce a fine plan.[1]

Advocating for the special expertise of the urbanist, the authors took a top-down perspective, with the plan d'ensemble as the central instrument for coordinating and controlling the building.[2] It conceptualized the city as a distinct object, with clear boundaries, parts, and systems. No longer were debates about measures or the map's composition important. This representational form had become thoroughly entrenched in the urban administrative and governance culture, and any contestations were emptied from the discourse around its use. Accordingly, as the nascent field of statistics immediately took up orthographic representations, maps could also spatialize social data. The implicit argument was that social phenomena and geography were linked, and that the purview of the urbanist was to resolve social problems through spatial design.

Anticipating a French victory, the authors projected that the war's end would bring an opportunity to reorganize the urban fabric of French cities and the administrative and financial infrastructure that undergirded all urban decisions. The intention was not to rebuild the destroyed cities as they had been, but to seize the chance to improve them. The same text explained:

> Even prior to their evacuation by the enemy the destroyed towns should be provided with plans of improvement and extension, and . . . the Prefects should be invited to work them out through commissions appointed for the purpose, and . . . no town should be allowed to touch the sums which the State is to pay to it, in the name of national solidarity, without having presented a town plan seriously and rationally based.[3]

Titled "The necessity of overall plans [plans d'ensemble] for the organization and extension of urban and rural agglomerations," the authors made a claim in the first chapter for the coordinated management of the built environment through a drawn total view.

Proper hygienic improvements, improved circulation, and thoughtful aesthetic programs would ensure moral improvement in cities as a precondition for renewed French national solidarity.[4] The view was that the distinct work of shaping cities was an entirely new practice to emerge from the social and political circumstances exacerbated during the nineteenth century, which left many French cities—especially Paris—with traffic congestion, overcrowding (especially related to workers' housing), a dearth of open spaces, and a lack of centralized urban planning structures. This work was to be handled by an empowered elite, notably through the specific administrative structure of committee work rather than by any

individual figure. This last point is of course in direct reaction to the perceived failures of the Second Empire and particularly of the oversized character of Haussmann, who has a ghostly presence in this founding text on urbanism. He is only mentioned briefly in the preface and in the legislative appendix, with more attention given to Henri IV and Louis XIV, as well as the Plan des artistes. Haussmann is credited as merely executing the plan with some additional "strategic openings."[5] From the Third Republic perspective of these urbanists, the nineteenth century was to be understood as simply carrying out the revolutionary plans drawn from Edme Verniquet's plates.

Each successive administration effected erasures by not giving credit to previous work, paralleling the graphic erasures of the urban plan. At its core, urbanism always supposed a break from the past. To describe a city's environment in a geometric language was to present the city in an atemporal form that disavowed contingencies, histories, and society. All was translated and reduced to lines and the spaces between, and those absences and voids on paper provided the projective space to rebuild French cities as well as to build new colonial settlements. For these urbanists, social paternalism combined socio-scientific practices into formal analyses, securing urban planning as a technocratic practice and the master plan as central. The results of these rationalized actions would be most clearly and violently articulated in the French planning of colonial cities during World War I and afterward. So, while claiming to enact social reforms through the seemingly coherent and rational figures of the urban plan, they simultaneously denied the complexities omitted by its graphic lines. It was the legacy of orthography's thorough bureaucratic assimilation.

The aim of the analyses presented in this book has been to demonstrate the efforts that went into making the orthographic urban plan a legible image, one specifically associated with the values of rationality, objectivity, and science in the name of shaping the modern city. The visual—and cartographic—rhetoric was of course highly aestheticized. The graphic norms presenting Paris as a product of technical processes and functional decisions were not assumed, and were contingent upon their alignment with a number of social and epistemological values related to scientific objectivity. It was not merely that the plan was accurate; it had to appear precise, with cartouches, measures, and tables composed to signify fidelity.

There are many unanswered questions in the study

offered here, but the primary point is this: in using these orthographic maps, administrators—among them, engineers and architects—were able to define a territory on which they could project their plans. The analysis of the plans themselves, their representational mechanics, and how the planning practices related to a public all demonstrate how cartographic images were potent, how they sustained an ideology of objectivity, and how they materially affected and constituted urban experiences. The most banal of practices—setting a ruler on a table and drawing a line on a plan—were given authority through the symbolic power of the map, in the administrative and legal regulations that ensured the map's authority, in the social construction of its scientific values, as well as by political will.

If the myth of modernity suggests that there could be a singular, spectacular image of the city—perhaps an ur-map such as the one handed to Haussmann by Napoléon III—this book has hopefully dispelled that notion. There was no single image from which Paris's urban form emerged. There were numerous maps and plans, and there were enormous and sustained efforts to make those images work, to make them cohere to the terrain. Yet the lived experience of the city always exceeded what could be captured and fixed on a sheet of paper. That vast gap between urban life and its cartographic representation ultimately provided room for the emergence of forms of resistance by the city's inhabitants, despite efforts by the government to hide those possibilities in the blank spaces of their maps.

This tension, between a proposed totality and the apparent impossibility of representing everything at once, is what made Paris's relationship to the map specifically modern. It was the knowledge that there was no physical experience of the city that was not negotiated through representations, which created a simultaneously embodied and alienated sensation of space. However, this feeling of not knowing where one is in a city, not being able to imagine that city, returns us continually to the map. Its incompleteness forces us to cling harder to its presentation, until we cannot conceive of the city outside of it. Perhaps this is what Borges intended to evoke when the map and territory share the same scale. At a certain point, the inhabitants lived on the map as their terrain, and there was no world outside of it.

Efforts to map and quantify urban spaces and beyond have not abated since the eighteenth century. Cartographic representations drawn and printed on paper have given way to digital maps that are now ever-more detailed in their resolution and pervade almost every aspect of daily life and society. In this sense, maps do more than define space. They determine political representation, access to services and certain rights, infrastructure development, and our collective and individual identities. But they do so not because of some intrinsic quality of the map itself. It is because we have shaped our world to conform to the map. We have allowed our cities, our institutions, and even our social relations to be flattened and framed by its lines.

I wrote most of this book while under severe lockdown in Italy and France for over a year. In the midst of the COVID pandemic, space bent in strange ways as we were confined to our homes but instantaneously connected to distant places through our computer screens. Our lives and cities became subject to a new geography based on invisible contagion, and in which all actions were geared toward limiting circulation and increasing restriction. In this context, maps and plans became even more central as institutions tried to measure the number of people who could fit within a specific area, as officials tried to close existing borders and create new ones, and as people tried to stay safe distances from one another.

During these stressful moments of confinement, it became clear that the map ceased to represent or work in the ways that were needed. Tallies and spatial guidelines created more confusion, not order and clarity. Maps and plans potentially helped to determine a safe number of people for a theater, market, or library. But they proved wholly inadequate and impractical in helping distribute care equitably and in managing refugees crossing borders to find asylum, homeless people who were caught outside of curfews, and other communities so powerless that they were not even mapped and counted by the state. These people were already victims, but during the pandemic they were made even more vulnerable to the consequences of our modern mapping practices that have clearly centered on a kind of speculation for control and profit by a few. As we begin to emerge from this global crisis, from our confined worlds, I can only hope that we find opportunities for new spaces of human connection that cannot be measured on a map.

ÉDIFICES DE PARIS
CONSTRUITS DE 1871 À 1889

La Teinte Rouge indique les Édifices.
La Teinte Bleue indique les Établissements
Universitaires et Scolaires.

Notes

INTRODUCTION

Epigraph: Walter Benjamin, "Rigorous Study of Art," trans. Thomas Y. Levin, *October* 47 (1988): 89.

1. Merruau, *Souvenirs*, 365. Unless otherwise noted, all translations are my own.
2. Merruau, *Souvenirs*, 374.
3. This aspect is found in many of the writings by administrators, including notably the memoirs of Jean-Gilbert-Victor Fialin, duc de Persigny, who gave specific credit to Louis Philippe. See "Les travaux de Paris," in Persigny, *Mémoires*, 237–65.
4. The complexities of the Paris administration originated in its status as a municipality and capital marked by a struggle between the state's centralization efforts and the city's efforts to preserve its independence. See Say, *Études*; for an analysis, see Mead, *Making Modern Paris*, 67; Van Zanten, *Building Paris*, 46–73.
5. For a historiographic analysis of Haussmann's role in the modernization of Paris, see Darin, "Haussmann," 97–113; Van Zanten, "Paris Space," 179–210.
6. On the metaphor of the body for the city, see Vidler, "Scenes of the Street," 29–111.
7. Haussmann, *Mémoires*, 3:54. This passage was also cited by T. J. Clark to discuss the class war associated with Haussmann's plans. See Clark, *Painting of Modern Life*, 39.
8. For a history of the term, see Choay, *Modern City*.
9. Casselle, *Commission des embellissements*; Bowie, *La modernité avant Haussmann*; Papayanis, *Planning Paris before Haussmann*.
10. Two notable studies on the cartographic culture of the French administration are Chapel, *L'oeil raisonné;* and Olson, *Cartographic Capital*. However, both focus their analysis on the Third Republic, when statistical mapping prevailed in the public administration.
11. Frängsmyr, Heilbron, and Rider, *Quantifying Spirit*.
12. Kula, *Measure and Men;* Rose-Redwood, "With Numbers in Place," 295–319.
13. In their introductory essay, Jeremy Crampton and Stuart Elden explain that both quantitative calculations and qualitative measures composed a "model of rationality" that was connected to mathematical models. Crampton and Elden, "Space, Politics, Calculation," 681–85.
14. Alder, *Measure of All Things*.
15. Wise, *Values of Precision*.
16. Lee, "Objective Point of View," 11–32.
17. Vismann, *Files*.
18. Mead, *Making Modern Paris*, 67–101.
19. Picon, *Architectes et ingénieurs*, 99–120.
20. Van Zanten, *Building Paris*, 69.
21. Ian Hacking and Norton Wise most notably make this claim not through a technologist approach to quantification but by questioning the ways in which precision itself was made valuable through the use of instruments and those who used them. In this sense, they revise Thomas Kuhn's theory of technological change. See Kuhn, *Structure of Scientific Revolutions;* Porter, *Rise of Statistical Thinking;* Porter, "Objectivity and Authority," 245–65; Wise, *Values of Precision;* Hacking, *Social Construction of What?*
22. This would continue into the twentieth century and manifest in the rhetoric of architects such as Le Corbusier. For an analysis of the use of the term "functionalism" and its relation to modernism, see Anderson, "Fiction of Function," 18–31.
23. Robin Evans argues against Erwin Panofsky's reading of perspective. See Panofsky, *Perspective as Symbolic Form;* Evans, *Translations*.
24. Evans, *Projective Cast*.
25. See Blau, Kaufman, and Evans, *Architecture and Its Image*. This important catalogue on the history and theory of architectural images mostly addresses the individual architect's plan.
26. Harley, "Deconstructing the Map," 149–68.
27. Crampton, "Maps as Social Constructions," 235–52.
28. These scholars include Barbara Belyea, Martin Brückner, Max Edelson, Matthew Edney, Laura Hostetler, Claudio Saunt, and Jean-François Palomino. For historiographic context, see Belyea, "Image of Power," 1–9.
29. Godlewska describes maps produced during Napoléonic France and the French formation of the practice of geography. Her work has been integral in laying out the historical particularities of maps as products of the Enlightenment. See Godlewska, *Geography Unbound*.
30. This claim follows the work of media theorists and historians such as Kafka, *Demon of Writing;* Gitelman, *Paper Knowledge;* Latour, *Making of Law*.
31. Daston and Galison, "Image of Objectivity," 84.
32. Wood, "Pleasure in the Idea," 4.
33. While they do not write about these specific instances, Roberto Casati and Achille Varzi do explain the semantic structure of maps and how they preserve information. Casati and Varzi, *Parts and Places*, 187–96.
34. Wood, "Introducing," 207–19.
35. Crary, *Techniques of an Observer*.
36. Foucault, *Security, Territory, Population*, 109.
37. Dunlop, *Cartographia*.
38. Foucault, *Power/Knowledge*, 74–75.

39. Foucault, *Security, Territory, Population*, 109; Crampton, "Cartographic Calculations," 92–103.

40. Hannah, *Governmentality*.

41. Elden, "Governmentality, Calculation, Territory," 565. See also Gordon, "Governmental Rationality," 1–51.

42. Foucault, *Security, Territory, Population*, 46–47.

43. Sennett, *Flesh and Stone*.

44. Rabinow, *French Modern*.

45. For contrasting interpretations, see Brenner, "Foucault's New Functionalism," 679–709; Pløger, "Foucault's 'Dispositif,'" 51–70. For its application in the French context, see Rabinow, *Anthropos Today*, 50.

46. Foucault, *Power/Knowledge*, 194–95.

47. Lobsinger, "Architectural History," 136.

48. For an overview of the main references that include Friedrich Kittler and Bernard Stiegler, see Horn, "There Are No Media," 7–13.

49. Benjamin, *Writer of Modern Life*.

50. Buck-Morss, *Dialectics of Seeing*.

51. Benjamin, *Arcades Project*. For analysis, see Osborne, *Politics of Time*; Hanssen, *Walter Benjamin*; Weber, *Benjamin's -abilities*, 227–39.

52. Examples of histories of mapping within imperial and colonial contexts include: Akerman, *Imperial Map*; Bazzaz, Batsaki, and Angelov, *Imperial Geographies*; Bell, Butlin, and Heffernan, *Geography and Imperialism*. For French colonial examples, see de Rugy, *Imperial Borderlands*; Hannoum, *Invention of the Maghreb*. For British examples, see Edney, *Mapping an Empire*; Chattopadhyay, *Representing Calcutta*.

CHAPTER 1. MASTER PLANS

Epigraph: Alfred Korzybski, *Science and Sanity: An Introduction to Non-Aristotelian Systems and General Semantics* (New York: International Non-Aristotelian Library Publishing Company, 1933), 58.

1. For the term's origins, see Jordan, "Haussmann and Haussmannisation," 88.

2. For examples of the term in other contexts, see Benchimol, *Pereira Passos*; Schubert and Sutcliffe, "'Haussmannisation' of London?," 115–44; Van Loo, "L'Haussmannisation de Bruxelles," 39–49; Vernière, "Les oubliés," 5–23; Volait, "Making Cairo Modern," 17–51.

3. There was also a corollary phrase, "Paris of the ___." Buenos Aires was the "Paris of the South"; Beirut was the "Paris of the Middle East"; and many other cities made claims of being the Paris of their given region. For a brief overview, see Soppelsa, "Haussmann's

Hegemony," 35–51. For the reception of nineteenth-century Paris's modernization, see Wakeman, "Nostalgic Modernism," 115–44.

4. See Haussmann, *Mémoires*, 2:53.

5. Lameyre, *Haussmann*, 92. See also Houssaye, *Les confessions*, 4:93; Haussmann, *Mémoires*, 3:52–55.

6. Pierre Casselle's publication of the Siméon commission's report from 1853 contributes to the dismantling of this characterization. See Casselle, *Commission des embellissements*.

7. Merruau, *Souvenirs*.

8. Morizet, *Du vieux Paris*.

9. See Benevolo, *Origins of Modern Town Planning*, 132–33, citing Houssaye, *Les confessions*, quoted in Lameyre, *Haussmann*.

10. Examples include a variety of disciplines, from art history to architectural history to history *sans phrase*. From the nineteenth century, see Beaumont-Vassy, *Histoire intime du Second Empire*, 189–90; Merruau, *Souvenirs*, 364; Granier de Cassagnac, *Souvenirs*, 223. For more recent examples, see Clark, *Painting of Modern Life*; Lavedan, *Histoire de l'urbanisme*; Pinkney, *Napoléon III*; Benevolo, *Origins of Modern Town Planning*; Harvey, *Paris*. This brief list does not include the numerous biographies of Haussmann that also recount this scene.

11. Lavedan, *Histoire de l'urbanisme*.

12. In particular, see the section "Mapping, Claiming, and Reclaiming," which lays out the ways in which visual technologies, such as mapping, were entangled in colonial and imperial projects as part of an epistemic regime cultivated through industrialization. Jay and Ramaswany, *Empires of Vision*, 211–82.

13. As a contemporary of Haussmann, Jules Simon is an important example who was at first a critic of Haussmann's *grands travaux* and later his defender as minister. See Bertocci, *Jules Simon*.

14. There are significant parallels to the Spanish court, and diplomat and political theorist Diego Fajardo's writings, well known at Louis XIV's court, advocated maps as educational tools for the prince. See Escobar, "Map as Tapestry," 50–69.

15. "Portrait," in Denis and d'Alembert, *Encyclopédie*, 13:153.

16. See "Portrait," in *Encyclopédie méthodique*, 145.

17. *Dictionnaire portatif*, 2:205.

18. Hannah Williams contends that Testelin's painting was not a portrait but a history painting. Williams, *Académie Royale*, 66–68.

19. For context of how this particular group scene diverged from others in the *Histoire

du roi* series, see Williams, *Académie Royale*, 66–69.

20. Napoléon III never approved the painting, and the commission was cancelled, but Flandrin did finish the painting. Théophile Gautier praised the work as capturing Napoléon's complex personality, managing to depict his "distant and inscrutable" look. He affirmed, "This is without doubt the first 'true' portrait we have of His Majesty." Gautier continued, "Monsieur Flandrin has produced an ideal of perfection, 'the modern sovereign.' . . . The face has a resemblance which is both intimate and historical. . . . The clouded, dreamy eyes have a look which goes beyond the visible and can see the forms of the future invisible to the rest of us. On his calm lips hangs the smile of affable majesty: nothing could be truer, simpler or grander." The simplicity and "true-to-life" quality present in the painting made it appear much more realistic than the portrait of the emperor in his coronation robes by Franz Xaver Winterhalter, exhibited during the Salon of 1855, in which the attempt at nobility and idealization became unabashed flattery. However, the positive public acclaim for Flandrin's portrait, especially at the 1861 Salon, forced the emperor to accept it, and it was re-exhibited in the 1867 Exposition universelle. See Gautier, *Abécédaire du Salon*, 154–56.

21. Bordes, *Jacques-Louis David*, 119.

22. Harley, "Deconstructing the Map," 155.

23. Harley, "Deconstructing the Map," 156.

24. For the legacy of this canonical text, and a critique of its postmodernist claims, see Edney, "Cartography and Its Discontents," 9–13.

25. Jacopo de' Barbari's woodcut map of Venice from 1500 is considered the first known appearance of a bird's-eye view that diverged from perspectival views of the fifteenth century. See Ballon and Friedman, "Portraying the City," 687.

26. Ballon and Friedman, "Portraying the City," 690.

27. Picon, "Nineteenth-Century Urban Cartography," 136; Vergneault-Belmont, "Espace et société," 37. Related to describing the urban map as a portrait, humanist theory equated the physical attributes of the city to its character; see Friedman, "'Fiorenza,'" 56–77; Smith, *Architecture*.

28. Marin, *Utopics*, 233–38. For the etymology, see Vergneault-Belmont, "Espace et société," 37–38.

29. Ballon and Friedman, "Portraying the City," 691; Karrow, *Mapmakers of the Sixteenth Century*.

30. The original French terms are, respectively, "*la carte spectacle*" and "*la carte instrument.*" Palsky, *Des chiffres et des cartes*, 18.

31. Bender and Marrinan, *Culture of Diagram*, 23.

32. Bender and Marrinan, *Culture of Diagram*, 23.

33. Jack Goody writes about the epistemological performances that written language allows. See Goody, *La raison graphique*. For its extension to diagrams from a Piercian perspective, see Stjernfelt, *Diagrammatology*.

34. Haussmann, *Mémoires*, 3:15.

35. Jean-Luc Arnaud has suggested that it was the large scale used for drawing that pushed these images to be called "plans" rather than "maps." See Arnaud, *Analyse spatiale*, 64.

36. By the end of the nineteenth century, due to increasing techniques and technologies in mapping, the production of these relief maps was discontinued.

37. For the complex history of the ingénieurs géographes, see Berthaut, *Les ingénieurs géographes militaires*. For a brief description of the 1802 and then the revised 1807 procedures for the cadastre, see Konvitz, *Cartography in France*, 32–62.

38. Ernst Otto Innocenz von Odeleben, *A Circumstantial Narrative of the Campaign in Saxony* (London: J. Murray, 1820), 1:145–46, cited in Black, "Revolution in Military Cartography?," 67–68.

39. Siegfried, "Naked History," 235–58.

40. Jean-Gilbert-Victor Fialin, duc de Persigny, minister of the interior, to Henri Siméon, head of the commission for improvements, August 2, 1853, BAVP, ms 1782.

41. Sitte, *Der Städte-Bau.*

42. Sitte, *Der Städte-Bau*, 101.

43. Cited in Yates, *Selling Paris*, 15.

44. Augustin-Thierry, "Souvenirs," 844–49.

45. This painting was the first of two to document this event. This version was rejected for insufficiently emphasizing the prestige of the Conseil municipal. He thus created a second work that was lost in the Paris Commune of 1871.

46. Visconti submitted elevations, sections, and several other plans, including of the surrounding area, to the Conseil des bâtiments civils in February 1852.

47. Haussmann, *Mémoires*, 3:40.

48. McQueen, "Shaped to Suit a Nation," 216–50.

49. McQueen, "Women and Social Innovation," 176–93; McQueen, *Empress Eugénie and the Arts.*

50. Both of these plans are taken from photographic reproductions by Édouard Baldus, who had worked closely with Lefuel to document the construction of the new Louvre. See Daniel, *Photographs of Édouard Baldus*; Bresc-Bautier, *Le photographe et l'architecte.*

51. Bray, "Prose Constructions," 115–26.

52. Hautecoeur, *Histoire du Louvre.*

CHAPTER 2. TRIANGULATING THE CITY

Epigraphs: Cited in Uta Lindgren, "Land Surveys, Instruments, and Practitioners in the Renaissance," in *The History of Cartography*, vol 3: *Cartography in the European Renaissance* (Chicago: University of Chicago Press, 2007), 478; Lancelot, *Nouveau traité d'arpentage et de toisé* (Troyes: Chez Laloy, 1833), 8.

1. Carroll, *Sylvie and Bruno Concluded*, 169.

2. Borges, "On Exactitude in Science."

3. For a history of early modern maps of Paris and its connection to royal patronage of the city, see the final chapter on Paris maps in Ballon, *Paris of Henri IV*, 212–49.

4. Ballon, *Paris of Henri IV*, 226.

5. Harley, "Deconstructing the Map," 155.

6. Haussmann, *Mémoires*, 3:15.

7. Dilke, *Roman Land Surveyors*; Campbell, *Writings of the Roman Land Surveyors*; Guillaumin, *Les arpenteurs romains*; Chouquer and Favory, *Les arpenteurs romains.*

8 Ancient civilizations in Mexico and Japan were also known to have developed gridded irrigation systems.

9. Willebrord Snellius is often cited as the first surveyor to use triangulation. See Haasbroek, *Frisius, Brahe and Snellius*. The Flemish mathematician Gemma Frisius is also cited as the inventor of this method. See Pogo, "Gemma Frisius," 469–85. This article reproduces part of the *Libellus de locorum describendorum ratione* (Booklet concerning a way of describing places) (1533).

10. Neugebauer, *History of Ancient Mathematical Astronomy*, 772.

11. Keay, *Great Arc.*

12. The first triangulations for the first degree of the longitudinal line through Paris were completed between 1668 and 1670 by Picard, thereby establishing the Paris meridian. See Widmalm, "Accuracy, Rhetoric, and Technology," 180.

13. Edney, "Irony of Imperial Mapping," 41; Edney, "Reconsidering Enlightenment Geography," 165–98.

14. Delagrive, *Manuel de trigonométrie pratique*, 4. Jean Delgrive was a cartographer and an abbot with the congregation of Saint-Lazare, referred to as Lazarists. He was nominated as *géographe de Paris* and published a plan of the city in 1728, but his 1740 map, *Les environs de Paris relevés géométriquement*, was most influential in providing a broad view of Paris as well as the Seine. See Boutier, "Une tentative de relevé cadastral," 107–20.

15. Cassini de Thury, *La méridienne de l'Observatoire royal*. See also Gillispie, *Science and Polity*, 113–15.

16. The other commissioners were the architect Pierre Garrez, Sr., who would become a member of the Conseil des bâtiments civils from 1809 to 1819; Charles-François Callet, who was the architect of the city of Paris for thirty-two years, from 1796 to 1828; and George Galimard, who was possibly also an architect of gardens.

17. Moréri, *Le grand dictionnaire historique*, 390–91.

18. Cited in Pinon, Le Boudec, and Carré, *Les plans de Paris*, 58.

19. Jean Delagrive was included in English dictionaries because he was a member of the Society of British Geographers. See Chalmers, *General Biographical Dictionary*, 16:354.

20. *Journal de trévoux, ou mémoires pour l'histoire des sciences et des beaux-arts* 228 (October 1757): 2678–79.

21. Konvitz, *Cartography in France*, 109–10.

22. By ordinance from the Bureau de la Ville de Paris, February 24, 1744.

23. *Journal des géomètres*, 2nd ser., 7 (1864): 124.

24. Pronteau, *Edme Verniquet*, 311, 348.

25. Amy Wyngaard presents an alternative hypothesis based upon Pierre Choderlos de Laclos's letter of June 17, 1787, addressed to the *Journal de Paris*, that proposes a numbering system to order the labyrinth of Paris's streets. She locates the significance of this letter within the context of urban chaos and fear documented in a 1775 novel, *Le paysan perverti* (The perverted peasant), by Nicolas-Edme Rétif de la Bretonne, and the work of the Société royale de médecin to identify and control the Parisian masses, including the numbering of houses, the development of a central police record listing all residents, and the installation of oil-burning lanterns. See Wyngaard, "Libertine Space," 2–3.

26. Santana-Acuña, "From Manual Art to Scientific Profession," 121.

27. In March 1768 Louis XV declared that all cities and towns were required to number their houses for military purposes. In this time, it was common for soldiers to be housed in civilian quarters. Thus, the house numbering made it easier to keep track of the soldiers when dispersed throughout a city or town. Paris was exempted from this decree because there were official barracks

for soldiers. See Rose-Redwood, "Indexing," 199.

28. Garrioch, *Making of Revolutionary Paris*, 237.

29. The term "surveillance" relates to Michel Foucault's formulation. See Foucault, *Discipline and Punish*.

30. Pronteau, *Edme Verniquet*, 349.

31. It is equivalent to 0.027 of a meter. For a history of measures, see Kula, *Measures and Men*.

32. The number varies in different publications, and probably related to the budget and to the season, as the surveying work lasted several years. I am using the number offered in the *Journal des géomètres*, 2nd ser., 7 (1864): 124.

33. Lynch, "Discipline," 43.

34. Delagrive's trigonometric calculations were first published in Delagrive, *Manuel de trigonométrie pratique* in 1754, and reprinted again in 1806.

35. Picon, "Nineteenth-Century Urban Cartography," 136–37.

36. Léri, *"Le Marais."*

37. Louis Marin has argued that the two modalities present in Gomboust's map—planar and perspectival—represent two sociopolitical dimensions. Figurative representation is reserved for political relations (hence, Montmartre and the Louvre, as well as the buildings within the map), and schematic representation is devoted to exchange, commerce, and the bourgeoisie. See Marin, *Utopics*, 217–18.

38. Bullet, *Traité de l'usage du pantomètre.*

39. Pinon, Le Boudec, and Carré, *Les plans de Paris*, 50.

40. For a discussion of the ideological function of cartouches with a critical history of cartography, see Harley, "Maps, Knowledge, and Power," 277–312. For a study specifically addressing the eighteenth century, see Clarke, "Taking Possession," 455–74.

41. Wise, *Value of Precision*, 7–9.

42. Shapin, *Social History of Truth.*

43. Porter, "Objectivity as Standardization," 21. There is broad scholarship on the social history of the scientific method, including Kuhn, *Structure of Scientific Revolutions;* Latour, *Science in Action;* and a notable example of the social production of facts in Hacking, *Social Construction of What?*

44. For a contemporary model of this point, see Latour, "Circulating Reference," 24–79.

45. Alder, "Making Things the Same," 516.

46. Sewell, *Work and Revolution in France*, 21–25.

47. Santana-Acuña uses the term "'impersonal' objective scientific labor." See Santana-Acuña, "Making of a National Cadastre," 199.

48. Cited in Pinon, Le Boudec, and Carré, *Les plans de Paris*, 80.

49. Teyssèdre, *Nouveau manuel de l'arpenteur*, 144. The excerpt on the application of color was from Dupain de Montesson, *La science de l'arpenteur.*

50. See Mukerji, "Engineering and French Formal Gardens," 22–43; Lauterbach, "Faire céder l'art à la nature," 176–93.

51. Significant texts included a manual on trigonometry from 1754 by Delagrive, who served as the geographer to the king, as well as *Méthode de lever les plans et les cartes de terre et de mer* (Method of mapping plans and maps of the land and the sea) by Jacques Ozanam, a private mathematics tutor and member of the Académie royale des sciences, first printed in 1691 and reprinted in 1781.

52. Alain Manesson-Mallet was much more famous for his work *Les travaux de Mars, ou la fortification nouvelle* (The works of Mars, or the new fortification) (1671).

53. Remmert, "Art of Garden," 22.

54. Fortification architect Sébastien Le Prestre de Vauban kept a copy of Manesson-Mallet's volume at the Château de Bazoches. See Pujo, *Vauban*, 427.

55. Victoria Thompson has noted that during the second half of the eighteenth century, guidebooks to Paris doubled in number and were constantly updated and reissued. See Thompson, "Knowing Paris," 28.

56. Thompson, "Knowing Paris," 30.

57. Lindgren, "Land Surveys," 478. For nautical maps, see Blais, "Qui dresse la carte?," 25–29.

58. Branco, "Fieldwork," 201–28.

59. Santana-Acuña, "Making of a National Cadastre," 228–29.

60. Guiot, *L'arpenteur forestier*, ii.

61. Since surveying was an office bought directly from the monarch, and the number of surveyors per locality did not surpass more than two or three (except for big villages and cities), they did not form a corporation or guild. Santana-Acuña, "Making of a National Cadastre," 243.

62. de La Hire, *L'École des arpenteurs*, 1.

63. The title of "ingenieurs-géographes" had originally been founded in 1769, but was disbanded by the Directory in 1791. The law also established the École centrale des travaux publics (renamed the École polytechnique in 1795), École normale (renamed the École normale supérieure in 1845), and the Conservatoire national des arts et métiers.

64. Konvitz, *Cartography in France*, 140.

65. Santana-Acuña, "Making of a National Cadastre," 289–90.

66. This is an argument that Lorraine Daston has made regarding scientific thought in general: "It is in part the systematic erasure of these details in the service of extended sociability that creates the impression of the uniformity of nature. . . . The uniformity of nature presupposed universalism among scientists, rather than the reverse." See Daston, "Moral Economy of Science," 10.

67. Santana-Acuña, "From Manual Art to Scientific Profession," 121–36.

68. This claim suggests that these epistemic values associated with the academy were developed outside of these elite intellectual circles.

69. William Sewell, Jr., has argued that the *Encyclopédie*'s images of labor do not depict the *process* of production; there is no sequential organization to the steps that make up a particular task. This was an indication of the dissolution of a unity of trade labor. See Sewell, "Visions of Labor," 258–86.

70. "Art," in Diderot and d'Alembert, *Encyclopédie*, 1:717.

71. Godlewska, "Map, Text, Image," 5–28; Rabinow, *French Modern.*

72. See Dykstra, "French Occupation of Egypt," 113–38.

73. In 1799 seven students enrolled in the École des ingénieurs-géographes enlisted to join Napoléon in Egypt, and by 1802 six students were serving in Egypt and two had joined the Baudin expedition. Of the 151 who accompanied the soldiers, more than half were engineers or technicians, and they were relatively young. The average age was twenty-five years, and most were recent graduates of the École polytechnique and the École des ponts et chaussées. The only established scientists were Gaspard Monge, Claude-Louis Berthollet, Jean-Baptiste-Joseph Fourier, Étienne Geoffroy Saint-Hilaire, and Déodat de Dolomieu. See Gillispie and Dewachter, *Monuments of Egypt*, 6.

74. There are no surviving archival records that document at what stage and by whom the triangulated survey was chosen. Godlewska speculates that it was Napoléon himself, and that the decision was made in France. A report was apparently sent by Joseph-Jérôme Lalande and Gaspard Monge that advised the *Carte topographique de Corse* be engraved at 1:86,400 in order to link to the Cassini map of France, which was of the same scale. Godlewska argues that it was Monge who suggested that the Egyptian survey use the same scale so that all the surveys and maps of the entire

Mediterranean region could relate. See Godlewska, "Map, Text, Image," 5.

75. Gillispie, "Scientific Aspects of the French Egyptian Expedition," 467.

76. Chevalier, "La politique financière," 213–40. For a broader discussion, see Petry, *Cambridge History of Egypt*, 130. See also Shaw, "Landholding and Land-Tax Revenues," 91–103.

77. Godlewska, "Map, Text, Image," 8–9.

78. Minutes of the meeting for the cadastral commission, March 3, 1801, Paris, BNF, Mss Fr. 11275, Doc. 93, 90–101. Jacques-François Menou to Jean-Antoine Chaptal, minister of the interior, March 1, 1801, AN, F17A 1100 Dossier 1, doc. 14.

79. A map of Cairo was produced and used in the service of heavy bombardment in response to a rebellion.

80. The map that had guided the invasion had been compiled in 1765 by a geographe du cabinet, the chevalier d'Anville, who worked from manuscript maps and books.

81. The surveying data and drafts were delivered directly to the Dépôt de la Guerre. Maps were then drawn and printed from copper plates, prepared by twenty-three engravers. See Gillispie, "Scientific Aspects of the French Egyptian Expedition," 466.

82. There was also a fight between the two agencies over the profits in the sale of the maps. See Godlewska, "Map, Text, Image," 10.

83. Godlewska and Smith, *Geography and Empire*, 31–53.

84. Gillispie, "Scientific Aspects of the French Egyptian Expedition," 469.

85. Many of the men who participated in the scientific mission of the conquest would take on high administration positions in France, including Jomard, already mentioned in this chapter, as well as Fourier, who served as prefect of the Isère and director of the Bureau de la statistique, and Chabrol de Volvic, who would become prefect of Paris and would institute the first comprehensive census of the city. See Chapter 4.

86. The original scale of the Cassini map was based on the toise, and then was later translated into the metric measure. The resolution came to 1:86,400.

87. Mitchell, *Colonising Egypt*, 13.

88. Mitchell, *Colonising Egypt*, 2.

89. Jomard, *Mémoire sur le système métrique*, 699.

90. Edme-François Jomard to Pierre Jacotin, February 6, 1801, Sment, BNF, Mss., Fr. 11275, Doc. 133–36.

91. Edney, "Irony of Imperial Mapping," 16.

92. Gillispie, "Scientific Aspects of the French Egyptian Expedition," 467.

93. Jomard, *Considérations sur l'objet*, 9–10.

94. An earlier image had been published as the first plate in Alphand's *Les promenades de Paris*, published in 1867.

95. Block, *Administration*, 274–75.

96. Van Zanten, *Building Paris*, 69.

97. Haussmann, *Mémoires*, 3:3.

98. Haussmann, *Mémoires*, 3:2–5.

99. *Le moniteur universel*, August 17, 1861, 26–27.

100. The carton includes a list of citations related to the surveying towers. APo, EB 8, include notes of vandalism and unauthorized use of the towers that date to September 29, 1862.

101. *Le moniteur universel*, January 31, 1860.

102. "Nivellement," in Block, *Dictionnaire*, 274–81.

103. Rabinow, *French Modern*, 63–67.

104. Laussaudet, *Recherches*, 424.

105. The "spatial turn" in the humanities was a response to the thorough conquest of this Cartesian notion of space into all disciplines. The effort to account for the particularities of spaces, especially through the experience of the body and the contextualization of places, represented an attempt to account for not only site-specific analyses, but also the power relations embedded into conceptions of space. See Soja, *Postmodern Geographies*; Appadurai, "Production of Locality," 204–25; Lefebvre, *Rhythmanalysis*.

106. This incommensurability claim refers to different genres of a map, such as pictorial versus orthographic; within a map genre, there can be advancements in precision, but accuracy is a value whose meaning is not progressive but contingent. For the full argument, see Ingraham, *Architecture and the Burdens of Linearity*, 1–29.

CHAPTER 3. DRAWING GRIDS

Epigraph: Pierre Lavedan, *Histoire de l'urbanisme à Paris* (Paris: Hachette, 1975), 301.

1. Maynard, "Perspective's Places," 27.

2. Clark, *Farewell to an Idea*, 7 (quoting from Max Weber, quoting Schiller). The discussion of T. J. Clark's use of "disenchantment" is taken up by several critics, including Werckmeister, "Critique," 855–67.

3. Clark, *Painting of Modern Life*, 23–78.

4. See Lazare, *Les quartiers pauvres*.

5. There is no evidence of surveying being conducted in the press or in the police archives from this period. Nothing was reported in the *Journal des géomètres*, and considering that surveying activities were always a news item because of their threatening or speculative possibilities (as well as the disturbances caused by surveying activities), this seems to indicate that no

new ground surveys occurred. There are also no financial records of funds dispensed for surveying in the municipal budgets; surveying on this large scale would have required allocating significant finances, and therefore a string of legal authorizations. Moreover, because of the extensive costs related to surveying, engraving, and printing maps, the copying of maps and reuse of surveying data were economically viable and common approaches to the production of new maps. See Pedley, *Commerce of Cartography*, 40.

6. While the *cadastre napoléonien* from 1807 to 1821 covered the rest of France, it did not include the municipality of Paris.

7. The publication of the Paris cadastres ordered by Chabrol de Volvic were led by the Direction des contributions and drawn by architects Philibert Vasserot and J. S. Bellanger, who both worked at the finance ministry. The cadastral atlas comprised 240 sheets, of which only 155 were published. Importantly, each sheet pictured each quarter as a coherent and distinct figure, unlike the Jacoubet plans, whose map sheets interlocked to represent all of Paris's topography.

8. Gillispie, *Science and Polity in France*, 227. See also Nguyen, "Constructing Classicism"; Alder, "Revolution to Measure," 46.

9. The Assemblée constituante composed a commission to study the possibility of a unified system of measurement, and a final report was submitted on March 19, 1791. The members, including Jean-Charles de Borda, Joseph-Louis Lagrange, Pierre-Simon Laplace, Gaspard Monge, and Marie-Jean-Antoine-Nicolas de Caritat, marquis de Condorcet, recommended a measure taken from nature whose units would be based on a universal reference. The government adopted the report on March 26, 1791. It was the Convention nationale that adopted the metric system on August 1, 1793. After many political upheavals and scientific controversies, the final measured platinum meter was presented to the government in June 1799. See Alder, *Measure of All Things*.

10. Kula, *Measure and Men*, 29.

11. Zupko, *French Weights and Measures*.

12. Zupko, "Itinerary and Geographical Measures," 927.

13. Qualitative measures were not variable insofar as one could change a measure based on one's mood on a given morning. They were stable in their materiality and through the institutions that supported them. For example, land measures in a single parish in Normandy remained stable

over at least 750 years. See Kula, *Measure and Men*, 111.

14. As dated on June 22, 1799. Today, the true or invariable meter is defined as a length equal to 1,650,763.73 wavelengths of the orange light emitted by the krypton atom of mass 86 *in vacuo*.

15. Bousquet-Bressolier, "Survey of Meter," 965–66.

16. Alder, *Measure of All Things*.

17. In 1812 Napoléon returned France to the old standards, and then in the 1840s Louis Philippe reinstated the metric system.

18. Alder, "Making Things the Same," 499–545.

19. Condorcet, "L'arpentage," 567.

20. The consequences in terms of travel are analyzed in Schivelbusch, *Railway Journey*. In terms of objects and commodities, especially as they relate to the commodity fetish, see Marx, *Capital*, 165; Benjamin, "Capitalism as Religion," 288–91.

21. Picon, *Architectes et ingénieurs*, 97.

22. Konvitz, *Cartography in France*, 47.

23. Kula, *Measure and Men*, 115.

24. Gillispie, *Science and Polity in France*, 223–85.

25. For a distinction between precision and accuracy, see Wise, *Values of Precision*, 3–14.

26. In 1812, due to fierce popular resistance to its adoption and use, Napoléon issued a new law that revoked the metric measures for "*mesures usuelles*," or common measures. These were a hybrid of the traditional measures of the foot or pound, but translated into decimal units. In 1837 the Assemblée nationale under Louis Philippe reimposed the metric system as originally defined in 1799, and conformity was required by 1840.

27. Haüy, *Instruction*.

28. This was also the case for the *Atlas de Vasserot et Bellanger*, which was commercialized and presented for sale by subscription.

29. Kain and Baigent, *Cadastral Map*, 169–70.

30. Daston, "Description by Omission," 11–24.

31. Picon, *Architectes et ingénieurs*.

32. Boullée, *Architecture*, 49, cited in Picon, "From 'Poetry of Art' to Method," 23.

33. Picon, "Gestes," 132–47.

34. Pairault, *Gaspard Monge*. Descriptive geometry is now known as orthographic projection, and the graphical method is still used in technical and architectural drawing.

35. Monge, *Géométrie descriptive*, 5.

36. Pérez-Gómez and Pelletier, "Architectural Representation beyond Perspectivism," 34. For a fuller study, see Pérez-Gómez and Pelletier, *Architectural Representation and the Perspective Hinge*.

37. Jomard, *Considérations sur l'objet*, 15. For further studies of Jomard and the royal map collection, see Pelletier, "Jomard et le Département de cartes et plans," 18–27.

38. See Blau, Kaufman, and Evans, *Architecture and Its Image*, 28.

39. Durand, *Précis*; for analysis, see Pérez-Gómez, "Architecture as Drawing," 5.

40. Cited in Collins, "Origins of Graph Paper," 162. See also a reference to this use of gridded paper in Picon, "From 'Poetry of Art' to Method," 41.

41. Cited in Picon, "From 'Poetry of Art' to Method," 42.

42. Patte, *Monumens érigés*, plate 34.

43. Deming, "Une capitale et des ports," 51–66.

44. See Mosser and Rabreau, *Charles de Wailly*, 65–66.

45. In decrees from April 1793, this assembly, merged with the Commission des monuments, was named the Commission des artistes, which was created to manage and protect monuments and objects for the republic and the Commune des arts, also known as the Commune générale des arts. This commune was created by former *académiciens*, who during the revolutionary years denounced the Académie royale de peinture.

46. The commission was abolished by decree on March 31, 1797. See Léri, "La Commission et le plan des artistes," 152–59.

47. Léri, "Edme Verniquet," 208.

48. Léri, "Edme Verniquet," 209.

49. *Le journal*, May 8, 1899, reported that the Verniquet plan was lost.

50. Léo Taxil was a *nom de plume*. His given name was Marie Joseph Gabriel Antoine Jogand-Pagès.

51. Le Moël, Descat, and de Andia, *L'urbanisme Parisien*, 205–17.

52. Pronteau, *Edme Verniquet*, 358–59.

53. Papayanis, *Planning Paris before Haussmann*, 62–128.

54. The town was called Napoléonville instead of Pontivy from 1805 to 1814, and then from 1848 to 1871.

55. Papayanis, *Planning Paris before Haussmann*, 81; Godlewska, *Geography Unbound*, 212.

56. Supplemental notes to plans describing the drawn projections by Chabrol de Volvic can be found in AN, F13 1757B.

57. Morachiello and Teyssot, "State Town," 35.

58. Harouel, *L'embellissement des villes*.

59. Bowie, *La modernité avant Haussmann*.

60. Decree of February 24, 1811, articles 36–38, and Decree of May 19, 1811, articles 1–3, *Almanach impérial* (1811).

61. Van Zanten, *Building Paris*, 223–25.

62. *Agrandissement et construction des Halles Centrales d'approvisionnement, Rapport fait au Conseil municipal*, February 28, 1845, 32.

63. Plans were dated February 28 and July 15, 1844: *Agrandissement et construction des Halles Centrales*, 10–11. Lahure's first name is unknown—not to be confused with Louis-Auguste Lahure, city councilor and adjunct mayor of the fourth district.

64. *Agrandissement et construction des Halles Centrales*, 41.

65. *Agrandissement et construction des Halles Centrales*, 44.

66. Jacques-Antoine Dulaure, *Histoire physique, civile et morale de Paris depuis les premiers temps historiques jusqu'à nos jours* (Paris, 1824), 9:121.

67. Plans included in the Fonds Victor Baltard also show this alignment. ANCP, 332AP/7.

68. Baltard and Callet, *Monographie*.

69. Scholars have argued that the plan of the building is from the interior outward into the city. See Van Zanten, *Building Paris*, 218; Lemoine, *Les Halles de Paris*, 63–86.

70. Mead, *Making Modern Paris*, 144–225.

71. The proposals of François Debret, official state architect of the Opéra, and Edme-Jean-Louis Grillon suggested an orientation along the rue de la Grange-Batelière. Hector Horeau's counterproposals situated the building on the boulevard des Italiens. In 1846 Charles Rohault de Fleury replaced Debret as official architect of the Opéra, and summarized the possibilities at the Palais Royal and the rue de Rivoli; two different sites at rue Grange-Batelière; the Hôtel du Ministère des Affaires Étrangères; the Menus Plaisirs; the Bains Chinois; the boulevard des Capucines; the Palais Royal and the Louvre; and Louvois. Even after Haussmann's appointment, the proposals continued. In 1853 Max Berthelin and Louis Viguet situated their proposal according to Rohault de Fleury's preferences on the boulevard des Capucines, but oriented the building to the rue de la Chaussée-d'Antin. H. Barnout's suggestion in 1856–57 located the Opéra on the boulevard des Italiens, with a proposed rue Impériale connecting the building to the Louvre. The reasoning behind these propositions varied between Horeau's emphasis on public space and Rohault de Fleury's economic concerns. See Mead, *Charles Garnier's Paris Opéra*, 53.

72. Mead argues that this decree included no provisions for rues Scribe, Meyerbeer, and Gluck, nor for avenue Napoléon, because their existence was implicit. See Mead, *Charles Garnier's Paris Opéra*, 55.

73. For analysis of Rohault de Fleury's design, see Mead, *Charles Garnier's Paris Opéra*, 54–59.

74. A deadline of only one month was given for

proposals. One hundred and seventy-one were submitted for the first round, of which five were retained for a second-round competition. Charles Garnier was unanimously chosen as the winner and appointed as the official architect of the project on June 6, 1861. The first stone of the building was set on July 21, 1862, and the building was inaugurated on January 5, 1875.

75. Archived at the Bibliothèque nationale de France (BNF). There is another at the Archives nationales.

76. A different edition of the same Jacoubet plates located at the Art Institute of Chicago indicates even earlier sketches dated circa 1852 by P.-F.-L. Fontaine. Fontaine's freehand graphite sketches for the rue de Rouen on the west side of the Opéra site of the Art Institute plate and the drawings on the east side of the Opéra on the BNF plate attest to the cartographic grid as a departure point for new percements. Van Zanten, *Building Paris*, 16.

77. Castex, "Les origines du quartier," 42. One methodological problem with how he treats maps is that he considers plans within a history of greater and greater precision, without considering how the value of precision and its graphic expression have contingent histories of their own.

78. Castex, "Les origines du quartier," 48.

79. Loyer, *Paris XIXe siècle*.

80. Van Zanten, *Building Paris*, 26.

81. For a fuller discussion of Charles Marville's compositions, see Lee, "Constructing Nineteenth-Century Paris," 100–101.

82. Meynadier, *Paris sous le point*, 2.

83. Meynadier, *Paris sous le point*, 6.

84. "Plans de Paris.—Alignement," *Gazette municipale*, April 1843.

85. Grillon, Callou, and Jacoubet, *Études d'un nouveau système*, 14.

86. The pamphlet was published right after the February revolution of 1848, and thus issues of security were urgent and significant.

87. Grillon, Callou, and Jacoubet, *Études d'un nouveau système*, 23.

88. Moreover, their argument for order and security was no doubt influenced by the February and June revolutions of 1848, as Jacoubet had been one of the defenders of the order in June. See Maitron, *Dictionnaire biographique*, 2:366.

89. Laugier, *Essay on Architecture*, 129. For the original passage, see Laugier, *Essai sur l'architecture*, 224.

90. Haussmann, *Mémoires*, 3:15.

91. "Because of financial measures ratified by the Chambre in July 1848, I proposed to the Council a vast plan of eighty million, to be carried out in five years. It included Les Halles Centrales, the town halls of the third, eleventh, and twelfth arrondissements, the church of Sainte-Clotilde, the completion of the quays and bridges, the new opera, the place of the Palais-Royal, the extension of the rue de Rivoli, paving, sewers, sanitation, etc." See Rambuteau, *Mémoires*, 293.

92. Casselle, "Les travaux de la Commission des embellissements," 648.

93. Jean-Gilbert-Victor Fialin, duc de Persigny, minister of interior, to Henri Siméon, August 2, 1853, head of the commission for improvements, BAVP, ms 1782.

94. Manuscripts signed by Henri Siméon outlining the committee's process, September 2, 1853, BAVP, MS 1782, ff 22–25. In the end, while the Jacoubet plates might have been used for the study, the final report included several plans of varying scales, including two plans presenting an overview of new roadways and six others illustrating current and new circumscriptions.

95. Manuscripts signed by Siméon, BAVP, ms 1780.

96. Georges Haussmann, *Mémoire présenté par M. le préfet de la Seine au Conseil municipal de Paris* (Paris: Imprimerie impériale, 1858), 5.

97. Haussmann, *Mémoire présenté*, 10.

98. Vidler, *Architectural Uncanny*.

99. Frisby, *Fragments of Modernity*.

100. Benjamin, *Arcades Project*, 839.

101. Darin, *La comédie urbaine*.

CHAPTER 4. THE BUREAUCRACY OF PLANS

Epigraph: Cited in Georges Poisson, *Eugène Viollet-le-Duc: 1814–1879* (Paris: Picard, 2014), 287.

1. Kamea and Krygier, *Bureaucracy*.

2. Soll, *Information Master*.

3. James Cortada defines an ecosystem as "a body of information [that] circulates within some institution or group of individuals who communicate frequently with each other (in our case, diplomats and those with whom they worked) and their employers (in our example, diplomatic ministries). This ecosystem consists of documents and publications that circulate within a community, but it also includes shared knowledge, worldview, and the values its members have accumulated through experience and education." See Cortada, "Information Ecosystems," 224. For its spatial and corporeal values, see Rule and Trotter, *World of Paper*; Himmelfarb, "Versailles," 235–92.

4. Weber, *Economy and Society*, 957.

5. Gitelman, *Paper Knowledge*, 5.

6. Guillory, "Memo and Modernity," 113.

7. Guillory, "Memo and Modernity," 113.

8. Gitelman, *Paper Knowledge*, 3.

9. In anthropological scholarship it is referred to as "scheme of legibility." See Scott, *Seeing Like a State*.

10. Guillory, "Memo and Modernity," 126.

11. An important exception is Hull, *Government of Paper*.

12. This specific citation refers to a project at the Hôpital Saint-Louis and the opening and enlargement of roads around the building sessions of September 7, 1854, and May 3, 1855, AN, F21 2539. This phrasing, however, is not unique to this particular project and is found in almost every session of the Conseil des bâtiments civils.

13. Kafka, *Demon of Writing*, 21. In the first chapter, he argues for "the emergence of a radical new ethics of paperwork" specific to the revolutionary years in France.

14. Comte, "Evolution, Standardization, and Diffusion,'" 179–92; Comte, "King's Feet to Republican Metres," 281–92.

15. Comte, "King's Feet to Republican Metres," 284.

16. Harouel, "Les fonctions de l'alignement," 135.

17. Davenne, *Traité pratique de voirie urbaine*, 37.

18. Colbert, "Instruction," 27.

19. Following its conclusion in March 1790, Charles-Maurice de Talleyrand and Marie-Jean-Antoine-Nicolas de Caritat, marquis de Condorcet offered legislation for a national standard for measures, which was eventually adopted in 1793 by the revolutionary government. In 1812 Napoléon returned France to the old standards, and then in the 1840s, Louis Philippe reinstated the metric system.

20. In September 1785, Louis XVI agreed to establish and fund an official plan of the city and its surroundings, and Verniquet was charged with its official production. See Pronteau, *Edme Verniquet*.

21. Davenne, *Recueil méthodique*, 101 (emphasis mine).

22. Van Zanten, *Building Paris*, 285n29. Van Zanten discusses a case of the rue des Poulies at the rue de Rivoli.

23. Davenne, *Recueil méthodique*, 46.

24. Davenne, *Recueil méthodique*, 44.

25. In article 17 of the declaration of August 26, 1789, private property was elevated as a basic human right, but, as Jean-Louis Harouel writes, this law was only an illusory departure from the past. The procedures to administer properties developed under the monarchy were still enforced. See Harouel, *Histoire de l'expropriation*, 31.

26. Monnier, "La notion d'expropriation"; Mestre, "L'expropriation face à la propriété"; Harouel, "L'expropriation dans l'histoire du droit français."

27. Davenne, *Recueil méthodique*, 229.
28. Husson, *Traité de la législation des travaux publics*, 889–90.
29. Articles 1 and 2 from instruction of October 2, 1815. See Husson, *Traité de la législation des travaux publics*.
30. Article 5 from instruction of October 2, 1815. See Husson, *Traité de la législation des travaux publics*.
31. Articles 10 and 14 from instruction of October 2, 1815. See Husson, *Traité de la législation des travaux publics*.
32. Chabrol de Volvic, *Souvenirs*, 86–87.
33. See articles 1–4, 11, and 13 from instruction of October 2, 1815. See Husson, *Traité de la législation des travaux publics*.
34. Articles 5–10 from instruction of October 2, 1815. See Husson, *Traité de la législation des travaux publics*.
35. Law of July 18, 1837, article 19, section 7; article 31, section 18, cited in Block, *Administration*, 276.
36. Block, *Administration*, 275–77.
37. Davenne, *Recueil méthodique*, 85–86.
38. Van Zanten, *Building Paris*, 20.
39. Five articles of the law from May 3, 1841, specifically addressed the production and distribution of plans. For contemporaneous commentary, see Daffry de la Monnoye, *Les lois de l'expropriation*.
40. Daffry de la Monnoye, *Les lois de l'expropriation*, 237.
41. Daffry de la Monnoye, *Théorie et pratique*.
42. *Recueil des actes administratifs*, 3:80–81.
43. Hull, "Documents and Bureaucracy," 251–67.
44. Kain and Baigent, *Cadastral Map*, 218.
45. With the exception of Provence and Languedoc.
46. There were exceptions—for example, the commune of Saint-Geniès des Mourgues in the diocese of Montpellier, Cambon-lès-Lavaur, about twenty communities in the region of Toulouse and Languedoc, and the area of Chaussadenches in the Vivarais. See Fougères, "Les plans cadastraux," 54–69; Frêche, "Compoix, propriété foncière," 321–53.
47. This decoupling is distinct from the arguments of art historians such as Barbara Maria Stafford and Hector Reyes, who see in the ever-growing reliance on visual images that began in the eighteenth century a logocentrism that is still beholden to written or propositional language. The value of description within the French administration was aligned to scientific culture, not rhetorical tradition, that argued the synoptic quality of images was incorruptible, denying any aesthetic conditions and history that might have mediated their production. See Stafford, *Good Looking*; Reyes, "Rhetorical Frame," 287–302.
48. Poix de Fréminville, *La pratique universelle*, 121. See also Bloch, "Les plans parcellaires," 66.
49. Poix de Fréminville, *La pratique universelle*, 117.
50. Clergeot, *Recueil méthodique*, 5.
51. Article 1, November 23 to December 1, 1790. See Clergeot, *Recueil méthodique*, 5.
52. Clergeot, *Recueil méthodique*, 6.
53. Kain and Baigent, *Cadastral Map*, 220–21.
54. Aubry-Dubochet, *Exécution du cadastre general*, 497.
55. Clergeot, *Recueil méthodique*, 7.
56. Santana-Acuña, "Making of a National Cadastre."
57. Hennet, *Recueil méthodique des lois*, 61, article 176.
58. See Hennet, *Recueil méthodique*, 12, article 19.
59. Maurin, *Le cadastre en France*, 30–40.
60. Scott, *Seeing Like a State*.
61. Foucault, *Birth of Biopolitics*, 2.
62. Lemke, *Foucault, Governmentality, and Critique*, 9–24.
63. Burchell, Gordon, and Miller, *Foucault Effect*.
64. Foucault, *Power/Knowledge*, 173–82.
65. For a history of the ichnographic plan since the Renaissance, see Pinto, "Ichnographic City Plan," 35–50.
66. Pinto, "Ichnographic City Plan," 38.
67. Ballon, *Paris of Henri IV*, 218–19.
68. Lee, "Objective Point of View," 11–32.
69. Chapel, *L'oeil raisonné*.
70. The gradual shift from bird's-eye to planimetric views in the seventeenth century was not without some experimentation, and these two categories were often not separable. For example, when bird's-eye views dominated cartographic representations, the *Plan de Gomboust* (1652) represented streets and many lots orthogonally, but large edifices and monuments were represented at an angled aerial view, with the Seine oriented vertically. Conversely, almost a century later, when bird's-eye views were less common, the *Plan de Turgot* (1739) took a perspectival view oriented to the northwest with depictions of the buildings' façades, while simultaneously attempting to meet the synoptic advantages of planimetric modes by gradually angling the surface upward as the eye looked toward the horizon line.
71. Picon, "Nineteenth-Century Urban Cartography," 135–49.
72. Palsky, *Des chiffres et des cartes*, 18.
73. "L'esprit géométrique" was a term first used by Blaise Pascal in *De l'esprit géométrique* (1657). It was then picked up by Bernard Le Bovier de Fontenelle, who used the term "l'esprit géométrique" in the preface to his 1699 *Histoire de l'académie royal des sciences*. Tore Frängsmyr, J. L. Heilbron, and Robin E. Rider translated this term as "the quantifying spirit" for their edited collection, *The Quantifying Spirit in the Eighteenth Century* (1990). However, the graphic notation of quantification and its philosophical roots in geometry are lost in that translation, which I believe to be important in spatial understandings during the eighteenth and nineteenth centuries, especially with regard to mapmaking as a form of knowledge production.
74. See Porter, *Trust in Numbers*; Alder, "Making Things the Same," 499–545.
75. Vérin, *La gloire des ingénieurs*, 196.
76. Wood, Fels, and Krygier, *Rethinking the Power of Maps*, 86.
77. Fougères, "Les plans cadastraux," 56–58.
78. Kain and Baigent, *Cadastral Map*, 221.
79. Goubert, "En rouergue," 382–86. As surveying and mapmaking increasingly became structured and standardized in the Ponts et chaussées, the position of the ingénieurs-géographes became correspondingly unnecessary. The position was officially eliminated in 1831.
80. *Dictionnaire de l'Académie française*, 5th ed., 2:623.
81. Aubry-Dubochet, *L'exécution du cadastre générale*; Babeuf and Audiffred, *Projet de cadastre perpétuel*.
82. Gitelman, *Paper Knowledge*, 5.
83. Guillory, "Memo and Modernity."
84. Alliès, *L'invention du territoire*, 157.
85. Quesnay's tableau was printed in Versailles in 1758 in a run of four copies. There was a second edition of the *Tableau économique* published in the spring of 1759 with a print run of three copies, and, at the end of the same year, a third edition was published. Additionally Mirabeau included a version of this table in *L'ami des hommes* (1760) and in his *Philosophie rurale* (1763) and *Éléments de la philosophie rurale* (1767). The basic structure remains the same in all of these. Only one published in Quesnay's *Analyse du table économique* differs, and is often referred to as the *Tableau abrégé* to signal its abbreviated form for a general audience.
86. Vardi, *Physiocrats*, 79–82. She cites a letter dated 1759 from Quesnay to Victor Riqueti, marquis de Mirabeau.
87. Vardi, *Physiocrats*, 74.
88. Larrère, "L'arithmétique des physiocrates," 7.
89. Larrère, "L'arithmétique des physiocrates," 14.

90. Marx and Engels, *Marx and Engels Collected Works*, 239–40.

91. See Jomard, *Considérations sur l'objet*, 9–10. For studies of Jomard and the royal collections of maps, see Pognon, "Les collections du Département des cartes et plans," 195–204; Pelletier, "Jomard et le Département de Cartes et Plans," 18–27.

92. Even as early as the 1790s, Jomard was articulating the specificity of the cartographic medium when seeking to strengthen the case for including topographic maps in the *Description de l'Égypte* against objections by the ministry of war. See Godlewska, *Geography Unbound*, 199.

93. Godlewska, *Geography Unbound*, 210–12.

94. Ozouf Marignier, "Entre tradition et modernité," 747–62; Ozouf-Marignier, "Administration," 19–39.

95. Papayanis, *Planning Paris before Haussmann*, 75.

96. Chabrol de Volvic, *Recherches statistiques*, vol. 1, pt. 3, 10.

97. Porter, *Trust in Numbers*, 12.

98. See Margairaz, "La géographie des administrateurs," 185–216; Godlewska, *Geography Unbound*, 215–20.

99. Chabrol de Volvic, *Souvenirs*, 94.

100. Haussmann, *Mémoires*, 3:15.

101. Haussmann, *Mémoires*, 2.53.

102. See Jomard, *Considérations sur l'objet*, 9.

103. Bellanger, "Town Clerks in the Paris Region," 103–23. Specifically regarding the hierarchy among the bureaucrats, senior municipal employees included heads of surveying, architecture, tax collection, and octrois (municipal customs) as opposed to the *secrétaires de mairie*, who were often not formally trained. Bellanger, "Town Clerks in the Paris Region," 104.

104. Mazzarella, "Internet X-Ray," 476.

105. O'Connell, "Redefining the Past," 208; Van Zanten, "Paris Space."

106. Van Zanten, "Paris Space," 179–210.

107. Under his control, the Service du pavé became the Service de la voie publique et du nettoiement. The Service des eaux et égouts was separated from the Service du pavé and led by Eugène Belgrand; and the Service des promenades et plantations, a wholly new entity, was headed by Adolphe Alphand.

108. Haussmann, *Mémoires*, 3:3.

109. Lemoine and Mimram, *Paris d'ingénieurs*, 21.

110. The combination of these two types of mapping—both interior and exterior to the lot—was part of one agency, and was made possible by the 1852 requirement that obligated all builders to submit a plan and two sections of projected constructions, which in effect outsourced to private property owners a task that had been within the domain of cadastral surveyors.

111. Two versions of two different scales are known—one at a larger scale in 1864, marked heavily and located in the BNF; the other at a smaller scale, printed and more widely distributed in 1867.

112. Manuscript outlining the agency's organization plan, AN, F21 2023/Règlements des palais nationaux, 1851.

113. Beauregard, *Planning Matter*, 14–35.

114. Söderström, "Paper Cities," 261.

115. Ledgers function as lists, and Cornelia Vismann has theorized that "files are governed by lists. . . . Lists with tasks to be performed govern the inside of the file world, from the initial compilation to their final storage." See Vismann, *Files*, 7.

116. Vismann, *Files*, 106.

117. Bender and Marrinan, *Culture of Diagram*.

118. Gitelman, *Paper Knowledge*, 30.

119. Gitelman, *Paper Knowledge*, 31.

120. "Bureaucracy develops the more perfectly, the more it is 'dehumanized,' the more completely it succeeds in eliminating from official business love, hatred, and all purely personal, irrational, and emotional elements which escape calculation." Weber, *Economy and Society*, 975.

121. Kafka, *Demon of Writing*, 117.

122. Dossier related to the expropriation around the Butte des Moulins, AP, VO11 0452: Butte des Moulins, 1877.

123. *Journal des géomètres* (May 1847): 1.

124. *Journal des géomètres* (1881): 101–6. Reprinted from July 21, 1874, and signed by Adolphe Alphand as director of the Travaux de Paris.

125. For a discussion about shadow conventions in maps, see Imhof, *Cartographic Relief Presentations*. Alphand would have taken his cues from conventions in technical and cartographic drawings in which a forty-five-degree angle was established by the nineteenth century.

126. Alphand, *Les promenades de Paris*, xlvi.

127. Porter, *Trust in Numbers*, ix.

128. Deleuze and Guattari, *Thousand Plateaus*, 101.

129. Conseil des bâtiments civils, session 7, August 1854, AN, F21 2539.

130. Ferrier, *Notice sur l'Hôtel de Ville de Paris*, 87.

131. Le Roux de Lincy, *Histoire de l'Hôtel de Ville de Paris*, 101.

CHAPTER 5. CARTOGRAPHIC PRESENTATIONS
Epigraph: Louis Marin, *De la représentation* (Paris: Seuil/Gallimard, 1994), 206.

1. Gillispie, *Science and Polity in France*.

2. Foucault, *Sécurité, territoire, population*.

3. Lehning, *To Be a Citizen*, 1–13.

4. Lehning, *To Be a Citizen*, 2.

5. Alphand, *Rapport*.

6. What Kory Olson designates "the cartographic capital." See Olson, *Cartographic Capital*.

7. Conklin, *Mission to Civilize*, 11–37.

8. Söderström, "Paper Cities," 261.

9. Rearick, "Festivals in Modern France," 458n10.

10. Rearick, "Festivals in Modern France," 437.

11. Levin, "Inventing Modern Paris," 35–56.

12. Michelet, *Nos fils*, 419.

13. *La république française*, 1871–1924. See the year 1877 in particular.

14. Bernard, *Les fêtes célèbres*, 1.

15. Pearce, "Framing the Days," 24.

16. "Administration générale: Administration municipal," APo, EB 1, 1867.

17. Franklin, "Les anciens plans de Paris," 1:vi.

18. For a fuller discussion about the establishment of the archive, initiated by the French Revolution, that gave way in the long term not to the burning of files (as has received much emphasis), but rather to the preserving and launching of new archival projects, see Vismann, *Files*, 117–19.

19. "Administration générale: Administration municipal," APo, EB 1, 1867.

20. A bibliographic list of historical works was included with the publication of the *Atlas des anciens plans de Paris*.

21. Foucault, "Leçon du 29 mars 1978" and "Leçons du 5 avril 1978," in *Sécurité, territoire, population*, 319–70.

22. Birignani, "State, Civil Society, Architecture."

23. Alphand, *Atlas des anciens plans de Paris*, Table analytique.

24. Hobsbawm, "Mass-Producing Traditions," 272.

25. Alphand, *Atlas des anciens plans de Paris*, Table analytique.

26. Alphand, *Atlas des anciens plans de Paris*, Table analytique.

27. Lyotard, *Post-Modern Condition*, 25.

28. Lyotard, *Post-Modern Condition*, 25.

29. Lyotard, *Post-Modern Condition*, 26.

30. Denis Wood makes the same argument for atlases that privilege the borders of nation-states. See Wood, "Pleasure in the Idea," 31.

31. Alphand, *Atlas des anciens plans de Paris*, Table analytique.

32. Alphand, *Atlas des anciens plans de Paris*, Table analytique.

33. See Hobsbawm, "Mass-Producing Traditions," 263–307; Nora, "De la République à la nation," 651–59; Agulhon, *La République de 1880*; Ory, *L'Expo universelle, 1889*; Gérard, *La Révolution française*.

34. Eric Hobsbawm has argued that the

subject of history was avoided, insofar as political events since the French Revolution had been more divisive than united. See Hobsbawm, "Mass-Producing Traditions," 272.

35. Agulhon, *French Republic* and *La République*. Also see Néré, "French Republic," 300–322; Piñol, *Histoire de l'Europe urbaine*.

36. A decree dated July 28, 1886, nominated Alphand as director of the Travaux de l'exposition: Ministère du commerce, Commission de l'Exposition universelle, AN, F12 3757.

37. Michel Foucault, "Governmentality," in Burchell, Gordon, and Miller, *Foucault Effect*, 92.

38. Anthony Sutcliffe has argued that Alphand was able to gain as much power as Haussmann by reorganizing and consolidating all of the separate administrative agencies into the Ministère des travaux publics, of which he served as director. Sutcliffe, *Autumn of Central Paris*, 44.

39. Report presented by M. de Bouteiller, on behalf of the municipal commission of the exposition on the special exhibition of the city of Paris in 1889, AP, Tri Briand 255.

40. Notes from M. Fauve, *géomètre en chef* for the production of the *Atlas 1789–1889*, 8, AP, Tri Briand 255.

41. Report presented by M. de Bouteiller.

42. Friendly, "Golden Age of Statistical Graphics," 502–35; Lécuyer, "Statistician's Role in Society," 35–55; Funkhouser, "Historical Development of the Graphical Representation of Statistical Data," 269–405; Palsky, *Des chiffres et des cartes*.

43. La Bras, "Government Bureau of Statistics," 361–400.

44. Kory Olson makes a similar argument that the *Atlas* "transmits a message of progress to its audience." See Olson, *Cartographic Capital*, 86–129.

45. "Service du Plan," 1889, AP, Tri Brand 255.

46. James Akerman notes that the earliest atlas editors recognized the advantages of a uniform presentation of the images in a set of maps before the execution was possible. Abraham Ortelius observed these advantages in the introduction to his *Theatrum orbis terrarum* (Theater of the orb of the world) (first published in 1570), when he writes that one of his primary motivations for making his atlas was to provide maps in a uniformly small and convenient format. See Akerman, "On the Shoulders of a Titan," 101–2. In the case of the *Atlas*, the conception of the maps as a set was aided by reproduction technologies such as photography and photoengraving.

47. Akerman explains that the *Atlas* is merger of descriptive and pattern-seeking impulses. See Akerman, "On the Shoulders of a Titan," 134–35.

48. Margaret Wickens Pearce writes, "Narrative lay in the visual syntax between, rather than within, the maps in the atlas." See Pearce, "Framing the Days," 19.

49. Daston and Galison, *Objectivity*, 115–90.

50. Gross, "Temporality and the Modern State," 54.

51. Gross, "Temporality and the Modern State," 66–67.

52. Blanchard, Boëtsch, and Snoep, *Human Zoos*; Blanchard, Bancel, Boëtsch, Deroo, Lemaire, and Forsdick, *Human Zoos*.

53. Schneider, *Empire for the Masses*.

54. Persell, *French Colonial Lobby*.

55. The most oppressive and violent, the most invested and treasured, was Algeria, conquered by the French in 1830. See Çelik, *Urban Forms and Colonial Confrontations*.

56. Colonial pavilions and displays were part of earlier expositions beginning in 1867; however, in 1889 their elaboration and investment made the centennial exposition patently different from earlier displays. See Çelik, *Displaying the Orient*.

57. Delort de Gléon, *La Rue du Caire à l'Exposition universelle*.

58. Çelik, *Displaying the Orient*, 95–138.

59. Ferguson, *Paris as Revolution*.

60. The new buildings are located in the arrondissements 20, 19, 18, 17, 16, and 13, but also in two central zones, the fourth and the Latin Quarter.

61. Olson, *Cartographic Capital*, 47–48.

62. Olsen, *Cartographic Capital*, 94–95.

63. Andrews, "Early Life of Paul Vidal de la Blache," 174–82.

64. A key figure who defined a new direction was Paul Vidal de la Blache, who in 1883 published a geography textbook for secondary school curricula, *La terre. Géographie physique et économique. Histoire sommaire des découvertes* (The earth: Physical and economic geography. Brief history of discoveries), which included numerous illustrations.

65. Broc, *Regards sur la géographie française*; Claval, *Histoire de la géographie française*.

66. Cited in Broc, *L'établissement de la géographie en France*, 546.

67. Commission de l'Exposition universelle de 1889, minutes of a meeting of Tuesday, March 29, 1887, AP, carton 122.

68. Jacob, *Sovereign Map*, 67.

69. Levin, "Democratic Vistas," 82–108.

70. It was originally published in 1877, but the 1884 edition introduced maps with the text. For a detailed study of the text and its role

in national identity, see Ozouf and Ozouf, "Le Tour de la France," 291–321.

71. Olson, *Cartographic Capital*, 47–85.

72. For the difficulty of studying the reception of maps, see Jacob, "Towards a Cultural History of Cartography," 192.

73. Olson, *Cartographic Capital*, 115.

74. Vernes, "Au jardin comme à ville," 16.

75. Nominated on May 27, 1871, by Thiers as director of the Travaux de Paris, a post that included direction of roadways, public streets, promenades, the plan of Paris, and architectural works. With Belgrand's death in 1898, the direction of water and sewage was also placed under the direction of the Travaux de Paris. Alphand replaced Haussmann's seat as a member of the Académie des Beaux-Arts.

76. Personnel file, AN, F14 11459 Adolphe Alphand.

77. *Recueil. Dossiers biographiques Boutillier du Retail. Documentation sur Adolphe Alphand*, BNF. The Bibliothèque historique de la Ville de Paris (BHVP) holds a folder with all the obituaries printed upon his death in 1891. Both dossiers include several statements and lectures on Alphand, by a member of the Académie des Beaux-Arts, the president of the Conseil générale, by an official at the Association polytechnique, the assistant director of the Travaux de Paris, and the inspector general of Ponts et chaussées, all of whom repeat that he was a paragon of service to the city.

78. Jordan, "Haussmann and Haussmannisation," 88; Van Zanten, "Paris Space," 179–210.

79. As Anthony Sutcliffe summarizes, "The very comprehension of Haussmann's plan seems to have caused officials and councilors to stop thinking about its relevance to the city's problems—they sought only to carry it out." See Sutcliffe, *Autumn of Central Paris*, 326.

80. Haussmann, *Mémoires*, 1:213; 3:127.

81. *Rapport sur Exposition universelle 1889*, Saturday, November 29, 1885, AN, F12 3757. See also Mangin, "Les Expositions universelles," 538–40; *Bulletin municipal de la Ville de Paris*, November 6, 1884, 1904–5.

82. Association française pour l'avancement des sciences, fusionnée avec l'association scientifique de France, session of February 18, 1888, M. Berger, director general of the Exploitation de l'Exposition de 1889, 8–9, BHVP.

83. See Picard, *Rapport général*, 1:8.

84. For an analysis of the relationship between the formation of the Third Republic and the Exposition universelle of 1889, see Nelms, *Third Republic*.

85. The idea of prints being understood as "fact" because of their synoptic quality is referred to by several scholars; see Levin, "Democratic Vistas," 82–108, but Levin does not address maps. The ontology of "facts" is specifically addressed in Poovey, *History of the Modern Fact*, but it does not address graphic media.

86. *Le journal*, May 8, 1899, reported that Verniquet's map was moved to the ground floor room of the Bibliothèque imperiale during the First Empire, and in 1798 it was transferred to the Arts et métiers. Then on November 17, 1808, the minister of the interior ordered its transport to a small building, the Hôtel de Coti on the rue de Grenelle. Other reports say that from the Couvent des Cordeliers, in 1797, the plates were moved to the Hôtel de Ville, where all but two original plates were subsequently lost in 1871. See Léri, "Edme Verniquet," 204–5.

87. For specifics on the photoengraving process, see Lecuyer, *Histoire de la photographie;* Nadeau, *Encyclopedia.*

88. The photoengraving also lent itself to architectural subjects. The photoengraving process, first invented by Nicéphore Niépce in France, was advanced by Hippolyte-Louis Fizeau, whose first true photoengravings were of the Hôtel de Ville de Paris and a sculptural relief panel on Notre-Dame. When Fizeau first outlined his patented process at the Académie des sciences on February 13, 1843, the photo-engraved prints he distributed were of the church of Saint-Sulpice. See Jammes et al., *De Niépce à Stieglitz.* For a more detailed study of the relationship between photography, engineering, and architecture in the nineteenth century, see Weiss, "Engineering, Photography, and Construction."

89. Alphand, *Atlas des anciens plans de Paris,* Table analytique.

90. Benjamin, "Work of Art."

91. Adorno and Horkheimer, *Dialectic of Enlightenment,* 128.

92. Adorno, "Culture Industry Reconsidered," 12.

93. For a fuller discussion, see Krakauer, *Disposition of the Subject.*

94. For a discussion about "francization," see Weber, *Peasants into Frenchman,* 67–94.

95. Smith, "Idea of the French Hexagon," 145.

96. There was also a rival shape under consideration: the octagon.

97. Eugen Weber explains that the term "Hexagon" became popular in the 1960s. See Weber, "L'Hexagone," 1171–90. However, Todd Shepard critiques this interpretation as ignoring the term's history in relation to France's colonial and specifically Algerian occupations. See Shepard, "Birth of the Hexagonal," 53–72.

CONCLUSION. THE TOTAL VIEW OF THE CITY

Epigraph: T. S. Eliot, *The Waste Land, Prufrock, the Hollow Men and Other Poems* (New York: Dover Publications, 2022), 51.

1. Agache, Aubertin, and Redont, *Comment reconstruire,* 5.

2. Specifically listed were (1) an overall plan at 1/5,000; (2) an overall plan of the city's underground infrastructure; (3) an overall plan of the park systems; and (4) an overall plan of public transport. See Agache, Aubertin, and Redont, *Comment reconstruire,* 13.

3. Agache, Aubertin, and Redont, *Comment reconstruire,* xiv.

4. Agache, Aubertin, and Redont, *Comment reconstruire,* ix–xvi.

5. Agache, Aubertin, and Redont, *Comment reconstruire,* xi.

Bibliography

ARCHIVES

AN/ANCP   Archives nationales/Archives nationales, Cartes et plans
AP   Archives de Paris
APo   Archives de la police
BAVP   Bibliothèque administrative de la Ville de Paris
BHVP   Bibliothèque historique de la Ville de Paris
BNF   Bibliothèque nationale de France

PUBLISHED TEXTS

ACKERMAN, JAMES, AND WOLFGANG JUNG, EDS. *Conventions of Architecture Drawing: Representation and Misrepresentation.* Self-published, 2000.

ADORNO, THEODOR. "Culture Industry Reconsidered." *New German Critique* 6 (1975): 12–19.

ADORNO, THEODOR, AND MAX HORKHEIMER. *Dialectic of Enlightenment.* London: Verso, 1997.

AGACHE, HUBERT-DONAT-ALFRED, JACQUES AUBERTIN, AND ÉDOUARD REDONT. *Comment reconstruire nos cités détruites.* Paris: Librarie Armand, 1915.

AGRANDISSEMENT ET CONSTRUCTION DES HALLES CENTRALES D'APPROVISIONNEMENT: RAPPORT FAIT AU CONSEIL MUNICIPAL. February 28, 1845.

AGREST, DIANA. *Practice: Architecture, Technique and Representation.* London: Routledge, 2000.

AGULHON, MAURICE. *The French Republic, 1879–1992.* Oxford: Blackwell, 1993.

——. *La République: De Jules Ferry à François Mitterrand (1880–1995).* Paris: Hachette, 1997.

——. *La République de 1880 à nos jours.* Paris: Hachette, 1990.

AKERMAN, JAMES R. "On the Shoulders of a Titan: Viewing the World of the Past in Atlas Structure." PhD diss., Pennsylvania State University, 1991.

——, ED. *The Imperial Map.* Chicago: University of Chicago Press, 2009.

ALDER, KEN. *Engineering the Revolution.* Princeton: Princeton University Press, 1999.

——. "Making Things the Same: Representation, Tolerance and the End of the Ancien Régime in France." *Social Studies in Science* 28, no. 4 (August 1998): 499–545.

——. *The Measure of All Things: The Seven-Year Odyssey and Hidden Error That Transformed the World.* New York: Free Press, 2002.

——. "A Revolution to Measure: The Political Economy of the Metric System in France." In *The Values of Precision,* edited by M. Norton Wise, 39–71. Princeton: Princeton University Press, 1995.

ALLIÈS, PAUL. *L'invention du territoire.* Grenoble: Presses universitaires de Grenoble, 1980.

ALMANACH IMPÉRIAL (OR ALMANACH ROYALE OR ALMANACH NATIONAL DE FRANCE). Paris: Testu, 1810–54.

ALPERS, SVETLANA. *The Art of Describing: Dutch Art in the Seventeenth Century.* Chicago: University of Chicago Press, 1983.

ALPHAND, JEAN-CHARLES-ADOLPHE. *Atlas des anciens plans de Paris.* Paris, 1880.

——. *Les promenades de Paris.* Paris: Rothschild, 1867–73.

——. *Les travaux de Paris, 1789–1889.* Paris, 1889.

——. *Rapport sur la proposition tenant à reproduire par la photogravure les anciens plans de Paris et à les réunir dans un atlas qui figurera à l'Exposition universelle de 1878.* Paris, 1878.

——. *Recueil de lois, ordonnances, décrets et réglements relatifs aux alignements, à l'expropriation pour la cause d'utilité publique, spéciale-ment dans le voies de Paris.* Paris, 1886.

——. *Recueil des lettres patentes, ordonnances royales, décrets et arrêtés pré-fectoraux concernant les voies publiques.* Paris, 1886–1902.

ANDERSON, KIRSTI. *The Geometry of an Art: The History of the Mathematical Theory of Perspective from Albert to Monge.* New York: Springer, 2007.

ANDERSON, STANFORD. "The Fiction of Function." *Assemblage 2* (February 1987): 18–31.

——, ed. *On Streets.* Cambridge, MA: MIT Press, 1978.

ANDIA, BÉATRICE DE. *Le Paris des polytechniciens: Des ingénieurs dans la ville: 1794–1994.* Paris: Délégation à l'action artistique de la Ville de Paris, 1997.

ANDREWS, HOWARD. "The Early Life of Paul Vidal de la Blache and the Makings of Modern Geography." *Transactions of the Institute of British Geographers* 11, no. 2 (1986): 174–82.

ANGENOT, MARC, ed. *Le centenaire de la Révolution 1889.* Paris: Direction de la documentation française, 1989.

APPADURAI, ARJUN. "Production of Locality." In *Counterworks: Managing the Diversity of Knowledge,* edited by R. Fardon, 204–25. London: Routledge, 1995.

ARNAUD, JEAN-LUC. *Analyse spatiale, cartographie et histoire urbaine.* Marseille: Éditions parenthèses, 2008.

ARTZ, FREDERICK B. *The Development of Technical Education in France, 1500–1850.* Cleveland: Society of the History of Technology, 1966.

AUBRY-DUBOCHET, PIERRE-FRANÇOIS. *L'exécution du cadastre générale de la France et d'un cadastre provisoire pour la répartition des impôts en 1791.* Paris, 1791.

AUGUSTIN-THIERRY, A. "Souvenirs d'un peintre militaire." *Revue des deux mondes* 17, no. 4 (1933): 844–49.

BABELON, JEAN-PIERRE, MYRIAM BACHA, AND BÉATRICE DE ANDIA, EDS. *Les Expositions universelles à Paris de 1855 à 1937.* Paris: Délégation action artistique de la Ville de Paris, 2005.

BABEUF, FRANÇOIS N., AND J. P. AUDIFFRED. *Projet de cadastre perpétuel.* Paris, 1789.

BAGROW, LEO. *History of Cartography,* revised and enlarged by R. A. Skelton. London: C. A. Watts, 1964.

BAIGRIE, BRIAN S. *Picturing Knowledge: Historical and Philosophical Problems Concerning the Use of Art in Science.* Toronto: University of Toronto Press, 1996.

BALLON, HILARY. *The Paris of Henri IV: Architecture and Urbanism.* Cambridge, MA: MIT Press, 1991.

BALLON, HILARY, AND DAVID FRIEDMAN. "Portraying the City in Early Modern Europe: Measurement, Representation, and Planning." In *The History of Cartography,* vol. 3, pt. 1, *Cartography in the European Renaissance,* edited by David Woodward, 680–704. Chicago: University of Chicago Press, 2007.

BALTARD, VICTOR, AND FÉLIX-EMMANUEL CALLET. *Monographie des Halles Centrales de Paris.* Paris, 1863.

BANN, STEPHEN. *Parallel Lines: Printmakers, Painters, and Photographers in Nineteenth-Century France.* New Haven: Yale University Press, 2001.

BARRET-KRIEGEL, BLANDINE, AND BRUNO FORTIER. *La politique de l'espace parisien: À la fin de l'Ancien Régime.* Paris: Comité de la rechercher et du développement en architecture, 1975.

BARTHES, ROLAND. *The Eiffel Tower, and Other Mythologies.* New York: Hill and Wang, 1979.

——. *Image–Music–Text.* New York: Hill and Wang, 1977.

BATES, DAVID. "Cartographic Aberrations: Epistemology and Order in the Encyclopedia Map." In *Using the Encyclopédie: Ways of Knowing, Ways of Reading,* edited by Daniel Brewer and Julia Candler Hayes, 1–20. Oxford: Voltaire Foundation, 2002.

BAXANDALL, MICHAEL. *Shadows and Enlightenment.* New Haven: Yale University Press, 1995.

BAZZAZ, SAHAR, YOTA BATSAKI, AND DIMITER ANGELOV, EDS. *Imperial Geographies in Byzantine and Ottoman Space.* Cambridge, MA: Blackwell, 1994.

BEAUMONT-VASSY, E. F. DE. *Histoire intime du Second Empire.* Paris, 1874.

BEAUREGARD, ROBERT A. *Planning Matter: Acting with Things.* Chicago: University of Chicago Press, 2015.

BECKER, PETER, AND RÜDIGER VON KROSIGK, EDS. *Figures of Authority: Contributions towards a Cultural History of Governance from the Seventeenth to the Twentieth Century.* Brussels: Peter Lang, 2008.

BELHOSTE, BRUNO. "Du dessin d'ingénieur à la géométrie descriptive: L'enseignment de Chastillon à l'École royale du génie de Mézières." In *Extenso* (June 1990): 103–35.

——. *Le Paris des polytechniciens: Des ingénieurs dans la ville, 1794–1994.* Paris: Délégation à l'action artistique de la Ville de Paris, 1994.

——. *Paris Savant: Capital of Science in the Age of the Enlightenment.* Translated by Susan Emanuel. Oxford: Oxford University Press, 2019.

BELHOSTE, BRUNO, AMY DAHAN-DALMEDICO, AND ANTOINE PICON. *La formation polytechnicienne: 1794–1994.* Paris: Dunod, 1994.

BÉLIER, CORINNE, BARRY BERGDOLL, AND MARC LE COEUR. *Henri Labrouste: Structure Brought to Light.* New York: Museum of Modern Art, 2012.

BELL, MORAG, ROBIN BUTLIN, AND MICHAEL HEFFERNAN, EDS. *Geography and Imperialism, 1820–1940.* Manchester, UK: Manchester University Press, 1995.

BELLANGER, EMMANUEL. "Town Clerks in the Paris Region: The Design of a Professional Identity in the Late Nineteenth and Early Twentieth Centuries." In *Municipal Services and Employees in the Modern City: New Historic Approaches,* edited by Michèle Dagenais, Irene Maver, and Pierre-Yves Saunier, 103–23. Hampshire, UK: Ashgate, 2003.

BELYEA, BARBARA. "Image of Power: Derrida/Foucault/Harley." *Cartographica: The International Journal for Geographic Information and Geovisualization* 29, no. 2 (Summer 1992): 1–9.

BENCHIMOL, JAIME L. *Pereira Passos: Um Haussmann tropical: A renovação urbana da cidade do Rio de Janeiro no início do século XX.* Rio de Janeiro: Prefeitura da Cidade do Rio de Janeiro, Secretaria Municipal de Cultura, Turismo e Esportes, Departamento Geral de Documentação e Informação Cultural, 1990.

BENDER, JOHN, AND MICHAEL MARRINAN. *The Culture of Diagram.* Stanford, CA: Stanford University Press, 2010.

——. *Regimes of Description: In the Archive of the Eighteenth Century.* Stanford, CA: Stanford University Press, 2005.

BENEVOLO, LEONARDO. *The Origins of Modern Town Planning.* Cambridge, MA: MIT Press, 1967.

BENJAMIN, WALTER. *The Arcades Project.* Translated by Rolf Tiedemann. Cambridge, MA: Belknap, 1999.

——. "Capitalism as Religion." In *Selected Writings,* vol. 1, edited by Marcus Bullock and Michael W. Jennings, 288–91. Cambridge, MA: Harvard University Press, 2004.

——. "The Work of Art in the Age of Its Technological Reproducibility." In *The Work of Art in the Age of Its Technological Reproducibility and Other Writings on Media,* edited by Michael W. Jennings, Brigid Doherty, and Thomas Y. Levin, 9–18. Cambridge, MA: Harvard University Press, 2008.

——. *The Writer of Modern Life: Essays on Charles Baudelaire*. Edited by Michael W. Jennings, translated by Howard Eiland, Edmund Jephcott, Rodney Livingstone, and Harry Zohn. Cambridge, MA: Harvard University Press, 2006.

BENOISTON DE CHÂTEAUNEUF, LOUIS-FRANÇOIS. *Extraits des recherches statistiques sur la Ville de Paris et le Département de la Seine: Recueil de tableaux*. Paris, 1824.

BERDOULAY, VINCENT, AND PAUL CLAVAL, EDS. *Aux débuts de l'urbanisme français*. Paris: L'Harmattan, 2001.

BERG, LAWRENCE D., AND JANI VUOLTEENAHO, EDS. *Critical Toponymies: The Contested Politics of Place Naming*. London: Routledge, 2009.

BERGDOLL, BARRY. *European Architecture, 1750–1890*. New York: Oxford University Press, 2000.

——. *Léon Vaudoyer: Historicism in the Age of Industry*. Cambridge, MA: MIT Press, 1994.

BERGER, GEORGES, AND JEAN-CHARLES-ADOLPHE ALPHAND. *Exposition universelle internationale de 1889 à Paris*. Paris, 1892–95.

BERGERON, LOUIS. *Paris, genèse d'un paysage*. Paris: Picard, 1989.

BERNARD, FRÉDÉRIC. *Les fêtes célèbres de l'antiquité, du moyen âge, et des temps modernes*. Paris, 1878.

BERTHAUT, HENRI-MARIE-AUGUSTE. *Les ingénieurs géographes militaires, 1624–1831: Étude historique*. 2 vols. Paris: Service géographique de l'armée, 1902.

BERTIN, JACQUES. *Semiology of Graphics: Diagrams, Networks, Maps*. Redlands, CA: Esri, 2011.

BERTOCCI, PHILIP. *Jules Simon: Republican Anti-Clericalism and Cultural Politics in France, 1848–1886*. Columbia: University of Missouri Press, 1978.

BERTY, ADOLPHE. *Topographie historique du vieux Paris*. 8 vols. Paris, 1866–97.

BESSE, JEAN-MARC. *Face au monde; Atlas, jardins, géoramas*. Paris: Desclée de Brouwer, 2003.

——. *Voir la terre: Six essais sur le paysage et la géographie*. Paris: Actes Sud/ENSP/Centre du paysage, 2000.

BIAGIOLI, MARIO, ED. *The Science Studies Reader*. New York: Routledge, 1999.

BIGGS, MICHAEL. "Putting the State on the Map: Cartography, Territory, and European State Formation." *Comparative Studies in Society and History* 41, no. 2 (1999): 374–405.

BIRIGNANI, CESARE. "State, Civil Society, Architecture: A Critique of the Representations of the Good City." PhD diss., Columbia University, 2012.

BLACK, JEREMY. *Maps and History: Constructing Images of the Past*. New Haven: Yale University Press, 1997.

——. "A Revolution in Military Cartography?: Europe 1650–1815." *Journal of Military History* 73, no. 1 (2009): 49–68.

BLAIS, HÉLÈNE. "Qui dresse la carte? La controverse entre savants et voyageurs au XIXe siècle." *Revue du comité français de cartographie*, no. 175 (March 2003): 25–29.

BLAIS, HÉLÈNE, AND ISABELLE LABOULAIS, EDS. *Géographies plurielles: Les sciences géographiques au moment de l'émergence des sciences humaines (1750–1850)*. Paris: L'Harmattan, 2006.

BLANCHARD, ANNE. *Les ingénieurs du roy de Louis XIV à Louis XVI: Étude du corps des fortifications*. Montpellier: Université Paul-Valéry, 1979.

BLANCHARD, PASCAL, NICOLAS BANCEL, GILLES BOËTSCH, ÉRIC DEROO, SANDRINE LEMAIRE, AND CHARLES FORSDICK, EDS. *Human Zoos: Science and Spectacle in the Age of Colonial Empires*. Liverpool: Liverpool University Press, 2008.

BLANCHARD, PASCAL, GILLES BOËTSCH, AND NANETTE JACOMIJN SNOEP, EDS. *Human Zoos: The Invention of the Savage*. Arles: Actes Sud, Musée du quai Branly, 2011.

BLAU, EVE, EDWARD KAUFMAN, AND ROBIN EVANS, EDS. *Architecture and Its Image: Four Centuries of Architectural Representation: Works from the Collection of the Canadian Centre for Architecture*. Montreal: Centre Canadien d'architecture, 1989.

BLOCH, MARC. "Les plans parcellaires: Allemagne, Angleterre, Danemark, France." *Annales d'histoire économique et sociale* 1 (1929): 60–70.

BLOCK, MAURICE. *Administration de la Ville de Paris et du Département de la Seine*. Paris: Guillaumin, 1884.

——. *Dictionnaire de l'administration française*. Paris: Berger-Levrault, 1891.

BLONDEL, JACQUES-FRANÇOIS, AND PIERRE PATTE. *Cours d'architecture*. 6 vols. Paris: Desaint, 1771–77.

BOOKER, P. J. "Gaspard Monge (1746–1818) and His Effect on Engineering Drawing and Technical Education." *Transactions of the Newcomen Society* 34, no. 1 (1961): 15–36.

BORDES, PHILIPPE. *Jacques-Louis David: Empire to Exile*. New Haven: Yale University Press, 2005.

BORGES, JORGE LUIS. "On Exactitude in Science." In *A Universal History of Infamy*. Translated by Norman Thomas de Giovanni, 131. London: Penguin, 1975.

BOSSELMANN, PETER. *Representation of Places: Reality and Realism in City Design*. Berkeley: University of California Press, 1998.

BOUDON, PHILIPPE. *Figuration graphique en architecture*. Paris: Atelier de recherche et d'études d'aménagement, 1974.

——. *Langages singuliers et partages de l'urbain*. Paris: L'Harmattan, 1999.

BOULLÉE, ÉTIENNE-LOUIS. *Architecture: Essai sur l'art*, edited by Jean-Marie Pérouse de Montclos. Paris: Hermann, 1968.

BOURDIEU, PIERRE. *Esquisse d'une théorie de la pratique*. Paris: Droz, 1972.

BOURILLON, FLORENCE. "Changer la ville: La question urbaine au milieu du 19e siècle." "Villes en crise?" Special issue, *Vingtième siècle*, no. 64 (October–December 1999): 11–23.

——. *Les villes en France au XIXe siècle*. Paris: Ophrys, 1992.

BOUSQUET-BRESSOLIER, CATHERINE. "Survey of Meter." *History of Cartography*, vol. 4, *Cartography in the European Enlightenment*, edited by Matthew H. Edney and Mary Spongberg Pedley, 965–66. Chicago: University of Chicago Press, 2019.

BOUTIER, JEAN. "Une tentative de relevé cadastral de Paris: Le plan de l'Abbé Jean Delagrive, 1735–1757." "Les plans de Paris du XVIe au XVIIIe siècles." Special issue, *Cahiers du CREPIF*, no. 50 (1995): 107–20.

BOWIE, KAREN, ED. *La modernité avant Haussmann: Formes de l'espace urbain à Paris, 1801–1853*. Paris: Recherches, 2001.

BRAHAM, ALLAN. *The Architecture of the French Enlightenment*. Berkeley: University of California Press, 1980.

BRANCO, RUI. "Fieldwork, Map-Making, and State Formation: A Case Study in the History of Science and Administration." In *Figures of Authority: Contributions towards a Cultural History of Governance from the Seventeenth to the Twentieth Century*, edited by. P. Becker and R. von Krosigk, 201–28. Brussels: Peter Lang, 2008.

BRAUDEL, FERNAND, AND ERNEST LABROUSSE. *Histoire économique et sociale de la France*. Paris: Presses universitaires de France, 1970–82.

BRAY, PATRICK M. "Prose Constructions: Nerval, Baudelaire, and the Louvre." *L'esprit créateur* 54, no. 2 (Summer 2014): 115–26.

BRENNER, NEIL. "Foucault's New Functionalism." *Theory and Society* 23 (1994): 679–709.

BRESC-BAUTIER, GENEVIÈVE. *Le photographe et l'architecte: Édouard Baldus, Hector-Martin Lefuel et le chanter du nouveau Louvre de Napoléon III*. Paris: Éditions de la réunion des musées nationaux, 1995.

BRESSANI, MARTIN. *Architecture and the Historical Imagination:*

Eugène-Emmanuel Viollet-le-Duc, 1814–1879. Burlington, VT: Ashgate, 2014.

BRETT, DAVID. "Drawing and the Ideology of Industrialization." *Design Issues* 3, no. 2 (Autumn 1986): 59–72.

BRIOT, CHARLES, AND CHARLES VACQUANT. *Arpentage, Levé des Plans et Nivellement.* Paris, 1859.

BROC, NUMA. "L'établissement de la géographie en France: Diffusion, institutions, projets (1870–1890)." *Annales de géographie* 83 (1974): 545–68.

——. "La pensée géographique en France au XIXe siècle. Rupture ou continuité?" *Revue géographique des Pyrénées et du Sud-Ouest* (1976): 225–46.

——. *Regards sur la géographie française de la Renaissance à nos jours.* Perpignan: Presses universitaires de Perpignan, 1995.

BRÜCKNER, MARTIN. *The Geographic Revolution in Early America: Maps, Literacy, and National Identity.* Chapel Hill: University of North Carolina Press, 2006.

BUCHLOH, BENJAMIN, SERGE GUILBAUT, AND DAVID H. SOLKIN, EDS. *Modernism and Modernity: The Vancouver Conference Papers.* Halifax: Press of the Nova Scotia College of Art and Design, 1983.

BUCK-MORSS, SUSAN. *The Dialectics of Seeing: Walter Benjamin and the Arcades Project.* Cambridge, MA: MIT Press, 1989.

BUISSERET, DAVID. *Envisioning the City: Six Studies in Urban Cartography.* Chicago: University of Chicago Press, 1998.

——. "Les ingénieurs du roi au temps de Henri IV." *Bulletin de la section de géographie* 77 (1964): 13–84.

——. *Monarchs, Ministers, and Maps: The Emergence of Cartography as a Tool of Government in Early Modern Europe.* Chicago: University of Chicago Press, 1992.

BULLET, PIERRE. *Traité de l'usage du pantomètre.* Paris, 1675.

BURCHELL, GRAHAM, COLIN GORDON, AND PETER MILLER, EDS. *The Foucault Effect: Studies in Governmentality.* Chicago: University of Chicago Press, 1991.

BURDEAU, FRANÇOIS. *Histoire de l'administration française du XVIIIe au XXe siècle.* Paris: Montchrestien, 1994.

CAMPBELL, BRIAN. *The Writings of the Roman Land Surveyors.* Journal of Roman Studies Monograph. London: Roman Society, 2000.

CARMONA, MICHEL. *Haussmann: His Life and Times, and the Making of Modern Paris.* Chicago: I. R. Dee, 2000.

CARON, FRANÇOIS, ED. *Paris et ses réseaux, naissance d'un mode de vie urbain XIXe–XXe siècles.* Paris: Bibliothèque historique de la Ville de Paris, 1990.

CARPO, MARIO. *The Alphabet and the Algorithm.* Cambridge, MA: MIT Press, 2011.

——. *Architecture in the Age of Printing: Orality, Writing, Typography and Printed Images in the History of Architectural Theory.* Cambridge, MA: MIT Press, 2001.

CARPO, MARIO, AND FRÉDÉRIQUE LEMERLE, EDS. *Perspective, Projections, and Design: Technologies of Architectural Representation.* London: Routledge, 2008.

CARROLL, LEWIS. *Sylvie and Bruno Concluded.* New York: Macmillan, 1894.

CARS, JEAN DES. *Haussmann, la gloire du Second Empire.* Paris: Librairie académique Perrin, 1978.

CARS, JEAN DES, AND PIERRE PINON. *Paris-Haussmann, "le pari d'Haussmann."* Paris: Pavillon de l'arsenal, 1991.

CASATI, ROBERTO, AND ACHILLE VARZI. *Parts and Places: The Structures of Spatial Representation.* Cambridge, MA: MIT Press, 1999.

CASSELLE, PIERRE. "Les travaux de la Commission des embellissements de Paris en 1853; Pouvait-on transformer la capitale sans Haussmann?" *Bibliothèque de l'École des chartes* 155, no. 2 (1997): 645–89.

——, ED. "Commission des embellissements de Paris: Rapport à l'empereur Napoléon III." Special issue, *Cahiers de la rotonde* 23 (2000).

——, ED. *Commission des embellissements de Paris: Rapport à l'empereur Napoléon III rédigé par le comte Henri Siméon (décembre 1853).* Paris: Commission du vieux Paris, 2000.

CASSINI DE THURY, CÉSAR-FRANÇOIS. *La méridienne de l'Observatoire royal de Paris: Verifiée dans toute l'étendue du royaume par de nouvelles observations.* Paris, 1744.

CASTEX, JEAN. *Formes urbaines; de l'ilot à la barre.* Paris: Dunod, 1977.

——. "Les origines du quartier." In *Autour de l'Opéra: Naissance de la ville moderne,* edited by François Loyer, 43–44. Paris: Délégation à l'action artistique de la Ville de Paris, 1995.

CASTEX, JEAN, JEAN-LOUIS COHEN, JEAN-CHARLES DEPAULE, AND DANIEL LE COUËDIC. *Histoire urbaine: Anthropologie de l'espace.* Paris: CNRS, 1995.

ÇELIK, ZEYNEP. *Displaying the Orient: Architecture of Islam at Nineteenth-Century World's Fairs.* Berkeley: University of California Press, 1992.

——. *Urban Forms and Colonial Confrontations; Algiers under French Rule.* Berkeley: University of California Press, 1997.

CERDÀ, ILDEFONSO. *La théorie générale de l'urbanisation.* Translated by Antonio Lopez de Aberasturi. Paris: Seuil, 1979.

CHABROL DE VOLVIC, GILBERT-JOSEPH-GASPARD. *Recherches statistiques sur la Ville de Paris et le Département de la Seine, publiées sous la direction du comte Chabrol de Volvic.* 4 vols. Paris, 1821–29.

——. *Souvenirs inédits de M. le comte Chabrol de Volvic.* Paris: Commission des travaux historiques, Ville de Paris, 2002.

CHALMERS, ALEXANDER. *The General Biographical Dictionary Containing an Historical and Critical Account of the Lives and Writing of the Most Eminent Persons in Every Nation; Particularly the British and Irish; from the Earliest Accounts to the Present Time.* Vol. 16. London, 1814.

CHAPEL, ENRICO. "Cartes et figure de l'urbanisme scientifique en France (1910–1943)." PhD diss., Université de Paris VIII, 2000.

——. "La statistique graphique dans l'urbanisme français avant 1945: Une pratique, des démarches." In *Langages singuliers et partagés de l'urbain,* edited by Philippe Boudon, 185–209. Paris: L'Harmattan, 1999.

——. "Les architectes et l'objectivation de l'urbain. La carte statistique au centre de l'action urbanistique." In *Figures de la ville et construction des savoirs: Architecture, urbanisme, géographie,* edited by Frédéric Pousin, 173–83. Paris: CNRS, 2005.

——. *L'oeil raisonné: L'invention de l'urbanisme par la carte.* Geneva: Métis, 2010.

CHAPMAN, BRIAN. *The Life and Times of Baron Haussmann, Paris in the Second Empire.* London: Weidenfeld and Nicolson, 1957.

CHARDON, C. A. *Cours pratique de de géométrie, d'arpentage, de dessin linéaire et d'architecture ouvrage à l'usage des Écoles, des pensions, des cours d'adultes, des apenteurs, des ouvriers, et de toutes les personnes qui s'occupent de dessin.* Paris, 1856.

CHARLTON, D. G. *Positivist Thought in France during the Second Empire, 1852–1870.* Oxford: Clarendon, 1959.

CHARTIER, ROGER. *Au bord de la falaise: L'histoire entre certitudes et inquiétude.* Paris: A. Michel, 1998.

CHATTOPADHYAY, SWATI. *Representing Calcutta: Modernity, Nationalism, and the Colonial Uncanny.* London: Routledge, 2005.

CHEVALIER, LOUIS. "La politique financière." *Cahiers d'histoire égyptienne* 8 (1956): 213–40.

CHOAY, FRANÇOISE. *La règle et le modèle.* Paris: Seuil, 1980.

——. *The Modern City: Planning in the Nineteenth Century.* Translated by M. Hugo and G. R. Collins. New York: George Braziller, 1969.

CHOUQUER, G., AND F. FAVORY. *Les arpenteurs romains: Théorie et pratique.* Paris: Errance, 1992.

CHU, PETRA TEN-DOESSCHATE, AND GABRIEL P. WEISBERG, EDS. *Popularization of Images: Visual Culture under the July Monarchy.* Princeton: Princeton University Press, 1994.

CILLEULS, ALFRED DES. *Histoire de l'administration parisienne au XIXe siècle.* Paris: Champion, 1900.

———. *L'administration parisienne sous le 3e République.* Paris: Picard fils, 1910.

———. *Origines et développement du régime des travaux publics en France.* Paris: Imprimerie nationale, 1895.

CLARK, T. J. *Farewell to an Idea: Episodes from a History of Modernism.* New Haven: Yale University Press, 1999.

———. *The Painting of Modern Life: Paris in the Art of Manet and His Followers.* New York: Knopf, 1985.

CLARKE, G. N. G. "Taking Possession: The Cartouche as Cultural Text in Eighteenth-Century American Maps." *Word and Image* 4, no. 2 (April–June 1988): 455–74.

CLAUSEN, MEREDITH. "The Department Store—Development of a Type." *Journal of Architectural Education* 39, no. 1 (Fall 1985): 20–29.

CLAVAL, PAUL. "From Michelet to Braudel: Personality, Identity and the Organization of France." In *The New Cambridge Modern History,* vol. 11, *Material Progress and World-Wide Problems, 1870–1898,* edited by F. H. Hinsley, 39–57. Cambridge: Cambridge University Press, 1979.

———. *Histoire de la géographie française de 1870 à nos jours.* Paris: Éditions Nathan, 1998.

———. "Qu'est-ce que la géographie?" *Geographical Journal* 133, no. 1 (March 1967): 33–39.

CLAYSON, HOLLIS. *Paris in Despair: Art and Everyday Life under Siege (1870–71).* Chicago: University of Chicago Press, 2002.

CLAYSON, HOLLIS, AND ANDRÉ DOMBROWSKI, EDS. *Is Paris Still the Capital of the Nineteenth Century? Essays on Art and Modernity, 1850–1900.* London: Routledge, 2018.

CLEARY, RICHARD. *The Place Royale and Urban Design in the Ancien Régime.* Cambridge: Cambridge University Press, 1999.

CLÉMENT, PIERRE, ED. *Lettres, instructions et mémoires de Colbert.* Paris: Imprimerie nationale, 1861–73.

CLERGEOT, PIERRE. *Recueil méthodique des lois, décrets, réglements, instructions et décisions sur le cadastre de la France approuvé par le ministre des finances.* Paris: Imprimerie impérial, 1811.

COHEN, STUART E. "History as Drawing." *Journal of Architectural Education* 32, no. 1 (September 1978): 2–3.

COLBERT, JEAN-BAPTISTE. "Instruction pour les maîtres des requêtes, commissaires départies dans les provinces" (September 1663). Reprinted in *Lettres, instructions et mémoires de Colbert,* vol. 4, edited by Pierre Clément. Paris: Imprimerie nationale, 1861–73.

COLLINS, GEORGE R. "Linear Planning throughout the World." *Journal of the Society of Architectural Historians* 18, no. 3 (October 1959): 74–93.

COLLINS, GEORGE R., AND CHRISTIANE CRASEMANN COLLINS. *Camillo Sitte and the Birth of Modern City Planning.* New York: Columbia University/Random House, 1965.

COLLINS, PETER. "The Origins of Graph Paper as an Influence on Architectural Design." *Journal of the Society of Architectural Historians* 21, no. 4 (December 1962): 159–62.

COMTE, BARBARA SHAPIRO. "The Evolution, Standardization, and Diffusion of Architects' Construction Drawing Models through Printed Sources, 1750s–1870s." In *La Construction savante: Les avatars de la littérature technique,* edited by Jean-Philippe Garric, Valérie Nègre, and Alice Thomine-Berrada, 179–92. Paris: Picard, 2008.

———. "King's Feet to Republican Metres: The Evolution of Construction Drawings, Paris, 1782–1876." *The Proceedings of the Fifth International Congress on Construction History,* vol. 3, edited by Donald Friedman, 281–92. Raleigh: Lulu, 2015.

CONDORCET, NICOLAS DE. "L'arpentage." In *Supplément à l'encyclopédie ou dictionnaire raisonné des sciences, des arts et des métiers,* 1:567. Amsterdam: Chez M. M. Rey, 1776–77.

CONKLIN, ALICE. *A Mission to Civilize: The Republican Idea of Empire in France and West Africa, 1895–1930.* Stanford, CA: Stanford University Press, 1997.

CONLEY, TOM. *The Self-Made Map: Cartographic Writing in Early Modern France.* Minneapolis: University of Minnesota Press, 2011.

COOK, KAREN SEVERUD. "From False Starts to Firm Beginnings: Early Colour Printing of Geological Maps." *Imago Mundi* 47 (1995): 155–72.

CORBOZ, ANDRÉ. "A Network of Irregularities and Fragments: Genesis of a New Urban Structure in the Eighteenth Century." *Diadolos* 34 (1989): 64–71.

CORTADA, JAMES. "The Information Ecosystems for National Diplomacy: The Case of Spain, 1815–1936." *Information and Technology* 48, no. 2 (Spring 2013): 222–59.

CORTESAO, ARMANDO. *History of Portuguese Cartography.* 2 vols. Coimbra: Junta de Investigações do Ultramar, 1969–71.

COSGROVE, DENIS. *Geography and Vision: Seeing, Imagining and Representing the World.* London: I. B. Tauris, 2008.

———. *The Iconography of Landscape: Essays on the Symbolic Representation, Design and Use of Past Environments.* Cambridge: Cambridge University Press, 1988.

———. *Mappings.* London: Reaktion, 1999.

COSTA MEYER, ESTHER DA. *Dividing Paris: Urban Renewal and Social Inequality, 1852–1870.* Princeton: Princeton University Press, 2022.

COURAL, NATALIE, HÉLÈNE GROLLEMUND, AND DOMINIQUE CORDELLIER, EDS. *Le papier à l'oeuvre: Exposition au Musée du Louvre, 9 juin–5 septembre 2011.* Paris: Hazan, 2011.

COYECQUE, ERNEST. *Les plans cadastraux de la Ville de Paris aux archives nationales.* Paris, 1909.

CRAMPTON, JEREMY W. "Cartographic Calculations of Territory." *Progress in Human Geography* 35 (2011): 92–103.

———. "Maps as Social Constructions: Power, Communication and Visualization." *Progress in Human Geography* 25, no. 2 (2001): 235–52.

CRAMPTON, JEREMY W., AND STUART ELDEN. "Space, Politics, Calculation: An Introduction." *Social and Cultural Geography* 7 (2006): 681–85.

CRAMPTON, JEREMY W., AND STUART ELDEN, EDS. *Space, Knowledge and Power: Foucault and Geography.* Hampshire, UK: Ashgate, 2007.

CRARY, JONATHAN. *Techniques of an Observer: On Vision and Modernity in the Nineteenth Century.* Cambridge, MA: MIT Press, 1992.

CROSLAND, MAURICE. "'Nature' and Measurement in Eighteenth-Century France." In *Studies on Voltaire and the Eighteenth Century,* edited by Theodore Besterman, 277–309. Banbury, UK: Voltaire Foundation, 1972.

CROWE, NORMAN, AND STEVEN W. HURTT. "Visual Notes and the Acquisition of Architectural Knowledge." *Journal of Architectural Education* 39, no. 3 (Spring 1986): 6–16.

DAFFRY DE LA MONNOYE, LÉON. *Les lois de l'expropriation pour la cause d'utilité publique expliquées par la jurisprudence.* Paris, 1859.

———. *Théorie et pratique de l'expropriation pour cause d'utilité publique.* Paris: Pedone-Lauriel, 1879.

DAINVILLE, FRANÇOIS DE. "Cartes et contestations au XVe siècle." *Imago Mundi* 24 (1970): 99–121.

——. "How Did Oronce Fine Draw His Large Map of France?," *Imago Mundi* 24 (1970): 49–55.

——. *Le langage des géographes, 1500–1800*. Paris: Picard, 1964.

DANIEL, MALCOLM. *The Photographs of Édouard Baldus*. New York: Metropolitan Museum of Art, 1994.

DARIN, MICHAËL. "Haussmann: Reconsidering His Role in the Transformation of Paris." In *Shapers of Urban Form: Explorations in Morphological Agency*, edited by Peter J. Larkham and Michael Conzen, 97–113. New York: Routledge, 2014.

——. *La comédie urbaine: Voir la ville autrement*. Gollion, Switz.: Infolio, 2009.

——. "Les grandes percées urbaines du XIXe siècle: Quatre villes de province." *Annales: Économies, sociétés, civilisations* 43, no. 2 (1988): 477–505.

DASTON, LORRAINE. "Description by Omission: Nature Enlightened and Obscured." In *Regimes of Description: In the Archive of the Eighteenth Century*, edited by John Bender and Michael Marrinan, 11–24. Stanford, CA: Stanford University Press, 2005.

——. "Enlightenment Calculations." *Critical Inquiry* 21, no. 1 (Autumn 1994): 182–202.

——. "The Moral Economy of Science." In "Constructing Knowledge in the History of Science." Special issue, *Osiris* 10 (1995): 2–24.

——. "Objectivity and the Escape from Perspective." *Social Studies of Science* 22 (November 1992): 597–618.

DASTON, LORRAINE, AND PETER GALISON. "The Image of Objectivity." *Representations*, no. 40 (Autumn 1992): 81–128.

——. *Objectivity*. Cambridge, MA: MIT Press, 2007.

DAVENNE, HENRI-JEAN-BAPTISTE. *Recueil méthodique et raisonné des lois et réglemens sur la voirie, les alignements et la police des construction contenant un résumé de la jurisprudence de Ministère de l'intérieur et du conseil d'état sur cette matière*. Paris, 1836.

——. *Traité pratique de voirie urbaine, ou législation et principes qui régissent cette branche de l'administration*. Paris, 1858.

DEBOFLE, PIERRE, ET AL., EDS. *L'administration de Paris (1789–1977)*. Geneva: Droz/Champion, 1979.

DEFORGE, YVES. *Le graphisme technique: Son histoire et son enseignement*. Seyssel: Champ Villon, 1981.

DELAGRIVE, JEAN. *Manuel de trigonométrie pratique*. Paris, 1754.

DELAMARE, NICOLAS. *Traité de la police*. 4 vols. Paris, 1719–22.

DELEUZE, GILLES, AND FÉLIX GUATTARI. *A Thousand Plateaus: Capitalism and Schizophrenia*. Translated by Brian Massumi. Minneapolis: University of Minnesota Press, 1987.

DELORT DE GLÉON, ALPHONSE. *La Rue du Caire à l'Exposition universelle de 1889*. Paris, 1889.

DEMANGEON, ALAIN, AND BRUNO FORTIER. *Les vaisseaux et les villes*. Brussels: Mardaga, 1978.

——. "The Politics of Urban Space: The City around 1800." *Architectural Design* 48 (1978): 8–13.

DEMING, MARK K. "Une capitale et des ports: Embellissement et planification urbaine à la fin de l'Ancien Régime." In *Les architectes de la liberté, 1789–1799*, edited by Annie Jacques and Jean-Pierre Mouilleseaux, 51–66. Paris: École nationale des Beaux-Arts, 1989.

DÉMY, ALFRED. *Essai Historique sur les Expositions Uiverselles de Paris*. Paris, 1907.

DÉRENS, JEAN. *À la découverte des plans de Paris, du XVIe au XVIIIe siècle, exposition à la Bibliothèque historique de la Ville de Paris (14 juin–25 septembre)*. Paris: Direction des affaires culturelles, 1994.

DESPORTES, MARC, AND ANTOINE PICON. *De l'espace au territoire: L'aménagement en France XVIe–XXe siècles*. Paris: Presses de l'École nationale des ponts et chaussées, 1997.

DICTIONNAIRE DE L'ACADÉMIE FRANÇAISE, 5TH ED. Paris, 1789.

DICTIONNAIRE PORTATIF DE PEINTURE, SCULPTURE ET GRAVURE. Paris, 1781.

DIDEROT, DENIS, AND JEAN LE ROND D'ALEMBERT, EDS. *Encyclopédie ou dictionnaire raisonné des sciences, des arts, et des métiers*. Paris, 1751–80.

DILKE, O. A. W. *The Roman Land Surveyors: An Introduction to the Agrimensores*. Newton Abbot, UK: David and Charles, 1971.

DORRIAN, MARK. "The Agency of Specifications, Contracts and Technical Literature." *Arq* 16, no. 3 (2012): 201–4.

——. "On Google Earth." In *Scales of the Earth*, edited by E. H. Jazairy, 164–70. New Geographies 4. Cambridge, MA: Harvard University Press, 2011.

DUBY, GEORGES, ED. *Histoire de la France urbaine*, vol. 4, *La ville à l'âge industriel*. Paris: Seuil, 1980.

DUFFY, CHRISTOPHER. *Siege Warfare: The Fortress in the Early Modern World, 1494–1600*. London: Routledge/Kegan Paul, 1979.

DULAURE, JACQUES-ANTOINE. *Histoire physique, civile et morale de Paris depuis les premiers temps historiques jusqu'à nos jours*. Paris, 1824.

DUMAS, JEAN-PHILIPPE. "Représentation et description des propriétés à Paris au XIXe siècle. Cadastre et plan parcellaire." *Mélanges de l'École française de Rome, Italie et Méditerranée* 111, no. 2 (1999): 779–93.

DUNLOP, CATHERINE TATIANA. *Cartographia: Maps and the Search for Identity in the French-German Borderland*. Chicago: University of Chicago Press, 2015.

DUPUY, G. *L'urbanisme des réseaux. Théories et méthodes*. Paris: Dunod, 1991.

DURAND, JEAN-NICOLAS-LOUIS. *Précis of the Lectures on Architecture: With Graphic Portion of the Lectures on Architecture*. Introduction by Antoine Picon, translated by David Britt. 1823–25. Reprint, Los Angeles: Getty Research Institute, 2000.

DYKSTRA, DARRELL. "The French Occupation of Egypt, 1798–1801." In *The Cambridge History of Egypt*, vol. 2, *Modern Egypt, from 1517 to the End of the Twentieth Century*, edited by M. W. Daly, 113–38. Cambridge: Cambridge University Press, 1998.

EDGERTON, SAMUEL Y. *The Renaissance Rediscovery of Linear Perspective*. New York: Basic, 1975.

EDNEY, MATTHEW. "Cartography and Its Discontents." *Cartographica: The International Journal for Geographic Information and Geovisualization* 50, no. 1 (2015): 9–13.

——. *Cartography in the European Enlightenment*. Chicago: University of Chicago Press, 2020.

——. "Cartography without 'Progress': Reinterpreting the Nature and Historical Development of Mapmaking." *Cartographica* 30 (1993): 54–68.

——. "The Irony of Imperial Mapping." In *The Imperial Map*, edited by James Akerman, 11–46. Chicago: University of Chicago Press, 2009.

——. *Mapping an Empire: The Geographical Construction of British India, 1765–1843*. Chicago: University of Chicago Press, 1997.

——. "Reconsidering Enlightenment Geography and Map Making: Reconnaissance, Mapping, Archive." In *Geography and Enlightenment*, edited by David N. Livingstone and Charles W. J. Withers, 165–98. Chicago: University of Chicago Press, 1999.

——. "Theory and the History of Cartography." *Imago Mundi* 48 (1996): 185–91.

ELDEN, STUART. "Governmentality, Calculation, Territory." *Environment and Planning D: Society and Space* 25 (2007): 562–80.

ELKINS, JAMES. "Art History and Images That Are Not Art." *Art Bulletin* 77, no. 4 (December 1995): 553–71.

——. *The Domain of Images.* Ithaca: Cornell University Press, 1999.

ENCYCLOPÉDIE MÉTHODIQUE PAR ORDRE DE MATIÈRES. Paris, 1788.

ESCOBAR, JESÚS. "Map as Tapestry: Science and Art in Pedro Teixeira's 1656 Representation of Madrid." *Art Bulletin* 96, no. 1 (March 2014): 50–69.

ESTIVALS, ROBERT, AND JEAN-CHARLES GAUDY. *La bibliologie graphique: L'évolution graphique des plans de Paris, 1530–1798.* Paris: Société de bibliologie et de schématisation, 1983.

EVANS, ROBIN. *The Projective Cast: Architecture and Its Three Geometries.* Cambridge, MA: MIT Press, 1995.

——. *Translations from Drawing to Building and Other Essays.* Cambridge, MA: MIT Press, 1997.

EVENSON, NORMA. *Paris: A Century of Change.* New Haven: Yale University Press, 1979.

EXPOSITION UNIVERSELLE DE PARIS EN 1889, CATALOGUE DE L'EXPOSITION SPÉCIALE DE LA VILLE DE PARIS ET DU DÉPARTEMENT DE LA SEINE. Paris, 1889.

FAURE, ALAIN. "Spéculation et société: Les grands travaux à Paris au XIXe siècle." *Histoire, économie et société* 23, no. 3 (2004): 433–48.

FAURE, F. "The Development and Progress of Statistics in France." In *The History of Statistics: Their Development and Progress in Many Countries,* edited by J. Koren, 218–329. New York: Macmillan, 1918.

FÉLIX, MAURICE. *Le régime administrative de la Ville de Paris et du Département de la Seine.* 4 vols. Paris: La documentation française, 1957–59.

FERGUSON, EUGENE S. "The Mind's Eye: Non-Verbal Thought in Technology." *Science* 197 (1977): 827–36.

FERGUSON, PRISCILLA PARKHURST. *Paris as Revolution: Writing the Nineteenth-Century City.* Berkeley: University of California Press, 1994.

FERRIER, A. *Notice sur l'Hôtel de Ville de Paris.* Paris, 1855.

FIERRO, ALFRED, AND BÉATRICE DE ANDIA, EDS. *Patrimoine parisien, 1789–1799: Destructions, créations, mutations.* Paris: Délégation à l'action artistique de la Ville de Paris, 1989.

FISCHER, HUBERTUS, VOLKER R. REMMERT, AND JOACHIM WOLSCHKE-BULMAHN, EDS. *Gardens, Knowledge and the Sciences in the Early Modern Period.* Basel: Birkhauser, 2016.

FORTIER, BRUNO. *La métropole imaginaire: Un atlas de Paris.* Liège: Mardaga, 1989.

——. "La politique de l'espace parisien à la fin de l'Ancien Régime." In *Rapport de recherche.* Paris: CORDA, 1975.

——. *Un atlas des tracés parisiens. L'idée de la ville.* Paris: Picard, 1984.

FOUCART, B., AND V. NOEL-BOUTON. "Les projets d'église pour Napoléonville (1802–1809)." *Bulletin de la société de l'histoire de l'art français* (1971): 235–52.

FOUCAULT, MICHEL. *The Birth of Biopolitics: Lectures at the Collège de France, 1978–79.* Translated by Michel Senellart. New York: Palgrave Macmillan, 2008.

——. *Discipline and Punish: The Birth of the Prison.* Translated by Alan Sheridan. New York: Vintage, 1995.

——. *Les mots et les choses: Une archéologie des sciences humaines.* Paris: Gallimard, 1966.

——. *Power/Knowledge: Selected Interviews and Other Writings 1972–1977.* Edited by Colin Gordon, translated by Colin Gordon, Lelio Marshall, John Mepham, and Kate Soper. New York: Pantheon, 1980.

——. *Sécurité, territoire, population. Cours au Collège de France. 1977–1978.* Paris: Gallimard Seuil, 2004.

——. *Security, Territory, Population: Lectures at the Collège de France, 1977–1978.* Translated by Graham Burchell. New York: Picador, 2007.

FOUGÈRES, MARC. "Les plans cadastraux de l'Ancien Régime." *Mélanges d'histoire sociale, Annales d'histoire sociale* 3 (1945): 54–69.

FOURNEL, VICTOR. *Ce qu'on voit dans les rues de Paris.* Paris, 1858.

FOX, ROBERT, AND GEORGE WEISZ. *The Organization of Science and Technology in France, 1808–1914.* Cambridge: Cambridge University Press, 1980.

FRÄNGSMYR, TORE. "The New Academies and the Scientific Climate of the Eighteenth Century." *Proceedings of the American Philosophical Society* 143, no. 1 (March 1999): 109–15.

FRÄNGSMYR, TORE, J. L. HEILBRON, AND ROBIN E. RIDER, EDS. *The Quantifying Spirit in the Eighteenth Century.* Berkeley: University of California Press, 1990.

FRANKLIN, ALFRED. *Les anciens plans de Paris. Notices historiques et topographiques.* Paris: Léon Willem, Librairie-Éditeur, 1878.

FRÈCHE, GEORGES. "Compoix, propriété foncière, fiscalité, et demographie historique en pays de taille réelle (XVIe–XVIIIe siècles)." *Revue d'histoire moderne et contemporaine* 18 (1971): 321–53.

FRIEDMAN, DAVID. "'Fiorenza': Geography and Representation in the Fifteenth-Century City View." *Zeitschrift für Kunstgeschichte* 64 (2001): 56–77.

FRIENDLY, MICHAEL. "The Golden Age of Statistical Graphics." *Statistical Science* 32, no. 4 (2008): 502–35.

FRISBY, DAVID. *Fragments of Modernity: Theories of Modernity in the Work of Simmel, Kracauer and Benjamin.* Cambridge: Polity, 1985.

FULTON, ROBERT. "Crafting a Site of State Information Management: The French Case of the Dépôt de la Guerre." *French Historical Studies* 40, no. 2 (April 2017): 215–40.

FUNKHOUSER, H. G. "Historical Development of the Graphical Representation of Statistical Data." *Osiris* 3 (1937): 269–405.

GAILLARD, JEANNE. *Paris, la Ville, 1852–1870: L'urbanisme parisien à l'heure d'Haussmann.* Paris: H. Champion, 1977.

GALISON, PETER. *Einstein's Clocks, Poincaré's Maps.* New York: W. W. Norton, 2003.

GALLET, MICHEL. *Les architectes parisiens du XVIIIe siècle: Dictionnaire biographique et critique.* Paris: Mengès, 1995.

GARRIOCH, DAVID. *The Formation of the Parisian Bourgeoisie, 1690–1830.* Cambridge, MA: Harvard University Press, 1996.

——. *The Making of Revolutionary Paris.* Berkeley: University of California Press, 2002.

GAUTIER, THÉOPHILE. *Abécédaire du Salon de 1861.* Paris, 1861.

GAY, RECTEUR JEAN. *Histoire de l'administration de la Ville de Paris, et études diverses sur l'organisation municipal avant et après la Révolution.* Nancy: Presses Universitaires de Nancy, 2011.

GEISON, GERALD. *Professions and the French State, 1700–1900.* Philadelphia: University of Pennsylvania Press, 1984.

GELL, ALFRED. "How to Read a Map: Remarks on the Practical Logic of Navigation." *Man* 20, no. 2 (June 1985): 271–86.

GÉRARD, ALICE. *La Révolution française, mythes et interprétations, 1789–1970.* Paris: Flammarion, 1970.

GERBINO, ANTHONY, AND STEPHEN JOHNSTON. *Compass and Rule: Architecture as Mathematical Practice in England 1500–1750.* New Haven: Yale University Press, 2009.

GILLISPIE, CHARLES COULSTON. *Science and Polity in France: The Revolutionary and Napoléonic Years.* Princeton: Princeton University Press, 2004.

——. *Science and Polity in France at the End of the Old Regime.* Princeton: Princeton University Press, 1980.

——. "Scientific Aspects of the French Egyptian Expedition, 1798–1801."

Proceedings of the American Philosophical Society 133, no. 4 (December 1989): 447–74.

GILLISPIE, CHARLES COULSTON, AND MICHEL DEWACHTER, EDS. *Monuments of Egypt. The Napoléonic Edition: The Complete Archeological Plates from La Description de l'Égypte*. 2 vols. Princeton: Princeton Architectural Press, 1987.

GIRVEAU, BRUNO, ED. *Charles Garnier: Un architecte pour un empire*. Paris: École nationale supérieure des Beaux-Arts, 2010.

GITELMAN, LISA. *Paper Knowledge: Towards a Media History of Documents*. Durham: Duke University Press, 2014.

GODLEWSKA, ANNE MARIE CLAIRE. *Geography Unbound: French Geographic Science from Cassini to Humboldt*. Chicago: University of Chicago Press, 2000.

———. "Map, Text, Image: The Mentality of Enlightened Conquerors: A New Look at the *Description de l'Égypte*." *Transactions of the Institute of British Geographers* 20 (1995): 5–28.

———. *The Napoléonic Survey of Egypt: A Masterpiece of Cartographic Compilation and Early Nineteenth-Century Fieldwork*. Toronto: Winters College, York University, 1988.

———. "Traditions, Crises, and New Paradigms in the Rise of the Modern French Discipline of Geography, 1760–1850." *Annals of the Association of American Geographers* 79, no. 2 (1989): 192–213.

GODLEWSKA, ANNE MARIE CLAIRE, AND NEIL SMITH, EDS. *Geography and Empire*. Oxford: Oxford University Press, 1994.

GOFFART, WALTER. *Historical Atlases: The First Three Hundred Years, 1570–1870*. Chicago: University of Chicago Press, 2003.

GOLDSTEIN, JAN. *The Post-Revolutionary Self: Politics and Psyche in France, 1750–1850*. Cambridge, MA: Harvard University Press, 2008.

GOMBRICH, E. H. *The Image and the Eye: Further Studies in the Psychology of Pictorial Representation*. Ithaca: Cornell University Press, 1982.

GOOCH, G. P. *History and Historians in the Nineteenth Century*. London: Longmans, 1952.

GOODY, JACK. *La raison graphique: La domestication de la pensée sauvage*. Paris: Éditions de Minuit, 1979.

GORDON, COLIN. "Governmental Rationality: An Introduction." In *The Foucault Effect: Studies in Governmentality*, edited by Graham Burchell, Colin Gordon, and Peter Miller, 1–51. Chicago: University of Chicago Press, 1991.

GOUBERT, PIERRE. "En rouergue: Structure agraires et cadastres au XVIIIe siècle." *Annales: Histoire, sciences sociales* 9, no. 3 (July–September 1954): 382–86.

GRANIER DE CASSAGNAC, ADOLPHE. *Souvenirs du Second Empire*. Vol 2. Paris, 1879–82.

GREGORY, DEREK. "Interventions in the Historical Geography of Modernity: Social Theory, Spatiality and the Politics of Representation." "Meaning and Modernity: Cultural Geographies of the Invisible and the Concrete." Special issue, *Geografiska Annaler* 73, no. 1, (1991): 17–44.

GRILLON, M., G. CALLOU, AND THÉODORE JACOUBET. *Études d'un nouveau système d'alignemens et de percemens de voies publiques faites en 1840 et 1841*. Paris: Chez Chaillou, 1848.

GROSS, DAVID. "Temporality and the Modern State." *Theory and Society* 14, no. 1 (January 1985): 53–82.

GUILLAUME, EUGÈNE. *Traité pratique de la voirie urbaine*. Paris, 1876.

GUILLAUMIN, JEAN-YVES. *Les arpenteurs romains*. Vols. 1–3. Paris: Les belles lettres, 2005–14.

GUILLERME, ANDRÉ. *Bâtir la ville: Révolutions industrielles dans les matériaux de construction: France–Grande-Bretagne (1760–1840)*. Seyssel: Champ Vallon, 1995.

———. *Corps à corps sur la route: Les routes, les chemins et l'organisation des services au XIXe siècle*. Paris: Presses de l'École nationale des ponts et chaussées, 1994.

———. *Genèse du concept de réseau; Territoire et génie en Europe de l'Ouest, 1760–1815*. Marne-la-Vallée: École nationale des ponts et chaussées, 1988.

———. "La formation des nouveau édiles: Ingénieurs des ponts et chaussées et architectes (1804–1815)." In *Villes et territoire pendant la période napoléonienne (France et Italie). Actes du colloque de Rome (3–5 mai, 1984)*, 35–57. Rome: École française de Rome, 1987.

———. "Réseau: Genèse d'une catégorie dans la pensée de l'ingénieur sous la restauration." *Flux 6* (1991): 5–17.

GUILLERME, JACQUES. *L'art du projet: Histoire, technique, architecture*. Wavre: Mardaga, 2008.

———. *Sur la politique architecturale*. Paris: Klincksieck, 1975.

GUILLORY, JOHN. "The Memo and Modernity." *Critical Inquiry* 31, no. 1 (Autumn 2004): 108–32.

GUIOMAR, JEAN-YVES. "Le tableau géographique de la France." In *Les lieux de mémoire*, vol. 1, *La République*, edited by Pierre Nora, 1073–98. Paris: Gallimard, 1984.

GUIOT. *L'arpenteur forestier; ou, Méthode nouvelle de mesurer, calculer et construire toutes sortes de figures, suivant les principes géométriques et trigonométriques, avec un traité d'arpentage appliqué à la réformation des forêts, trèsutile tant aux arpenteurs et géographes, qu'aux marchands et propriétaires des bois*. Paris: Saillant, Nyon, et Desaint, 1770.

HAASBROEK, N. D. *Gemma Frisius, Tycho Brahe and Snellius and Their Triangulation*. Delft: Netherlands Geodetic Commission, 1968.

HACKING, IAN. *The Social Construction of What?* Cambridge: Cambridge University Press, 1990.

HALBWACHS, MAURICE. *La théorie de l'homme moyen: Essai sur quetelet et la statistique morale*. Paris: Alcan, 1913.

———. *Les expropriations et le prix des terrains à Paris*. Vol 1., *La population et les tracés de voies à Paris depuis un siècle*. Paris: Presses universitaires de France, 1928.

HALL, PETER, AND MARK TEWDWR-JONES. *Urban and Regional Planning*. London: Routledge, 2010.

HALL, THOMAS. *Planning Europe's Capital Cities: Aspects of Nineteenth-Century Urban Development*. London: E. and F. N. Spon, 1997.

HANNAH, MATTHEW G. *Governmentality and the Mastery of Territory in Nineteenth-Century America*. Cambridge: Cambridge University Press, 2000.

HANNOUM, ABDELMAJID. *The Invention of the Maghreb: Between Africa and the Middle East*. Cambridge: Cambridge University Press, 2021.

HANSSEN, BEATRICE, ED. *Walter Benjamin and the Arcades Project*. London: Continuum, 2006.

HARLEY, J. B. "Deconstructing the Map." In *The New Nature of Maps: Essays in the History of Cartography*, edited by Paul Laxton, 149–68. Baltimore: Johns Hopkins University Press, 2001.

———. "Maps, Knowledge, and Power." In *The Iconography of Landscape: Essays on the Symbolic Representation, Design and Use of Past Environments*, edited by Denis Cosgrove and Stephen Daniels, 277–312. Cambridge: Cambridge University Press, 1988.

———. "Silences and Secrecy: The Hidden Agenda of Cartography in Early Modern Europe." *Imago Mundi* 40 (1988): 57–76.

HARLEY, J. B., AND PAUL LAXTON, ED. *The New Nature of Maps: Essays in the History of Cartography*. Baltimore: Johns Hopkins University Press, 2001.

HARLEY, J. B., AND DAVID WOODWARD, EDS. *Art and Cartography: Six Historical Essays*. Chicago: University of Chicago Press, 1987.

HAROUEL, JEAN-LOUIS. *Histoire de l'expropriation*. Paris: Presses universitaires de France, 2000.

———. *L'embellissement des villes: L'urbanisme française au XVIIIe siècle*.
Paris: Picard, 1993.

———. "Les fonctions de l'alignement dans l'organisme urbain." *Dix-
huitième siècle*, no. 9 (1977): 135–49.

———. "L'expropriation dans l'histoire du droit français." In *L'expropriation*,
2:39–77. Recueils de la Société Jean Bodin pour l'histoire comparative des institutions 67. Brussels: De Boeck University, 2000.

HARVEY, DAVID. *Paris, Capital of Modernity*. New York: Routledge, 2003.

HAUSSMANN, GEORGES-EUGÈNE. *Les mémoires d'Haussmann*. Edited and
introduction by Françoise Choay. Paris: Seuil, 2000.

———. *Mémoires du baron Haussmann*. 3 vols. Paris: Victor-Havard,
1890–93.

HAUTECOEUR, LOUIS. *Histoire du Louvre: Le château—le palais—le musée,
des origines à nos jours, 1200–1940*. Paris: L'Illustration, 1940.

HAÜY, RENÉ-JUST. *Instruction sur les mesures déduites de la grandeur de la
terre, uniformes pour toute la République, et sur les calculs relatifs à leur
division décimale*. Paris: Imprimerie de Marchant, 1794.

HEADRICK, DANIEL R. *When Information Came of Age: Technologies of
Knowledge in the Age of Reason and Revolution, 1700–1850*. Oxford:
Oxford University Press, 2002.

HEILBRON, JOHN L. "The Politics of the Meter Stick." *American Journal of
Physics* 57 (1989): 988–92.

———. *Weighing Imponderables and Other Quantitative Sciences around 1800*.
Los Angeles: University of California Press, 1993.

HENNET, ALBERT-JOSEPH-ULPIEN. *Recueil méthodique des lois, décrets,
règlemens, instructions et décisions sur le cadastre de la France*. Paris:
Imprimerie impériale, 1811.

HERBERT, DANIEL M. "Graphic Processes in Architectural Study
Drawings." *Journal of Architectural Education* 46, no. 1 (September
1992): 28–39.

———. "Study Drawings in Architectural Design: Their Properties as a
Graphic Medium." *Journal of Architectural Education* 41, no. 2 (Winter,
1988): 26–38.

HEWITT, MARK. "Representational Forms and Modes of Conception:
An Approach to the History of Architectural Drawing." *Journal of
Architectural Education* 39, no. 2 (Winter 1985): 2–9.

HILAIRE-PÉREZ, LILIANE, VALÉRIE NÈGRE, DELPHINE SPICQ, AND KOEN
VERMEIR, EDS. *Le livre technique avant le XXe siècle: À l'échelle du
monde*. Paris: CNRS, 2017.

HIMMELFARB, HÉLÈNE. "Versailles: Function and Legends." In *Les lieux
de mémoire*, vol. 2, *La nation*, pt. 2, edited by Pierre Nora, 235–92.
Paris: Gallimard, 1984–86.

HIRE, PHILIPPE DE LA. *L'École des arpenteurs, où l'on ensigne toutes les
pratiques de la géométrie qui sont necessaire à un arpenteur*. Paris:
T. Moette, 1692.

HOBSBAWM, ERIC. "Mass-Producing Traditions: Europe, 1870–1914." In
The Invention of Tradition, edited by Eric Hobsbawm and Terence
Ranger, 263–308. London: Cambridge University Press, 1983.

HOOSON, DAVID, ED. *Geography and National Identity*. Cambridge, MA:
Blackwell, 1994.

HORN, EVA. "There Are No Media." *Grey Room* 29 (Fall 2007): 7–13.

HOUSSAYE, ARSÈNE. *Les confessions: Souvenirs d'un demi-siècle, 1830–1880*.
Paris, 1885.

HULL, MATTHEW. "Documents and Bureaucracy." *Annual Review of
Anthropology* 41 (2012): 251–67.

———. *Government of Paper: The Materiality of Bureaucracy in Urban
Pakistan*. Berkeley: University of California Press, 2012.

HUSSON, ARMAND. *Traité de la législation des travaux publics et de la voirie
en France*. Paris, 1850.

IMHOF, EDUARD. *Cartographic Relief Presentation*. Edited by H. J. Steward.
Berlin: Walter de Gruyter, 1965.

INGRAHAM, CATHERINE. *Architecture and the Burdens of Linearity*. New
Haven: Yale University Press, 1998.

IVINS, WILLIAM. *On the Rationalization of Sight*. New York: Da Capo, 1973.

———. *Prints and Visual Communications*. Cambridge, MA: Harvard
University Press, 1953.

JACOB, CHRISTIAN. *L'empire des cartes: Approche théorique de la cartographie à travers l'histoire*. Paris: Albin Michel, 1992.

———. *The Sovereign Map: Theoretical Approaches in Cartography throughout
History*. Translated by Tom Conley. Chicago: University of Chicago
Press, 2006.

———. "Towards a Cultural History of Cartography." *Imago Mundi* 48
(1996): 91–98.

JACQUES, ANNIE, AND JEAN-PIERRE MOUILLESEAUX, EDS. *Les architectes
de la liberté, 1789–1799*. Paris: École nationale supérieure des Beaux-
Arts, 1988.

JAMES, CARLO, AND MARJORIE B. COHN. *Old Master Prints and Drawings:
A Guide to Preservation and Conservation*. Amsterdam: Amsterdam
University Press, 1997.

JAMMES, ANDRÉ, ET AL. *De Niépce à Stieglitz: La photographie en taille-
douce*. Lausanne: Musée de l'Élysée, 1982.

JAY, MARTIN, AND SUMATHI RAMASWANY, EDS. *Empires of Vision: A Reader*.
Durham: Duke University Press, 2014.

JENKINS, LLOYD. "Utopianism and Urban Change in Perreymond's Plans
for the Rebuilding of Paris." *Journal of Historical Geography* 32 (2006):
336–51.

JENNINGS, MICHAEL W., ED. *Walter Benjamin: Selected Writings*, vol 1,
1913–1926. Cambridge, MA: Harvard University Press, 2004.

JOMARD, EDME-FRANÇOIS. *Considérations sur l'objet et les avantages d'une
collection spéciale consacrée aux cartes géographiques et aux diverses
branches de la géographie*. Paris, 1831.

———. *Introduction à l'atlas des monuments de la géographie*. Paris, 1879.

———. *Mémoire sur le système métrique des anciens Égyptiens: Contenant des
recherches sur leurs connaissances géométriques et sur les mesures des
autres peuples de l'antiquité*. Paris, 1817.

———. *Souvenirs sur Gaspard Monge et ses rapport avec Napoléon*. Paris, 1853.

JONES, COLIN. "Theodore Vacquer and the Archaeology of Modernity
in Haussmann's Paris." *Transactions of the Royal Historical Society* 17
(2007): 157–83.

JORDAN, DAVID. "Haussmann and Haussmannisation: The Legacy for
Paris." *French Historical Studies* 27, no. 1 (Winter 2004): 87–113.

———. *Transforming Paris: The Life and Labors of Baron Haussmann*. New
York: Free Press, 1995.

JOURNAL DES GÉOMÈTRES, 1847–66.

KAFKA, BEN. *The Demon of Writing: Powers and Failures of Paperwork*.
Cambridge, MA: MIT Press, 2012.

KAIN, ROBERT J. P., AND ELIZABETH BAIGENT. *The Cadastral Map in the
Service of the State*. Chicago: University of Chicago Press, 1992.

KAMEA, EUGENE, AND MARTIN KRYGIER. *Bureaucracy, the Career of a
Concept*. New York: St. Martin's, 1979.

KAPLAN, STEVEN LAURENCE, AND CYNTHIA J. KOEPP, EDS. *Work in France:
Representations, Meanings, Organization, and Practice*. Ithaca: Cornell
University Press, 1986.

KARROW, ROBERT W., ED. *Mapmakers of the Sixteenth Century and Their
Maps*. Chicago: Speculum Orbis, 1993.

KAZMIERCZAK, ELZBIETA. "Design as Meaning Making: From Making

Things to the Design of Thinking." *Design Issues* 19, no. 2 (Spring 2003): 45–59.

KEAY, JOHN. *The Great Arc: The Dramatic Tale of How India Was Mapped and Everest Was Named.* New York: HarperCollins, 2000.

KENNEL, SARAH, ED. *Charles Marville: Photographer of Paris.* Chicago: University of Chicago Press, 2013.

KEYLOR, WILLIAM R. *Academy and Community: The Foundation of the French Historical Profession.* Cambridge, MA: Harvard University Press, 1975.

KIELY, EDMOND RICHARD. *Surveying Instruments: Their History and Classroom Use.* New York: Bureau of Publications, Teachers College, Columbia University, 1947.

KIESS, WALTER. *Urbanismus im Industriezeitalter.* Berlin: Ernst und Sohn, 1991.

KITCHIN, ROB, AND MARTIN DODGE. "Rethinking Maps." *Progress in Human Geography* 31, no. 3 (2007): 331–44.

KOMARA, ANN E. "Concrete and the Engineered Picturesque: The Parc des Buttes Chaumont (Paris, 1867)." *Journal of Architectural Education* 58, no. 1 (September 2004): 5–12.

——. "Measure and Map: Alphand's Contours of Construction at the Parc des Buttes Chaumont, Paris 1867." *Landscape Journal* 28 (2009): 22–39.

KONVITZ, JOSEF. *Cartography in France, 1660–1848: Science, Engineering, and Statecraft.* Chicago: University of Chicago Press, 1987.

KRAKAUER, ERIC. *The Disposition of the Subject: Reading Adorno's Dialectic of Technology.* Evanston, IL: Northwestern University Press, 1998.

KRAUSS, ROSALIND. "Grids." *October* 9 (Summer 1979): 50–64.

KUHN, THOMAS. *The Structure of Scientific Revolutions.* Chicago: University of Chicago Press, 1970.

KULA, WITOLD. *Measure and Men.* Translated by R. Szreter. Princeton: Princeton University Press, 1986.

LABORDE, ALEXANDRE DE. *Projets d'embellissemens de Paris et de travaux d'utilité publique, concernant les Ponts et chaussées.* Paris, 1833.

LA BRAS, HERVÉ. "The Government Bureau of Statistics: La statistique générale de la France." In *Rethinking France: Les lieux de memoires,* vol. 1, *The State,* edited by Pierre Nora and David Jourdan, translated by Mary Seidman Trouille, 361–400. Chicago: University of Chicago Press, 1999.

LACROIX, SIGISMOND. *Rapport sur l'organisation municipale de la Ville de Paris.* Paris, 1880.

LAMEYRE, GÉRARD-NOËL. *Haussmann, "préfet de Paris."* Paris: Flammarion, 1958.

LANCELOT. *Nouveau traité d'arpentage et de toisé, avec des tables de conversions de mesures anciennes et nouvelles.* Paris, 1830.

LANDAU, BERNARD, CLAIRE MONOD, EVELYNE LOHR, AND GÉRALDINE RIDEAU. *Les grands boulevards: Un parcours d'innovation et de modernité.* Paris: Action artistique de la Ville de Paris, 2000.

LANSING, KENNETH M. "The Effect of Drawing on the Development of Mental Representations: A Continuing Study." *Studies in Art Education* 25, no. 3 (1984): 167–75.

LA RÉPUBLIQUE FRANÇAISE, 1871–1924.

LARKIN, PETER, AND MICHAEL CONZEN, EDS. *Shapers of Urban Form: Explorations in Morphological Agency.* New York: Routledge, 2014.

LARRÈRE, CATHERINE. "L'arithmétique des physiocrates: La mesure de l'évidence." *Histoire et mesure* 7 1/2 (1992): 5–24.

LATOUR, BRUNO. "Circulating Reference: Sampling the Soil in the Amazon Forest." In *Pandora's Hope: Essays on the Reality of Science Studies,* 24–79. Cambridge, MA: Harvard University Press, 1999.

——. *The Making of Law: An Ethnography of the Conseil d'État.* Cambridge, MA: Polity, 2009.

——. *Science in Action: How to Follow Scientists and Engineers through Society.* Cambridge, MA: Harvard University Press, 1987.

——. "Visualization and Cognition: Drawing Things Together." *Knowledge and Society: Studies in the Sociology of Culture and Present* 6 (1986): 1–40.

——. "Where Are the Missing Masses? The Sociology of a Few Mundane Artifacts." In *Shaping Technology/Building Society: Studies in Sociotechnical Change,* edited by Wiebe Bijker and John Law, 225–58. Cambridge, MA: MIT Press, 1992.

LATOUR, BRUNO, AND STEVE WOOLGAR. *Laboratory Life: The Social Construction of Scientific Facts.* Princeton: Princeton University Press, 1986.

LAUGIER, MARC-ANTOINE. *Essai sur l'architecture.* Paris: Chez Duchesne, 1753.

——. *An Essay on Architecture.* Translated by Wolfgang and Anni Herrmann. Los Angeles: Hennessy and Ingalls, 1977.

LAUSSAUDET, AIMÉ. *Recherches sur les instruments, les méthodes, et le dessin topographique.* Paris, 1898–1901.

LAUTERBACH, IRIS. "'Faire céder l'art à la nature': Natürlichkeit in der Gartenkunst um 1700." In *Neue Modelle im Alten Europa: Traditionsbruch und Innovation als Herausforderung in der Frühen Neuzeit,* edited by Christoph Kampmann et al., 176–93. Cologne: Böhlau, 2011.

LAVEDAN, PIERRE. *Histoire de l'urbanisme à Paris.* Paris: Assocation pour la publication d'une histoire de Paris/Diffusion Hachette, 1975.

——. *L'architecture française.* Paris: Librairie Larousse, 1944.

——. *La question de déplacement de Paris et du transfert des Halles au Conseil municipal sous la monarchie de juillet.* Paris: Sous-Commission de recherches d'histoire municipale contemporaine, 1969.

LAZARE, FÉLIX, AND LOUIS LAZARE. *Dictionnaire administrative et historique des rues et monuments de Paris.* Paris, 1844.

LAZARE, LOUIS. *Les quartiers pauvres de Paris: Le 20eme arrondissement.* Paris: Bureau de la Bibliothèque municipale, 1870.

LEBOIS, VALÉRIE. "Habitants et architectes: Des créatures d'images." *Labyrinthe* 15 (2003), http://labyrinthe.revues.org/470.

LÉCUYER, BERNARD-PIERRE. "The Statistician's Role in Society: The Institutional Establishment of Statistics in France." *Minerva* 25, nos. 1–2 (March 1987): 35–55.

LÉCUYER, RAYMOND. *Histoire de la photographie.* Paris: Baschet, 1945.

LEE, MIN KYUNG. "Constructing Nineteenth-Century Paris through Cartography and Photography." In *Piercing Time: Paris after Marville and Atget, 1865–2012,* by Peter Sramek, 100–111. Chicago: University of Chicago Press, 2013.

——. "An Objective Point of View: The Orthogonal Grid in Eighteenth-Century Plans of Paris." *Journal of Architecture RIBA* 17, no. 1 (2012): 11–32.

LEFEBVRE, HENRI. *Introduction à la modernité.* Paris: Éditions de Minuit, 1962.

——. *Rhythmanalysis; Space, Time and Everyday Life.* Translated by Stuart Elden and Gerald Moore. London: Continuum, 2004.

LEHNING, JAMES. *To Be a Citizen: The Political Culture of the Early French Third Republic.* Ithaca, NY: Cornell University Press, 2001.

LEMERCIER, ALFRED. *La lithographie français de 1786 à 1896 et les arts qui s'y rattachent, Manuel Pratique s'addressant aux artistes et aux imprimeurs.* Paris, 1896.

LEMKE, THOMAS. *Foucault, Governmentality, and Critique.* New York: Routledge, 2016.

LE MOËL, MICHEL, SOPHIE DESCAT, AND BÉATRICE DE ANDIA. *L'urbanisme parisien au siècle des Lumières, collection Paris et son patrimoine.* Paris: Action artistique de la Ville de Paris, 1997.

LEMOINE, BERTRAND. *Les Halles de Paris: L'histoire d'un lieu, les péripéties d'une reconstruction, la succession des projets, l'architecture d'un monument, l'enjeu d'une Cité.* Paris: L'Équerre, 1980.

LEMOINE, BERTRAND, AND MARC MIMRAM, EDS. *Paris d'ingénieurs.* Paris: Pavilon de l'arsenal, 1995.

LE MONITEUR UNIVERSEL, 1860–89.

LEPETIT, BERNARD. *Les villes dans la France moderne (1740–1840).* Paris: Albin Michel, 1988.

——. "L'évolution de la notion de ville d'après les tableaux et descriptions géographiques de la France (1650–1850)." *Urbi*, no. 2 (December 1979): 99–107.

——. *The Pre-Industrial Urban System: France, 1740–1840.* Cambridge: Cambridge University Press, 1994.

LÉRI, JEAN-MARC. "Edme Verniquet (1727–1804) cartographe du grand plan de Paris." In *Les architectes de la liberté, 1789–1799*, edited by Jean-Pierre Mouilleseaux and Annie Jacques, 202–9. Paris: École nationale supérieure des Beaux-Arts, 1989.

——. "La Commission et le plan des artistes." In *Patrimoine parisien, 1789–1799: Destructions, créations, mutations*, edited by Alfred Fierro and Béatrice de Andia, 152–59. Paris: Délégation à l'action artistique de la Ville de Paris, 1989.

——. *"Le Marais" par Jacques Gomboust, 1652.* Paris: Ateliers de Saint-Martin-de-Nigelle, 1983.

——. "Les travaux parisiens sous le préfet Rambuteau." *Cahiers du CREPIF*, no. 18 (March 1987): 203–13.

LE ROUX DE LINCY, ANTOINE. *Histoire de l'Hôtel de Ville de Paris.* Paris, 1846.

LE ROUX DE LINCY, ANTOINE, AND L. M. TISSERAND. *Paris et ses historiens aux XIVe et XVe siècles.* Paris, 1867.

LE SIÈCLE, 1871.

LEVIN, MIRIAM R. "Democratic Vistas–Democratic Media: Defining a Role for Printed Images in Industrializing France." *French Historical Studies* 18, no. 1 (Spring, 1993): 82–108.

——. "The Eiffel Tower Revisited." *French Review* 62, no. 6 (May 1989): 1052–64.

——. "Inventing Modern Paris: The Dynamic Relationship between Expositions, Urban Development and Museums." *Quaderns d'historia de l'enginyeria* 13 (2012): 35–56.

——. *Republican Art and Ideology in Late Nineteenth-Century France.* Ann Arbor: UMI Research Press, 1986.

LEVINE, NEIL. *Modern Architecture: Representation and Reality.* New Haven: Yale University Press, 2009.

LEY, DAVID, AND MARWYN SAMUELS, EDS. *Humanistic Geography: Prospects and Problems.* Chicago: Maaroufa, 1978.

LINDGREN, UTA. "Land Surveys, Instruments, and Practitioners in the Renaissance." In *The History of Cartography: Cartography of the European Renaissance*, vol. 3, pt. 1, edited by David Woodward, 477–508. Chicago: University of Chicago Press, 2007.

LIPSTADT, H. *Architecte et ingénieur dans la presse: Polémique, débat, conflit.* Paris: IERAU, 1980.

LIVINGSTONE, DAVID N., AND CHARLES W. J. WITHERS, EDS. *Geography and Enlightenment.* Chicago: University of Chicago Press, 1999.

LOBSINGER, MARY LOUISE. "Architectural History: The Turn from Culture to Media." *Journal of the Society of Architectural Historians* 75, no. 2 (June 2016): 135–39.

LORTIE, ANDRÉ, AND PIERRE SCHALL, EDS. *Parcs et promenades de Paris.* Paris: Éditions du demi-cercle/Pavillon de l'arsenal, 1988.

LOYER, FRANÇOIS. *Architecture of the Industrial Age, 1789–1914.* Geneva: Skira, 1983.

——. *Histoire de l'architecture française de la Révolution à nos jours.* Paris: Éditions Mengès, 1999.

——. *Paris Nineteenth Century: Architecture and Urbanism.* New York: Abbeville, 1988.

——. *Paris XIXe siècle: L'immeuble et la rue.* Paris: Hazan, 1978.

LOYER, FRANÇOIS, JEAN-FRANÇOIS PINCHON, ANNE-MARIE CHATELET, AND BÉATRICE DE ANDIA. *Autour de l'Opéra: Naissance de la ville moderne.* Paris: Délégation à l'action artistique de la Ville de Paris, 1995.

LOYSEL, JULIEN. *Essai sur l'organisation des ponts-et-chaussées.* Vannes, 1816.

LUCAN, JACQUES. *Composition, Non-Composition: Architecture et théories, XIXe–XXe siècles.* Lausanne: Presses polytechniques et universitaires romandes, 2009.

——, ED. *Paris des faubourgs: Formation, Transformation.* Paris: Pavillon de l'arsenal, 2005.

LYNCH, KEVIN. *The Image of the City.* Cambridge, MA: MIT Press, 1960.

LYNCH, MICHAEL. "Discipline and the Material Form of Images: An Analysis of Scientific Visibility." *Social Studies of Science* 15, no. 1 (February 1985): 37–66.

LYOTARD, JEAN-FRANÇOIS. *The Post-Modern Condition: A Report on Knowledge.* Translated by Geoff Bennington and Brian Massumi. Minneapolis: University of Minnesota Press, 1984.

MAINARDI, PATRICIA. *Art and Politics of the Second Empire: The Universal Expositions of 1855 and 1867.* New Haven: Yale University Press, 1987.

MAITRON, JEAN, ED. *Dictionnaire biographique du mouvement ouvrier français, 1789–1864.* Vol. 2. Paris: Editions ouvrières, 1965.

MALET, HENRI. *Le Baron Haussmann et la rénovation de Paris.* Paris: Éditions municipales, 1973.

MANESSON-MALLET, ALAIN. *Les travaux de Mars, ou la fortification nouvelle.* Paris, 1671.

MANGIN, ARTHUR. "Les Expositions universelles: L'exposition projétée pour 1889." *L'economiste français* (November 1, 1884): 538–40.

MARCUS, SHARON. *Apartment Stories: City and Home in Nineteenth-Century Paris and London.* Los Angeles: University of California Press, 1999.

MAREY, ÉTIENNE-JULES. *La méthode graphique, dans les sciences expérimentales et principalement en physiologie et en médecine.* Paris, 1878.

MARGAIRAZ, DOMINIQUE. "La géographie des administrateurs." In *Géographie plurielles: Les sciences géographique au moment de l'émergence des sciences humaines, 1750–1850*, edited by Hélène Blais and Isabelle Laboulais-Lesage, 185–216. Paris: L'Harmattan, 2006.

MARIN, LOUIS. *De la représentation.* Paris: Seuil Gallimard, 1994.

——. *Le portrait du roi.* Paris: Éditions de Minuit, 1981.

——. "Narrative Theory and Piero as History Painter." *October* 65 (Summer, 1993): 106–32.

——. *Politiques de la représentation.* Paris: Kimé, 2005.

——. *Utopics: Spatial Play.* Translated by Robert A. Vollrath. London: Macmillan, 1984.

MARTIN, LÉON. ed. *Encyclopédie municipale de la Ville de Paris.* Paris: G. Roustan, 1904.

MARTIN, LESLIE. "The Grid as Generator." *Arq: Architectural Research Quarterly* 4, no. 4 (2000): 309–20.

MARX, KARL. *Capital: Critique of the Political Economy.* 1867. Reprint, London: Penguin Classics, 1990.

——. *The Eighteenth Brumaire of Louis Bonaparte.* New York: International, 1963.

MARX, KARL, AND FREDERICK ENGELS. *Marx and Engels Collected Works*, vol. 31, *1861–1863.* London: Lawrence and Wishart, 1989.

MAURIN, ANDRÉ. *Le cadastre en France: Histoire et rénovation.* Paris: Éditions du Centre national de la recherche scientifique, 1990.

MAYNARD, PATRICK. "Perspective's Places." *Journal of Aesthetics and Art Criticism* 54, no. 1 (Winter 1996): 23–40.

MAZZARELLA, WILLIAM. "Internet X-Ray: E-Governance, Transparency, and the Politics of Immediation in India." *Public Culture* 18, no. 3 (2006): 473–505.

MCLEAN, MATTHEW ADAM. *The Cosmographia of Sebastian Münster: Describing the World in the Reformation.* Hampshire, UK: Ashgate, 2007.

MCQUEEN, ALISON. *Empress Eugénie and the Arts: Politics and Visual Culture in the Nineteenth Century.* London: Ashgate, 2011.

——. "Shaped to Suit a Nation: Mid-Nineteenth-Century Representations of the Last Empress of the French." In *Empresses and Queens in the Courtly Public Sphere from the Seventeenth to the Twentieth Century,* edited by Marion Romberg, 216–50. Leiden: Brill, 2021.

——. "Women and Social Innovation during the Second Empire: Empress Eugénie's Patronage of the Fondation Eugène Napoléon." *Journal of the Society of Architectural Historians* 66, no. 2 (2007): 176–93.

MCQUIRE, SCOTT. *The Media City: Media, Architecture and Urban Space.* London: Sage, 2008.

MEAD, CHRISTOPHER CURTIS. *Charles Garnier's Paris Opéra: Architectural Empathy and the Renaissance of French Classicism.* Cambridge, MA: MIT Press, 1991.

——. *Making Modern Paris: Victor Baltard's Central Markets and the Urban Practice of Architecture.* University Park: Pennsylvania State University Press, 2012.

——. "Urban Contingency and the Problem of Representation in Second Empire Paris." *Journal of the Society of Architectural Historians* 54, no. 2 (June 1995): 138–74.

MEISNER ROSEN, CHRISTINE. "Infrastructural Improvement in Nineteenth-Century Cities: A Conceptual Framework and Cases." *Journal of Urban History* 12, no. 3 (May 1986): 211–56.

MERRUAU, CHARLES. *Souvenirs de l'Hôtel de Ville, 1848–1852.* Paris: Plon, 1875.

MESTRE, JEAN-LOUIS. "L'expropriation face à la propriété." *Droits* 1 (1985): 51.

MEYNADIER, HIPPOLYTE. *Paris sous le point de vie pittoresque et monumental.* Paris, 1843.

MICHELET, JULES. *Nos fils.* Paris, 1870.

MIDDLETON, ROBIN, ED. *The Beaux-Arts and Nineteenth-Century French Architecture.* Cambridge, MA: MIT Press, 1982.

MITCHELL, TIMOTHY. *Colonising Egypt.* Berkeley: University of California Press, 1991.

MONGE, GASPARD. *Application de l'analyse à la géométrie: À l'usage de l'École impériale polytechnique.* 4th ed. Paris, 1809.

——. *Géométrie descriptive, leçons données aux écoles normales, l'an 3 de la République.* Paris: Baudouin, 1799.

MONMONIER, MARK S. *How to Lie with Maps.* Chicago: University of Chicago Press, 1991.

MONNIER, FRANÇOIS. "La notion d'expropriation au XVIIIe siècle d'après l'exemple de Paris." *Journal des savants,* nos. 3–4 (July–December 1984): 223–58.

MONTESSON, LOUIS-CHARLES DUPAIN DE. *La science de l'arpenteur.* Paris: Chez le S. Jaillot, 1766.

MONTIGNY, GILLES. *De la ville à l'urbanisation: Essai sur la genèse des études urbaines françaises en géographie, sociologie et statistique sociale.* Paris: L'Harmattan, 1992.

MORACHIELLO, PAOLO, AND GEORGES TEYSSOT. "State Town: Colonization of the Territory during the First Empire." *Lotus,* no. 24 (1979): 24–29.

MORÉRI, LOUIS. *Le grand dictionnaire historique, ou mélange curieux de l'histoire sacrée et profane.* Paris, 1759.

MORIZET, ANDRÉ. *De vieux Paris au Paris moderne; Haussmann et ses prédécesseurs.* Paris: Hachette, 1932.

MOSSER, MONIQUE, AND DANIEL RABREAU. *Charles de Wailly: Peintre architecte dans l'Europe des Lumières.* Paris: Caisse nationale des monuments historiques et des sites, 1979.

MUKERJI, CHANDRA. "Engineering and French Formal Gardens in the Reign of Louis XIV." In *Tradition and Innovation in French Garden Art,* edited by John Dixon Hunt and Michel Conan, 22–43. Philadelphia: University of Pennsylvania Press, 2002.

——. *Territorial Ambitions and the Gardens of Versailles.* Cambridge: Cambridge University Press, 1997.

MURPHY, KEVIN, AND SALLY O'DRISCOLL. "The Art/History of Resistance: Visual Ephemera in Public Space." *Space and Culture* 18, no. 4 (2015): 328–57.

NADEAU, LUIS. *Encyclopedia of Printing, Photographic, and Photomechanical Processes.* Fredericton, CA: Luis Nadeau, 1994.

NAGAÏ, NOBUHITO. *Les conseillers municipaux de Paris sous la IIIe République (1871–1914).* Paris: Sorbonne, 2002.

NAGEL, THOMAS. *The View from Nowhere.* Oxford: Oxford University Press, 1986.

NASR, JOE, AND MERCEDES VOLAIT, EDS. *Urbanism: Imported or Exported?* Chichester, UK: Wiley, 2003.

NELMS, BRENDA. *The Third Republic and the Centennial of 1789.* New York: Garland, 1987.

NÉRÉ, J. "The French Republic." In *The New Cambridge Modern History,* vol. 11, *Material Progress and World-Wide Problems, 1870–1898,* edited by F. H. Hinsley, 300–322. Cambridge: Cambridge University Press, 2008.

NEUGEBAUER, OTTO. *A History of Ancient Mathematical Astronomy, Part 2.* New York: Springer, 1975.

NGUYEN, JASON. "Constructing Classicism: Architectural Theory, Practice, and Expertise in Paris, c. 1670–1720." PhD diss, Harvard University, 2017.

NORA, PIERRE. "De la République à la nation." In *Les lieux de mémoire,* vol. 1, *La République,* edited by Pierre Nora, 651–59. Paris: Gallimard, 1984.

NOUVELLES ANNALES DES PONTS ET CHAUSSÉES, 1875–90.

NUTI, LUCIA. "The Perspective Plan in the Sixteenth Century: The Invention of a Representational Language." *Art Bulletin* 76 (1994): 105–28.

O'CONNELL, LAUREN M. *Architecture and the French Revolution: Change and Continuity under the Conseil des Bâtiments Civils, 1795–99.* PhD diss., Cornell University, 1989.

——. "Redefining the Past: Revolutionary Architecture and the Conseil des Bâtiment Civils." *Art Bulletin* 77 (June 1995): 207–24.

OGLE, VANESSA. *The Global Transformation of Time: 1870–1950.* Cambridge, MA: Harvard University Press, 2015.

OLSEN, DONALD. *The City as a Work of Art: London, Paris, Vienna.* New Haven: Yale University Press, 1986.

OLSON, KORY. *The Cartographic Capital: Mapping Third Republic Paris, 1889–1934.* Liverpool: Liverpool University Press, 2018.

ORILLARD, CLÉMENT. "Contrôler l'image de la ville." *Labyrinthe* 15 (2003), http://labyrinthe.revues.org/472.

ORY, PASCAL. *L'Expo universelle, 1889.* Brussels: Éditions complexes, 1989.

OSBORNE, PETER. *The Politics of Time: Modernity and the Avant-Garde.* London: Verso, 1995.

OZOUF, JACQUES, AND MONA OZOUF. "Le Tour de la France par deux enfants: Le petit livre rouge de la République." In *Les lieux de mémoire,* vol. 1, *La République,* edited by Pierre Nora, 291–321. Paris: Gallimard, 1984.

OZOUF-MARIGNIER, MARIE-VIC. "Administration, statistique aménagement du territoire: L'intinéraire du préfect Chabrol de Volvic (1773–1843)." *Revue d'histoire moderne et contemporaine* 44, no. 1 (January–March 1997): 19–39.

——. "Entre tradition et modernité: Les recherches statistiques sur la Ville de Paris (1821)." In *Mélanges de l'École française de Rome, Italie et Méditerrannée* 111, no. 2 (1999): 747–62.

PACCOUD, ANTOINE. "Planning Law, Power, and Practice: Haussmann in Paris (1853–1870)." *Planning Perspectives* 31, no. 3 (2016): 341–61.

PAIRAULT, FRANÇOIS. *Gaspard Monge: Le fondateur de Polytechnique.* Paris: Tallandier, 2000.

PALMOWSKI, JAN. "Travels with Baedeker: The Guidebook and the Middle Classes in Victorian and Edwardian England." In *The History of Leisure,* edited by Rudy Koshar, 105–30. New York: Berg, 2002.

PALOMINO, JEAN-FRANÇOIS. *Une carrière de géographe au siècle des Lumières: Jean-Baptiste d'Anville.* Oxford: Voltaire Foundation, 2018.

PALSKY, GILLES. "The Debate on the Standardization of Statistical Maps and Diagrams (1857–1901)." *Cybergeo* 65 (1999), http://cybergeo.revues.org/148.

——. *Des chiffres et des cartes: Naissance et développement de la cartographie quantitative française au XIX siècle.* Paris: Ministère de l'enseignement supérieur et de la recherche, Comité des travaux historiques et scientifiques, 1996.

PANOFSKY, ERWIN. *Perspective as Symbolic Form.* New York: Zone, 1991.

PAPAYANIS, NICHOLAS. *Planning Paris before Haussmann.* Baltimore: Johns Hopkins University Press, 2004.

PAQUOT, THIERRY, AND CHRIS YOUNÈS, EDS. *Géométrie, mesure du monde: Philosophie, architecture urbain.* Paris: Éditions la Découverte, 2005.

PARROCHIA, DANIEL. "Quelques aspects historiques de la notion de réseau." *Flux,* no. 62 (October–December 2005): 10–20.

PATTE, PIERRE. *Monumens érigés en France à la gloire de Louis XV.* Paris, 1765.

PEARCE, MARGARET WICKENS. "Framing the Days: Place and Narrative in Cartography." *Cartography and Geographic Information Science* 35, no. 1 (2008): 17–32.

PEARSON, KAREN S. "The Nineteenth-Century Colour Revolution: Maps in Geographical Journals." *Imago Mundi* 32 (1980): 9–20.

PEDLEY, MARY SPONBERG. *The Commerce of Cartography: Making and Marketing Maps in Eighteenth-Century France and England.* Chicago: University of Chicago Press, 2005.

——. "The Map Trade in Paris, 1650–1825." *Imago Mundi* 33 (1981): 33–45.

PEDLEY, MARY SPONBERG, AND MATTHEW EDNEY, EDS. *The History of Cartography,* vol. 4, *Cartography in the European Enlightenment.* Chicago: University of Chicago Press, 2020.

PELLETIER, ALPHONSE. *Notes sur l'administration des services et établissement municipaux compris dans la direction de l'administration générale.* Paris, 1879.

PELLETIER, MONIQUE. *Cartographie de la France et du monde de la Renaissance au siècle des lumières.* Paris: Bibliothèque nationale de France, 2001.

——. "Cartography and Power in France during the Seventeenth and Eighteenth Centuries." *Cartographica* 35, nos. 3/4 (Autumn/Winter 1998): 41–53.

——. "Jomard et le Département de cartes et plans." *Bulletin de la Bibliothèque nationale* 4 (1979): 18–27.

——. *La carte de Cassini: L'extraordinaire aventure de la carte de France.* Paris: Presses de l'École nationale des ponts et chaussées, 1990.

PELLETIER, MONIQUE, AND HENRIETTE OZANNE. *Portraits de la France: Les cartes, témoins de l'histoire.* Paris: Hachette, 1995.

PERCIN, J. *Principes d'arpentage et de nivellement.* Nancy, 1848.

PÉREZ-GÓMEZ, ALBERTO. "Architecture as Drawing." *Journal of Architectura; Education* 26, no. 2 (Winter 1982): 2–7.

PÉREZ-GÓMEZ, ALBERTO, AND LOUISE PELLETIER. *Architectural Representation and the Perspective Hinge.* Cambridge, MA: MIT Press, 1997.

——. "Architectural Representation beyond Perspectivism." *Perspecta* 27 (1992): 20–39.

PERITON, DIANA. "The 'Coupe Anatomique': Sections through the Nineteenth-Century Parisian Apartment Block." *Journal of Architecture* 22, no. 5 (2017): 933–48.

PERROT, J. C. "L'âge d'or de la statistique régionale française (an IV–1804)." *Annales historiques de la Révolution française* 224 (1976): 215–76.

PERSELL, STUART M. *The French Colonial Lobby, 1889–1938.* Stanford, CA: Hoover Institution, 1983.

PERSIGNY, JEAN-GILBERT-VICTOR FIALIN, DUC DE. *Mémoires du duc de Persigny.* Paris: Plon, 1896.

PERTUSIER, CHARLES. *La fortification ordonnée d'après les principes de la stratégie et de la ballistique modernes.* Paris, 1820.

PETERS, JEFFREY N. *Mapping Discord: Allegorical Cartography in Early Modern French Writing.* Newark: University of Delaware Press, 2004.

PETOT, JEAN. *Histoire de l'administration des Ponts et Chaussées (1599–1815).* Paris: Librairie M. Riviere, 1958.

PETRY, CARL F., ED. *The Cambridge History of Egypt.* Cambridge: Cambridge University Press, 2008.

PICARD, ALFRED. *Rapport général. Exposition universelle internationale de 1889 à Paris.* Vol. 1. Paris, 1891–92.

PICON, ANTOINE. *Architectes et ingénieurs au siècle des lumières.* Marseille: Éditions parenthèses, 1988.

——. "De la composition urbaine au 'génie urbain': Les ingénieurs des Ponts et chaussées et les ville françaises au XIXe siècle." In *Les langages de la ville,* edited by Bernard Lamizet and Pascal Sanson, 169–77. Paris: Éditions parenthèses, 1997.

——. "From 'Poetry of Art' to Method: The Theory of Jean-Nicolas-Louis Durand." In *Précis of the Lecture on Architecture: With Graphic Portion of the Lectures on Architecture,* by Jean-Nicholas-Louis Durand, 1–68. Los Angeles: Getty Research Institute, 2000.

——. "Gestes, ouvriers, opérations et processus techniques: La vision du travail des encyclopédistes." *Recherches sur Diderot et sur l'Encyclopédie,* no. 13 (October 1992): 132–47.

——. *L'art de l'ingénieur: Constructeur, entrepreneur, inventeur.* Paris: Centre Georges Pompidou/Le moniteur, 1997.

——. "Nineteenth-Century Urban Cartography and the Scientific Ideal, the Case of Paris." *Osiris* 18 (2003): 135–49.

PICON, ANTOINE, AND ALESSANDRA PONTE, EDS. *Architecture and the Sciences: Exchanging Metaphors.* New York: Princeton Architectural Press, 2003.

PICON, ANTOINE, JEAN-PAUL ROBERT, AND ANNA HARTMANN, EDS. *Le dessus des cartes: Un atlas parisien.* Paris: Pavillon de l'arsenal, 1999.

PINKNEY, DAVID. *Napoléon III and the Rebuilding of Paris.* Princeton: Princeton University Press, 1958.

PIÑOL, JEAN-LUC. *Histoire de l'Europe urbaine*, vol. 2, *De l'ancien régime à nos jours: expansion et limite d'un modèle*. Paris: Seuil, 2003.

PINON, PIERRE. *Atlas du Paris Haussmannien: La ville en héritage du Second Empire à nos jours*. Paris: Parigramme, 2002.

PINON, PIERRE, BERTRAND LE BOUDEC, AND DOMINIQUE CARRÉ. *Les plans de Paris: Histoire d'une capitale*. Paris: Bibliothèque nationale de France, 2004.

PINTO, JOHN A. "Origins and Development of the Ichnographic City Plan." *Journal of the Society of Architectural Historians* 35, no. 1 (March 1976): 35–50.

PLØGER, JOHN. "Foucault's 'Dispositif' in the City." *Planning Theory* 7, no. 1 (March 2008): 51–70.

POËTE, MARCEL. *Exposition de la Bibliothèque et des travaux historique de la Ville de Paris*. Paris: Paul Dupont, 1910–13.

——. *Introduction à l'urbanisme: L'évolution du plan des villes, la leçon de l'histoire, l'antiquité*. 2nd ed. Paris: Anthropos, 1967.

POGNON, EDMOND. "Les collections du Département des cartes et plans de la Bibliothèque National de Paris." In *The Map Librarian in the Modern World*, edited by Helen Wallis and Lothar Zögnar, 195–204. Munich: K. G. Saur, 1979.

POGO, A. "Gemma Frisius, His Method of Determining Differences in Longitude by Transporting Timepieces (1530) and Treatise on Triangulation (1538)." *Isis* 22 (1935): 469–85.

POIX DE FRÉMINVILLE, CLAUDE EDME DE LA. *La pratique universelle pour la rénovation des terriers et des droits seigneuriaux*. Paris, 1752.

POLLACK, MARTHA. "Architecture and Cartography in the Design of the Early Modern City." In *Envisioning the City: Six Studies in Urban Cartography*, edited by David Buisseret, 109–24. Chicago: University of Chicago Press, 1998.

POOVEY, MARY. *A History of the Modern Fact: Problems of Knowledge in the Sciences of Wealth and Society*. Chicago: University of Chicago Press, 1998.

PORTER, THEODORE. "Objectivity and Authority: How French Engineers Reduced Public Utility to Numbers." *Poetics Today* 12, no. 2 (Summer 1991): 245–65.

——. "Objectivity as Standardization: The Rhetoric of Impersonality in Measurement, Statistics, and Cost-Benefit Analysis." *Annals of Scholarship* 9 (1992): 19–59.

——. *The Rise in Statistical Thinking, 1820–1900*. Princeton: Princeton University Press, 1986.

——. *Trust in Numbers: The Pursuit of Objectivity in Science and Public Life*. Princeton: Princeton University Press, 1997.

POUSIN, FRÉDÉRIC. *Figures de la ville et construction des savoirs: Architecture, urbanisme, géographie*. Paris: CNRS, 2005.

PROCTOR, ROBERT. "Constructing the Retail Monument: The Parisian Department Store and Its Property, 1855–1914." *Urban History* 33, no. 3 (December 2006): 393–410.

PRONTEAU, JEANNE. *Edme Verniquet, 1727–1804: Architecte et auteur du "Grand Plan de Paris."* Paris: Commission des travaux historiques, 1986.

PUJO, BERNARD. *Vauban*. Paris: Albin Michel, 1991.

RABINOW, PAUL. *Anthropos Today: Reflections on Modern Equipment*. Princeton: Princeton University Press, 2003.

——. *French Modern: Norms and Forms of the Social Environment*. Cambridge, MA: MIT Press, 1989.

RAMBUTEAU, CLAUDE-PHILIBERT BARTHELOT. *Mémoires de Rambuteau*. Paris: Calmann-Lévy, 1905.

RATTENBURY, KESTER, ED. *This Is Not Architecture: Media Constructions*. London: Routledge, 2002.

REARICK, CHARLES. "Festivals in Modern France: The Experience of the Third Republic." *Journal of Contemporary History* 12, no. 3 (July 1977): 435–60.

RECHERCHES STATISTIQUES SUR LA VILLE DE PARIS ET LE DÉPARTEMENT DE LA SEINE. 6 vols. Paris: Imprimerie royale, 1821–60.

RECUEIL DES ACTES ADMINISTRATIFS. Paris, 1876.

RECUEIL DES ACTES ADMINISTRATIFS DE LA PRÉFECTURE DE LA SEINE, 1859.

REES, RONALD. "Historical Links between Cartography and Art." *Geographical Review* 70 (1980): 1980.

REMMERT, VOLKER R. "The Art of Garden and Landscape Design and the Mathematical Sciences in the Early Modern Period." In *Gardens, Knowledge and the Sciences in the Early Modern Period*, edited by Hubertus Fischer, Volker R. Remmert, and Joachim Wolschke-Bulmann, 9–28. Basel: Birkhauser, 2016.

REVUE DE L'ARCHITECTURE ET DES TRAVAUX PUBLICS, 1853–63.

REYES, HECTOR. "The Rhetorical Frame of Poussin's Theory of the Modes." *Intellectual History Review* 19, no. 3 (2009): 287–302.

ROBINSON, ARTHUR H. *Early Thematic Mapping in the History of Cartography*. Chicago: University of Chicago Press, 1982.

RONCAYOLO, MARCEL. *La ville et ses territoires*. Paris: Gallimard, 1990.

——. *Lecture de villes: Formes et temps*. Marseille: Éditions parenthèses, 2002.

RONCAYOLO, MARCEL, AND THIERRY PAQUOT. *Ville et civilisation urbaine XVIIIe–XXe siècle*. Paris: Larousse, 1992.

ROSE-REDWOOD, REUBEN. "Indexing the Great Ledger of the Community: Urban House Numbering, City Directories, and the Production of Spatial Legibility." In *Critical Toponymies: The Contested Politics of Place Naming*, edited by Lawrence D. Berg and Jani Vuolteenaho, 199–225. London: Routledge, 2009.

——. "With Numbers in Place: Security, Territory, and the Production of Calculable Space." *Annals of the Association of American Geographers* 102 (2012): 295–319.

ROSS, REBECCA. "All Above: Visual Culture and the Professionalization of City Planning, 1867–1931." PhD diss., Harvard University, 2012.

ROULEAU, BERNARD. *Le tracé des rues de Paris: Formation, typologie, fonctions*. Paris: CNRS, 1967.

RUGY, MARIE DE. *Imperial Borderlands: Maps and Territory-Building in the Northern Indochinese Peninsula (1885–1914)*. Leiden: Brill, 2022.

RULE, JOHN C., AND BEN S. TROTTER. *A World of Paper: Louis XIV, Colbert de Torcy, and the Rise of the Information State*. Montreal: McGill-Queen's University Press, 2014.

RYKWERT, JOSEPH. "Translation and/or Representation." *RES: Anthropology and Aesthetics*, no. 34 (Autumn 1998): 64–70.

SAALMAN, HOWARD. *Haussmann: Paris Transformed*. New York: G. Braziller, 1971.

SABOYA, MARC. *Presse et architecture au XIXe siècle: César Daly et la Revue générale de l'architecture et des travaux publics*. Paris: Picard, 1991.

SANTANA-ACUÑA, ALVARO. "From Manual Art to Scientific Profession: Technical Books on Land Surveying in Eighteenth-Century France." In *Le livre technique avant le XXe siècle: À l'échelle du monde*, edited by Liliane Hilaire-Pérez, Valérie Nègre, Delphine Spicq, and Koen Vermeir, 121–36. Paris: CNRS, 2017.

——. "The Making of a National Cadastre (1763–1807): State Uniformization, Nature Valuation, and Organizational Change in France." PhD diss., Harvard University, 2014.

SAVIGNAT, J.-M. *Dessin et architecture du moyen-âge au XVIIIe siècle*. Paris: École nationale supérieure des Beaux-Arts/Ministère de la culture et de la communication, 1980.

SAY, HORACE. *Études sur l'administration de la Ville de Paris et du Département de la Seine.* Paris: Guillaumin, 1846.

SCHENKER, HEATH MASSEY. "Parks and Politics during the Second Empire in Paris." *Landscape Journal* 14, no. 2 (Fall 1995): 201–19.

SCHIVELBUSCH, WOLFGANG. *The Railway Journey: The Industrialization of Time and Space in the Nineteenth Century.* Berkeley: University of California Press, 2014.

SCHNEIDER, WILLIAM H. *An Empire for the Masses: The Image of West Africa in Popular French Culture, 1870–1900.* Westport, CT: Greenwood, 1980.

SCHUBERT, DIRK, AND ANTHONY SUTCLIFFE. "The 'Haussmannisation' of London?: The Planning and Construction of Kingsway-Aldwych, 1889–1935." *Planning Perspectives* 11, no. 2 (1996): 115–44.

SCHULZ, JUERGEN. "Jacopo de' Barbari's View of Venice: Map Making, City Views, and Moralized Geography before the Year 1500." *Art Bulletin* 60 (1978): 425–74.

SCOTT, JAMES C. *Seeing Like a State: How Certain Schemes to Improve the Human Condition Have Failed.* New Haven: Yale University Press, 1998.

SENNETT, RICHARD. *Flesh and Stone: The Body and the City in Western Civilization.* New York: W. W. Norton, 1994.

SEWELL, JR., WILLIAM. "Visions of Labor: Illustrations of the Mechanical Arts before, in and after Diderot's *Encyclopédie*." In *Work in France: Representations, Meanings, Organization, and Practice,* edited by Steven Laurence Kaplan and Cynthia J. Koepp, 258–86. Ithaca: Cornell University Press, 1986.

——. *Work and Revolution in France: The Language of Labor from the Old Regime to 1848.* Cambridge: Cambridge University Press, 1980.

SEXTON, F. M. "The Adoption of the Metric System in the Ordnance Survey." *Geographical Journal* 134, no. 3 (September 1968): 328–36.

SHAPIN, STEVEN. *The Social History of Truth: Civility and Science in Seventeenth-Century England.* Chicago: University of Chicago Press, 1994.

SHAW, STANFORD. "Landholding and Land-Tax Revenues in Ottoman Egypt." In *Political and Social Change in Modern Egypt,* edited by P. M. Holt, 91–103. Oxford: Oxford University Press, 1968.

SHEPARD, TODD. "The Birth of the Hexagonal: 1962 and the Erasure of France's Supranational History." In *Vertriebene and Pieds-Noirs in Postwar German and France: Comparative Perspectives,* edited by Manuel Borutta and Jan C. Jansen, 53–72. New York: Palgrave Macmillan, 2016.

SIEGFRIED, SUSAN LOCKE. "Naked History: The Rhetoric of Military Painting in Postrevolutionary France." *Art Bulletin* 75, no. 2 (June 1993): 235–58.

SITTE, CAMILLO. *Der Städte-Bau nach seinen künstlerischen Grundsätzen.* Vienna: K. Graeser, 1909.

SMITH, CHRISTINE. *Architecture in the Culture of Early Humanism: Ethics, Aesthetics, and Eloquence, 1400–1470.* New York: Oxford University Press, 1992.

SMITH, NATHANIEL B. "The Idea of the French Hexagon." *French Historical Studies* 6, no. 2 (Autumn 1969): 145.

SNYDER, JOHN P. *Flattening the Earth: Two Thousand Years of Map Projections.* Chicago: University of Chicago Press, 1993.

SÖDERSTRÖM, OLA. *Des images pour agir: Le visuel en urbanisme.* Lausanne: Éditions Payot, 2000.

——. "L'agir cartographique en urbanisme." *Politiques sociales,* nos. 1–2 (1999): 64–75.

——. "Paper Cities: Visual Thinking in Urban Planning." *Cultural Geographies* 3, no. 3 (1996): 249–81.

——. "Sélectionner et projeter: Les visualisations dans la pratique de l'urbanisme." *Espaces-temps,* nos. 62–63 (1996): 104–13.

SOJA, EDWARD W. *Postmodern Geographies: The Reassertion of Space in Critical Social Theory.* London: Verso, 1989.

SOLL, JACOB. *Information Master: Jean-Baptiste Colbert's Secret State Intelligence System.* Ann Arbor: University of Michigan Press, 2011.

SOPPELSA, PETER. "The Fragility of Modernity: Infrastructure and Everyday Life in Paris, 1870–1914." PhD diss., University of Michigan, 2009.

——. "How Haussmann's Hegemony Haunted the Early Third Republic." In *Is Paris Still the Capital of the Nineteenth Century? Essays on Art and Modernity, 1850–1900,* edited by Hollis Clayson and André Dombrowski, 35–51. London: Routledge, 2018.

SOUFFRIN, PIERRE. "La *Geometria pratica* dans les *Ludi rerum mathematicarum*." *Albertiana* 1 (1998): 87–104.

STAFFORD, BARBARA MARIA. *Good Looking: Essays on the Virtue of Images.* Cambridge, MA: MIT Press, 1996.

STIEBER, NANCY. "Microhistory of the Modern City: Urban Space and Its Use and Representation." *Journal of the Society of Architectural Historians* 58, no. 3 (September 1999): 382–91.

STIGLER, STEPHEN. *The History of Statistics: The Measurement of Uncertainty before 1900.* Cambridge, MA: Harvard University Press, 1986.

STJERNFELT, FREDERIK. *Diagrammatology.* Dordrecht: Springer, 2007.

STOICHITA, VICTOR. *L'Instauration du tableau: Métapeinture à l'aube des temps modernes.* Geneva: Droz, 1999.

STONE, JEFFREY C. "Imperialism, Colonialism and Cartography." *Transactions of the Institute of British Geographers* 13, no. 1 (1988): 57–64.

STROHMAYER, ULF. "Technology, Modernity, and the Restructuring of the Present in Historical Geographies." "The Structure of Space." Special issue, *Geografiska Annaler* 79, no. 3 (1997): 155–69.

SUTCLIFFE, ANTHONY. *The Autumn of Central Paris: The Defeat of Town Planning.* London: Edward Arnold, 1970.

——. *The Rise of Modern Urban Planning, 1800–1914.* New York: St. Martin's, 1980.

——. *Towards the Planned City: Germany, Britain, the United States, and France, 1780–1914.* New York: St. Martin's, 1981.

TANTNER, ANTON. *Addressing the Houses: The Introduction of House Numbering in Europe.* Paris: Éditions de l'EHESS, 2009.

TAXIL, LÉO. *Recueil d'actes administratifs et de conventions relatifs aux servitudes spéciales d'architecture.* Paris: Imprimerie nouvelle, 1905.

TAYLOR, KATHERINE FISCHER. *In the Theater of Criminal Justice: The Palais de Justice in Second Empire Paris.* Princeton: Princeton University Press, 1993.

TEXIER, SIMON, ED. *Les places et jardins dans l'urbanisme parisien, XIXe–XXe siècle.* Paris: Action artistique de la Ville de Paris, 2001.

THOILLET, FRANÇOIS. *L'art de lever les plans, arpentage, nivellement et lavis des plans.* Paris: Audin, 1834.

THOMAS, KATE LLOYD. "Specifications: Writing Materials in Architecture and Philosophy." *Arq* 8, nos. 3/4 (2004): 277–83.

THOMPSON, VICTORIA. "Knowing Paris: Changing Approaches to Describing the Enlightenment City." *Journal of Urban History* 37, no. 1 (2011): 28–42.

TOOLEY, ROBERT V. *Maps and Mapmakers.* London: Batsford, 1952.

TOPALOV, CHRISTIAN. "Cities of the Social Sciences: Seeing for Doing." In *Representing London,* edited by Martin Zerlang, 14–29. Copenhagen: Forlaget Spring, 2001.

——. *La ville des sciences sociales.* Paris: Belin, 1999.

——. "La ville 'terre inconnue,' l'enquête de Charles Booth et le peuple de Londres, 1886–1891." *Genèses*, no. 5 (1991): 5–34.

——. *Les divisions de la ville*. Paris: Éditions de la Maison des sciences de l'homme, 2002.

——, ed. *L'aventure des mots de la ville*. Paris: R. Laffont, 2010.

TUFTE, EDWARD R. *Envisioning Information*. Cheshire, CT: Graphics, 1990.

——. *The Visual Display of Quantitative Information*. Cheshire, CT: Graphics, 2001.

TULARD, JEAN. *Paris et son administration (1800–1830)*. Paris: Ville de Paris, Commission des travaux historiques, 1976.

TWYMAN, MICHAEL. *Lithography, 1800–1850: The Techniques of Drawing on Stone in England and France and Their Application in Works of Topography*. London: Oxford University Press, 1970.

VALANCE, GEORGES. *Haussmann le grand*. Paris: Flammarion, 2000.

VAN DEN BROECKE, MARCEL, PETER VAN DER KROGT, AND PETER MEURER, EDS. *Abraham Ortelius and the First Atlas: Essays Commemorating the Quadricentennial of His Death, 1598–1998*. Houten, Neth.: HES, 1998.

VAN LOO, ANNE. "L'Haussmannisation de Bruxelles: La construction des boulevards du centre, 1865–1880." *Revue de l'art* 106 (1994): 39–49.

VAN ZANTEN, DAVID. *Building Paris: Architectural Institutions and the Transformations of the French Capital, 1830–1870*. Cambridge: Cambridge University Press, 1994.

——. *Designing Paris: The Architecture of Duban, Labrouste, Duc, and Vaudoyer*. Cambridge, MA: MIT Press, 1987.

——. "Nineteenth-Century French Government Architectural Services and the Design of the Monuments of Paris." "Nineteenth-Century French Art Institutions." Special issue, *Art Journal* 48, no. 1 (Spring, 1989): 16–22.

——. "Paris Space: What Might Have Constituted Haussmannization." In *Manifestoes and Transformations in the Early Modernist City*, edited by Christian Hermansen Cordua, 179–210. Surrey: Ashgate, 2010.

VARDI, LIANA. *The Physiocrats and the World of the Enlightenment*. Cambridge: Cambridge University Press, 2012.

VELTMAN, KIM H. "Military Surveying and Topography: The Practical Dimension of Renaissance Linear Perspective." *Revista da Universidade de Coimbra* 27 (1979): 263–79.

VERGNEAULT-BELMONT, FRANÇOISE. "Espace et société: Un régard de cartographie sur les plans anciens de Paris." *história, histórias* 1, no. 2 (2013): 35–84.

VÉRIN, HÉLÈNE. *La gloire des ingénieurs: L'intelligence technique du XVIe au XVIIIe siècle*. Paris: Albin Michel, 1993.

VERNER, COOLIE. "Copperplate Printing." In *Five Centuries of Map Printing*, edited by David Woodward, 51–75. Chicago: University of Chicago Press, 1975.

VERNES, M. "Au jardin comme à ville 1855–1914, le style municipal." In *Parcs et promenades de Paris, 15–20*. Paris: Éditions du demi-cercle/Pavillon de l'arsenal, 1988.

VERNIÈRE, MARC. "Les oubliés de l'"Haussmannisation' Dakaroise: Crise du logement populaire et exploitation rationnelle des locataires." *L'espace géographique: Régions, environnement, aménagement* 6, no. 1 (1977): 5–23.

VIDLER, ANTHONY. *The Architectural Uncanny: Essays in the Modern Unhomely*. Cambridge, MA: MIT Press, 1992.

——. "The Scenes of the Street: Transformations in Ideal and Reality, 1750–1871." In *On Streets*, edited by Stanford Anderson, 29–111. Cambridge, MA: MIT Press, 1978.

——. *The Scenes of the Street and Other Essays*. New York: Monacelli, 2011.

VINEGAR, ARON. "Architecture under the Knife: Viollet-le-Duc's Illustrations for the 'Dictionnaire Raisonné' and the Anatomical Representation of Architectural Knowledge." PhD diss., McGill University, 1995.

VISMANN, CORNELIA. *Files: Law and Media Technology*. Translated by Geoffrey Winthrop-Young. Stanford, CA: Stanford University Press, 2008.

VOLAIT, MERCEDES. "Making Cairo Modern (1870–1950): Multiple Models for a 'European Style' Urbanism." In *Urbanism. Imported or Exported?*, edited by Joe Nasr and Mercedes Volait, 17–51. Chichester, UK: Wiley, 2003.

VON SCHÖNING, ANTONIA. *Die Administration der Dinge. Technik und Imagination im Paris des 19. Jahrhunderts*. Zurich: diaphanes, 2018.

WAKEMAN, ROSEMARY. "Nostalgic Modernism and the Invention of Paris in the Twentieth Century." *French Historical Studies* 27, no. 1 (2004): 115–44.

WEBER, EUGEN. "L'Hexagone." In *Les lieux de mémoire*, vol. 1, *La République*, edited by Pierre Nora, 1171–90. Paris: Gallimard, 1984.

——. *Peasants into Frenchmen: The Modernization of Rural France, 1870–1914*. Stanford, CA: Stanford University Press, 1976.

WEBER, MAX. *Economy and Society: An Outline of Interpretive Sociology*. Edited by Guenther Roth and Claus Wittisch. Berkeley: University of California Press, 1978.

WEBER, SAMUEL. *Benjamin's -abilities*. Cambridge, MA: Harvard University Press, 2008.

——. "Saussure and the Apparition of Language: The Critical Perspective." *MLN* 91, no. 5 (October 1976): 913–38.

WEISS, SEAN. "Engineering, Photography, and the Construction of Modern Paris, 1857–1911." PhD diss., Graduate Center, City University of New York, 2013.

——. "Making Engineering Visible: Photography and the Politics of Potable Water in Modern Paris." *Technology and Culture* 61, no. 3 (July 2020): 740–71.

WERCKMEISTER, O. K. "A Critique of T. J. Clark's *Farewell to an Idea*." *Critical Inquiry* 28, no. 4 (Summer 2002): 855–67.

WHITEMAN, JOHN, JEFFREY KIPNIS, AND RICHARD BURDETT, EDS. *Strategies in Architectural Thinking*. Cambridge, MA: MIT Press, 1992.

WIDMALM, SVEN. "Accuracy, Rhetoric, and Technology: The Paris-Greenwich Triangulation, 1784–88." In *The Quantifying Spirit in the Eighteenth Century*, edited by Tore Frängsmyr, J. L. Heilbron, and Robin E. Rider, 179–206. Berkeley: University of California Press, 1990.

WILFORD, JOHN NOBLE. *The Mapmakers*. New York: Knopf, 1981.

WILLIAMS, HANNAH. *Académie Royale: A History in Portraits*. Farnham, UK: Ashgate, 2015.

WILLIAMSON, J. H. "The Grid: History, Use, and Meaning." *Design Issues* 3, no. 2 (Autumn 1986): 15–30.

WINEARLS, JOAN, ED. *Editing Early and Historical Atlases*. Toronto: University of Toronto, 1995.

WISE, M. NORTON, ED. *The Values of Precision*. Princeton: Princeton University Press, 1995.

WITTMAN, RICHARD. *Architecture, Print Culture, and the Public Sphere in Eighteenth-Century France*. New York: Routledge, 2007.

WOLFE, MICHAEL. "Building a Bastion in Early Modern History." *Proceedings of the Western Society of French History* 25 (1998): 36–48.

WOOD, DENIS. "How Maps Work." *Cartographica* 29, nos. 3/4 (Autumn/Winter 1992): 66–74.

——. "Introducing the Cartography of Reality." In *Humanistic Geography: Prospects and Problems*, edited by David Ley and Marwyn Samuels, 207–19. Chicago: Maaroufa, 1978.

——. "Pleasure in the Idea/The Atlas as Narrative Form." *Cartographica* 24, no. 1 (Spring 1987): 24–46.

———. *The Power of Maps.* New York: Guilford, 1992.

WOOD, DENIS, WITH JOHN FELS AND JOHN KRYGIER. *Rethinking the Power of Maps.* New York: Guilford, 2010.

WOODWARD, DAVID, ED. *Five Centuries of Map Printing.* Chicago: University of Chicago Press, 1975.

———. *The History of Cartography,* vol. 3, pt. 1, *Cartography in the European Renaissance.* Chicago: University of Chicago Press, 2007.

———. "Maps and the Rationalization of Geographic Space." In *Circa 1492: Art in the Age of Exploration,* edited by Jay Levenson, 83–88. New Haven: Yale University Press, 1991.

———. "The Study of the History of Cartography: A Suggested Framework." *American Cartographer* 1 (1974): 101–15.

WYNGAARD, AMY. "Libertine Space: Anonymous Crowds, Secret Chambers, and Urban Corruption in Rétif de la Bretonne." *Eighteenth-Century Life* 22, no. 2 (1998): 104–22.

YATES, ALEXIA. "Developing Knowledge, the Knowledge of Development: Real Estate Speculators and Brokers in Late Nineteenth Century Paris." *Business and Economic History On-Line* 9 (2011), https://thebhc.org/sites/default/files/yates.pdf.

———. *Selling Paris: Property and Commercial Culture in the Fin-de-Siècle Capital.* Cambridge, MA: Harvard University Press, 2015.

ZUPKO, RONALD EDWARD. *French Weights and Measures before the Revolution: A Dictionary of Provincial and Local Units.* Bloomington: Indiana University Press, 1978.

———. "Itinerary and Geographical Measures." In *The History of Cartography,* vol. 4, *Cartography in the European Enlightenment,* edited by Matthew H. Edney and Mary Sponberg Pedley, 927–28. Chicago: University of Chicago Press, 2019.

ECHELE de 120 toises

5 10 20 30 40 50 60 70 80 90 100 110 120

Méridienne à 400 Toises

R. Trop va qui dure Ruë

Quai de la Megisserie

la Samaritaine

PONT AU CHANGE

Quai de l'Horloge

Tournelle grande Chambre

Morfondus Président Conciergerie grande Salle du Palais

Quai des S. Barthelemi

cour de Lamoignon Ruë St Barthelemi

Salle neuve Salle Dauphine

Henri IV Ruë de Harlay traverse quai des Prisonniers Salle aux Merciers

PLACE DAUPHINE cour neuve du 1er S. Et

Quai des Hôtel grande St Chapelle Barnabites

PONT chambre des comptes Cour du Palais

Orfeures fontaine S. Michel+ Dauphine

Ruë St Louis PONT St MICHEL boucher

Quai des Augustins dit la Vallée Ruë de Hurepoix M

PLAN DETAILLÉ
DE LA CITÉ

Dedié à Messire Louis Basile DE BERNAGE

Conseiller d'Etat Prevôt des Marchands

et a Messieurs les Echevins de la Ville de Paris

Par M. l'Abbé DELAGRIVE Geographe ordinaire de la Ville.

1754.

35423

Bourgoin le jeune Sculpsit

Index

Note: Page numbers in *italic* type indicate illustrations.

Académie des Beaux-Arts, 3
Académie royale des sciences, 7, 15, 45, 81
accuracy: in cartographic representation, 1, 4–5, 15, 19, 21, 22, 29–30, 35–36, 39–41, 59, 61, 63, 118, 123, 154; meanings of, 3, 14, 30, 61; as a value, 68. *See also* maps and plans: fidelity of; precision
administration. *See* bureaucracy
Adorno, Theodor, 151
Agache, Donat Alfred, *Comment reconstruire nos cités détruites* (How to rebuild our destroyed cities), 153–54
Agrandissement et construction des Halles Centrales d'approvisionnement (Enlargement and construction of the central markets for provisioning), 90, *91*
Agulhon, Maurice, 135
Akerman, James, 166n46, 166n47
Alberti, Leon Battista, 112
Alder, Ken, 40, 113
Alembert, Jean le Rond d'. See *Encyclopédie*
Algeria, 84
alignment, in urban plans, 22, 33, 53, 87, *88*, *93*, 94, *95*, 97, 102, 103, 154; in regulatory and legal rules, 107–9
Alliès, Paul, 114
Alphand, Adolphe, 77, 109, 120–23, 129, 144, 147–148, 150–51, 166n38, 166n77; *Atlas des anciens plans de Paris* (Atlas of old plans of Paris), *126* (detail), 128–30, *131–34*, 132–35, 150; *Atlas des travaux de Paris, 1789–1889* (Atlas of the works of Paris, 1789–1889), 77, 79, 80–81, *80*, *81*, 122–23, *124–25*, 128–29, 137, *138–43*, 139–47, *146*, 150, *152* (detail), *156* (detail), *168* (detail); *Les promenades de Paris, 104* (detail), 121–23, *122*, 147, *148–49*, 150
Alsace-Lorraine, 144, 147, 151
Amelot, Antoine-Jean, 76
ancien régime, 31, 43, 66–67

architecture: bureaucracy and, 117–18; descriptive geometry and, 74; drawing's role in, 74; engineering vs., 4, 74; plan as basis of, 4, 7, 25–26, 76, 107–9; principles and practices of, 73–74; standardization in, 117; urbanism in relation to, 7
Argenson, Marc-René d', 132
Arnaud, Jean-Luc, 159n35
arpenteurs (land surveyors), 43
Assemblée constituante, 111
Assemblée nationale, 130
atlases, 5, 128. *See also specific atlases*
Aubertin, Jean, *Comment reconstruire nos cités détruites* (How to rebuild our destroyed cities), 153–54

Baldus, Édouard, 159n50; *Plan for the Nouveau Louvre by L. Visconti*, from *Du Louvre et des Tuileries*, *27*; *Plan of the Louvre and Its Surrounding around 1830 by Charles Vasserot*, from *Du Louvre et des Tuileries*, *26*
Ballon, Hilary, 18, 30, 112
Baltard, Victor, 85–87; *Monographie des Halles Centrales de Paris*, 90, *91*
Balzac, Honoré de, 26
Barbari, Jacopo de', 158n25
Barnout, H., 162n71
Bastille, 79
Bâtiments du roi, 39, 117
Baudelaire, Charles, 7
Beauregard, Robert, 118
Béguillet, Edme, *Description historique de Paris*, 42–43
Bellanger, J. S., 66, 94, 161n7
Bender, John, 21
Benevolo, Leonardo, 12
Benjamin, Walter, 1, 7, 103, 151
Berger, M., 150
Bernard, Frédéric, 129
Berthelin, Max, 162n71

Bertin, Henri-Léonard-Jean-Baptiste, 110, 111
Berty, Adolphe, 130, 132
Bibliothèque du roi, 47, 115
Bibliothèque nationale de France, 130, 144
bird's-eye views, 18–20, 113, 158n25, 164n70
Blancard, Hippolyte: *Hôtel de Ville et le pont d'Arcole, 125*; *Universal Exposition of 1889: The Pavilion for the Préfecture of the Seine, City of Paris, 136*
Bloch, Maurice, 109
blocks: creation, revision, and destruction of, 79–80, 90, 108–9; plans and maps of, 3, 23, 34, 53, 84, 91, 93–94, 96, 103, 109
Blondel, François, 39, 75; *Plan de Paris*, ii–iii (detail), *38*, 39, *39* (detail)
Bois de Boulogne, 121, 139, 141, 144, 147
Bois de Vincennes, 121, 139, 141, 147
Bonnardot, Alfred, 135
Borda, Jean-Charles de, 161n9
Bordes, Philippe, 16
Borges, Jorge Luis, "On Exactitude in Science," 29, 61, 155
Boullée, Étienne-Louis, 74
Bouvard, Joseph-Antoine, 136
Broc, Numa, 145
Buffon, Georges-Louis Leclerc, comte de, 33
Bullet, Pierre, 39; *Plan de Paris*, *38*, 39, *39* (detail)
bureaucracy, 105–23; and architecture, 117–18; documents' role in, 1, 5, 23, 105–7, 109, 114, 118–20, 128, 129–30; emergence and growth of, 3, 23, 105–7, 109; and historiography, 129–30, 135–44; limitations of, 113–14; maps' and plans' value for, 1, 7–8, 17, 23, 30, 103, 105–23; and modernity, 5, 24, 106; and objectivity, 119–20; power and authority of, 112, 117; quantification as component of, 3–4, 107, 113, 115–16, 123, 127; standardization as component of, 107, 111–12, 118–20, *118*, *119*, 123; totalizing goals of, 113–14; urban planning and, 107–23. *See also* governance; governmentality

Bureau de géomètre du cadastre, 117
Bureau de la voirie (Office of the roadways), 117
Butte des Moulins, 120, *120*, *121*

cadastral maps, 33–34, 48, 66, 94, 110–13, 117, 161n7
cadastre napoléonien, 112
Cahier classique sur le cours de construction à l'usage des élèves de l'École royale de l'artillerie et du génie (Classic workbook on the course of construction for the students of the Royal School of Artillery and Engineering), 74–75
Callet, Charles-François, 76
Callet, Félix-Emmanuel, *Monographie des Halles Centrales de Paris*, 90, *91*
Callou, G., *Études d'un nouveau système d'alignemens et de percemens de voies publiques faites en 1840 et 1841* (Studies for a new system of alignments and openings for public roadways built from 1840 to 1841), 102
Cambault (architect), 77
"Canevas trigonométrique," from Deschamps's *Plan général de la Ville de Paris*, viii (detail), *28* (detail), 59, *61*
Carroll, Lewis, 29, 61
Carruthers, Mary, 106
Carte de France, 32, 49, 51, 77
Carte de l'Égypte, 49, 51
cartography: in education curricula, 4, 145–47; fidelity of, 29–31; grids used in, 63–64, *64*; ideological factors in, 5; labor involved in, 43; photoengraving technology for, 151; political uses of, 3, 128; quantification in, 20, 30, 40, 51, 64, 113; standardization of, 108. *See also* geography; maps and plans
cartouches, 4, 30, 35–36, 39, 40, 59, 68, 140, 154
Casselle, Pierre, 3
Cassini, Jacques, Paris meridian, 31, *31*
Cassini, Jean-Dominique, 32; Paris meridian, 31, *31*
Cassini de Thury, César-François, 32
Cassini family, 49, 143
Castex, Jean, 94
census of Paris (1821–1860), 115–16
Certeau, Michel de, 106
Chabouillé, Marie-Jean, 77
Chabrol de Volvic, Gilbert-Joseph-Gaspard, comte de, 22, 47, 65, 84, 108, 115–16, 144, 161n7
Champeaux, Alfred de, 129
Charles X, 109
circulation: comprehensive plans and, 102; embellissements and, 75, 79; grid as aid in, 84; military mapping and, 22; monuments in relation to, 97; urban planning centered on, 6, 75, 84, 90, 97, 103. *See also* streets
Clark, T. J., 64–65
Claval, Paul, 145
Colbert, Jean-Baptiste, 15, 105, 107, 143
colonialism: during and after World War I, 154; Exposition universelle (1889) and, 144–45, 166n56; and national unity, 128; and surveys, 47–51, 53, 144

commissaires de la voirie (road commissioners), 33–34, 159n16
Commission de la Description de l'Égypte, 48
Commission des artistes, 76–77, 81, 84, 162n45
Commission des embellissements de Paris (Commission for the improvements of Paris), 2–3, 22, 27, 84, 102–3
Commission des Halles, *Document à étudier, no. 2. Plan de la Nouvelle Halle*, 85, *85*
Commission des recherches sur l'histoire de Paris (Research commission on the history of Paris), 129, 135
Commission municipale des travaux historiques (Municipal commission for historical work), 129, 135
Comte, Barbara Shapiro, 107
Condillac, Étienne Bonnot de, 74
Condorcet, Marie-Jean-Antoine-Nicolas de Caritat, marquis de, 67, 161n9, 163n19
Conseil des bâtiments civils (Council for civic buildings), 3, 77, 105, 117, 123
Conseil général de la Seine, 22, 102
Conseil général des ponts et chaussées (Council of bridges and roadways), 77, 109, 148, 164n79
Conseil municipal de Paris, 23, 85–86, 103, 129, 135–36, 159n45
Cortada, James, 163n3
Cousin, Jules, 129
Crampton, Jeremy, 157n13
Crary, Jonathan, 5–6

Daffry de la Monnoye, Léon, 109
d'Alembert, Jean le Rond. *See Encyclopédie*
Daston, Lorraine, 5, 142, 160n66
Davenne, Henri-Jean-Baptiste, 108
David, Jacques-Louis, *The Emperor Napoléon in His Study at the Tuileries*, 16–17, *17*, 23
Davioud, Gabriel, 97
Debret, François, 162n71
Degas, Edgar, 65
Delagrive, Jean, 31, 33–34, 36, 159n14; *Manuel de trigonométrie pratique*, 40; *Plan détaillé de la Cité* (Plan Detailing the City) 33, *34*, *186* (detail); *Plan détaillé du quartier de Sainte-Geneviève* (Plan Detailing the Neighborhood of Sainte-Geneviève), 33, *33*
Delamare, Nicolas, *Traité de la police* (Treatise of the police), 130, *131*, 132
Delambre, Jean-Baptiste-Joseph, 67, 143
Deleuze, Gilles, 123
Delort de Gléon, Alphonse, 145
Dépôt de la Guerre, 43, 48, 161n81
Derrida, Jacques, 5
Descartes, René, 74
Deschamps, Eugène, 7, 54, 84, 117; *Atlas administratif des 20 arrondissements de la Ville de Paris*, 52–53, *53*, 59, *60*; *Plan général de la Ville de Paris et de ses environs comprenant les bois de Bologne et de Vincennes*, 59, *61*
descriptive geometry, 74, 162n34

diagrammatic culture, 23, 113
Diderot, Denis. *See Encyclopédie*
Direction des travaux de Paris, 117
Direction du service d'architecture, 117
dispositif (apparatus), 6
Division des travaux publics municipaux (Division of public municipal works), 116
documents, in modern bureaucracies, 1, 5, 23, 105–7, 109, 114, 118–20, 128, 129–30
Doyenné neighborhood, 26
Dupain de Montesson, Louis-Charles, *La science de l'arpenteur dans toute son étendue* (The science of surveying in all of its scope), 42, *42*, 45
Durand, Jean-Nicolas-Louis, 74–75, 76, 80, 84, 147; *Précis des leçons d'architecture données à l'École polytechnique* (Summary of the architectural lessons delivered at the École polytechnique), 76, 78, 80, *82–83*
Dürer, Albrecht, *Draughtsman Making a Perspective Drawing of a Reclining Woman*, 63, *64*

École des Beaux-Arts, 3, 4
École des géographes du cadastre, 111
École des ingenieurs-géographes, 43
École des ponts et chaussées (School of bridges and roadways), 4, 84, 111
École polytechnique, 4, 47, 59, 74, 75, 84, 115
educational reform, 145–47
Egypt, 7, 22, 47–51, 67, 115, 145, 160n73, 160n74
Elden, Stuart, 6, 157n13
elevations, 55, 57, 59, 68, 75–76
embellissements (improvements), 75–76, 79, 84, 101, 108
eminent domain, 91, 108–9
empiricism, 14, 18, 22, 40, 41, 45, 74
Encyclopédie (Diderot and d'Alembert), 3, 14, 21, *44*, 45, 67, 74
engineering, 4, 74
Enlightenment, 3, 5, 8, 40, 45, 74
epistemological value of maps and documents, 2, 4, 7, 17–18, 20–22, 30, 36, 40, 51, 113–14, 154
Eugénie (empress), 24–25
Evans, Robin, 4
Exposition coloniale internationale (Paris, 1931), 144
Expositions universelles, 6, 128, 135, 150
Exposition universelle (Paris, 1867), 150
Exposition universelle (Paris, 1878), 8, 77, 128–30, 147, 150
Exposition universelle (Paris, 1889), x (detail), 8, 50–51, 122, 128, 135–37, *135–38*, 144, 146–47, 150

Fauve, L., 129
Fer, Nicolas de, 130
Ferry, Jules, 145
Fizeau, Hippolyte-Louis, 167n88
Flandrin, Jean-Hippolyte, *Napoléon III, in Uniform as Division General in His Official Office at the*

Tuileries, 16, *16*, *16* (detail), 158n20
Fleury, François-Louis Joly de, 34
Fontaine, P.-F.-L., 163n76
fortification plans, 21
Foucault, Michel, 5, 6, 112, 136
Fouillée, Augustine, *Le tour de France par deux enfants* (The tour of France by two children), 147
Fourier, Jean-Baptiste-Joseph, 116
Franco-Prussian War, 12, 127, 145, 153
Franklin, Alfred, 130
French Revolution, 65, 77, 79, 122, 129, 130, 135, 139, 144, 165n18
Friedman, David, 18
Friedrich Wilhelm of Prussia, 12
Frisius, Gemma, 159n9

Gaëte, Martin-Michel-Charles Gaudin, duc de, 112
Galimard, George, 77
Galison, Peter, 5, 142
Gambetta, Léon, 129
gardens, 42
Garnier, Charles, Opéra, 26, 93, 162n74
Garrez, Pierre, 76–77
Gautier, Théophile, 26, 158n20
Gazette municipale, 102
géographes de cabinet (office geographer), 43
geography, 4, 5, 51, 74, 128, 145–47. *See also* cartography
geometry: architectural profession and, 74; authority ascribed to, 3, 59, 113–15, 134, 164n73; as basis of maps' fidelity, 30; descriptive, 74, 162n34; in education curricula, 4; governance role of, 117; and orthographic plans, 21, 74, 113; as underlying logic of spatial representation, 2–4, 20; urban planning based on, 23. *See also* quantification; science
Gibrat, Joseph, *Traité de la géographie moderne*, 145
Gitelman, Lisa, 106, 113–14, 119
Godlewska, Anne Marie Claire, 5, 48, 49, 160n74
Gomboust, Jacques, *Lutetia Paris*, iv (detail), 19, *19*, 37, *37*, 160n37, 164n70
governance: documents' role in, 5; Exposition universelle (1889) as representation of, 136–37; geometry's role in, 117; maps' role in, 8, 27, 29–30, 84, 127; orthographic plans' role in, 13, 112–14. *See also* bureaucracy; governmentality
governmentality, 6, 112, 127, 136. *See also* bureaucracy; governance
Grand Plan de Jouvin de Rochefort, 133
graph paper, 74–75
grids: cartographic use of, 63–64, *64*, 68; design function of, 73–81, 84, 87, 90, 93–94, 96–97; drawn on top of maps, *64*, 87; Haussmann and, 84; importance of, 7, 64; meanings associated with, 64; modernity associated with, 64–65, 84, 103; objectivity associated with, 63; in Paris, 84, 103; planet conceived according to, 31–32, 73; projection made possible by, 76; space

organized by means of, 75, 76; as totalizing systems, 65; in urban planning, 84, 97; uses of, 63
Grillon, Edme-Jean-Louis, 85, 162n71; *Études d'un nouveau système d'alignemens et de percemens de voies publiques faites en 1840 et 1841* (Studies for a new system of alignments and openings for public roadways built from 1840 to 1841), 102
Gross, David, 143
Guattari, Félix, 123
guidebooks, 42
Guillory, John, 106

Hacking, Ian, 157n21
Harley, J. B., 5, 7, 17–18
Harouel, Jean-Louis, 163n25
Harvey, David, 12
Haussmann, Georges-Eugène: Alphand's work under, 148, 150; and bureaucracy, 117, 147; demolitions ordered by, 24–25, 26; engineering emphasized by, 4; and the grid plan, 84; legacy of, 11, 99, 116, 147–48, 154; and maps/plans, 21, 30, 53, 84, 102–3, 116, 117; painting of, 23, *24*; and Place d l'Opéra, 91; as prefect of the Département du Seine, 2, 11, 27, 84; role of, in Paris's modernization, 1–2, 7, 11–12, 21, 26–27, 53, 102–3, 109, 148, 150, 166n79; and surveys, 53–54, 59; and synoptic perspectives, 116, 117
Haussmannization, 11, 148
Haüy, René-Just, *Instruction sur les mesures déduites de la grandeur de la terre, uniformes pour toute la République, et sur les calculs relatifs à leur division décimale* (Instruction on measures deduced from the size of the earth, uniform for the entire Republic, and on calculations relative to the decimal point, 68, 72, *72*
Henri IV, 26, 154
hexagon form, 147, 151
Histoire de Paris, 130
historiography: bureaucracy and, 129–30, 135–44; and constructing an image of Paris, 130–35
Hobsbawm, Eric, 165n34
Hochereau, Émile, 129
Horeau, Hector, 87, 162n71
Hôtel de Ville, 1, 12, 23, 27, 123, *125*, 127–28, 144
house numbering, 34, 159n25, 159n27
Houssaye, Arsène, 11–12, 14
Hugnin, Alexandre-François, 33; *Plan détaillé du quartier de Sainte-Geneviève* (Plan Detailing the Neighborhood of Sainte-Geneviève), 33, *33*; *Plan détaillé de la Cité* (Plan Detailing the City) 33, *34*, *186* (detail)

image-text relationship: decoupling of, 110, 113; images deemed superior, 23, 101–2, 110–16, 129; in tax matters, 110–12; texts deemed superior, 106, 164n47. *See also* vision/visuality
imperialism. *See* colonialism
Indochina, 84
ingénieurs-géographes (engineer-surveyors), 21, 113, 160n73, 164n79

interchangeability. *See* universality/neutrality/regularity/interchangeability

Jacotin, Pierre, 48
Jacoubet, Théodore, 65–66, 85–87, 102, 103, 144; *Atlas général de la ville, des faubourgs et des monuments de Paris*, *64*, 65–66, *65*, 68, *70–71*, *72–73*, *73*, 84, *86–90*, 87, 90, *92–96*, 94, 96–97, 163n76; *Études d'un nouveau système d'alignemens et de percemens de voies publiques faites en 1840 et 1841* (Studies for a new system of alignments and openings for public roadways built from 1840 to 1841), 102
Jaillot, Alexis-Hubert, 94
Jardin du roi (now Jardin des Plantes), 33
Jollain, Nicolas-René, le Vieux, *Louis XIV Holding a Plan of the Maison Royale de Saint-Cyr*, 16–17, *17*, 25
Jomard, Edme-François, 47–51, 53, 74, 115, 130; *Description de l'Égypte*, 47–51, *49–51*, 115, 144, 165n92; *Mémoire sur la construction de la Carte de l'Égypte*, 51; *Mémoire sur le système métrique*, 51
Journal des géomètres, 34, 120

Kafka, Ben, 120
knowledge. *See* epistemological value of maps and documents
Konvitz, Josef, 67
Korzybski, Alfred, 11
Kula, Witold, 66, 67

Labrouste, Henri, 54
La Caille, Jean de, *Plan de La Caille*, 132, *132*
Lacaille, Nicolas-Louis de, 32
Lagrange, Joseph-Louis, 161n9
Lahure (architect), 85, 162n63
Lalande, Joseph-Jérôme, 41, 135, 160n74
Lameyre, Gérard-Noël, 12
Lancelot, 29
Laplace, Pierre-Simon, 161n9
La Poix de Fréminville, Claude Edme de, 110
Larrère, Catherine, 114–15
Laugier, Marc-Antoine, 75, 102
Lavedan, Pierre, *Histoire de l'urbanisme*, 12, 63
law, and urban planning and administration, 108–12, 123
Lazare, Louis, 22, 65, 102
Le Bovier de Fontenelle, Bernard, 164n73
Lefuel, Hector-Martin, 25–26, 96
Legrand, H., 130
Leonardo da Vinci, 112
Le Prestre de Vauban, Sébastien, 21, 66, 160n54
Le Roux de Liney, A., 130
Les Halles Centrales, 84–90
leveling, 59, 68
Louis XIV, 15–17, *15*, *17*, 21, 25, 39, 42, 43, 105, 107, 112, 154
Louis XV, 75, 159n27
Louis XVI, 34, 163n20

Louis Philippe, 157n3, 162n17, 162n26, 163n19
Louvre, 24, 26, *26*, 79–80, 96, 100
Loyer, François, 96
Lynch, Michael, 35
Lyotard, Jean-François, 134

Mallet, Philippe, 42
Manesson-Mallet, Alain, *La géométrie pratique*, 42, *45, 45, 46, 66*
Manet, Édouard, 65
mapping: academic approaches to, 4, 30; cadastral, 33; in Egypt, 47–50; as expression of governmentality, 6, 111–13; of history, 128–35; ideological factors in, 5; and intermediality of architecture and urbanism, 7; military, 21–22; modernization of Paris shaped by, 8; scientific approaches to, 13, 51; unity as theme of, 128–29
maps and plans: administrative functions of, 1, 7–8, 17, 23, 30, 103, 105–23; architecture based on, 4, 7, 25–26, 76, 107–9; audiences for/users of, 5, 116, 133; color in, 108; commodification of space by, 23; diagrammatic character of, 3, 21, 23, 37, 40, 113; drawing on, 64, 79–80, 87, 90, 93–94, 96, 101, 103, 107–8; epistemological value of, 2, 4, 7, 17–18, 20–22, 30, 36, 40, 51, 106, 113–14, 154; in Exposition universelle (1889), 136–37; fidelity of, 39–40, 59, 61, 63, 151, 154 (*see also* accuracy; precision); historical situatedness of, 7, 8; institutional/discursive underpinning of, 3–8, 118; as instruments, 20, 22, 54; and legal issues, 108–12; mediation in creation and implementation of, 3, 12; military, 21–22, 153; modernization of Paris associated with, 1–3, 7, 8, 11–14, 53; Napoléon III's original Paris plan, 1–2, 11–12, *13*, 27, 151, 155; natural or given appearance of, 2, 5, 8, 51, 103, 115–17, 142, 144; nomenclature for, 159n35; objectivity ascribed to, 1, 3, 5–6, 27, 30, 35–36, 40, 61, 116, 123, 155; Paris shaped by, 61, 103, 107–8, 110, 117, 129, 155; personal and sociopolitical agency concealed by, 1–2, 4–5, 8, 17–18, 27, 35, 41, 44, 51, 61, 116–17, 122, 142; physical world in relation to, 29–30; planimetric, 3, 19–20, 109, 116, 164n70; popularization of, 147, 150–51; as portraits, 18–19; portraits of rulers featuring, 14–17; positivist conceptions of, 5, 13, 128, 130; power and authority attributed to, 1–3, 8, 14, 15, 17, 22, 27, 29–31, 35–36, 39–40, 59, 116, 118, 153, 155; projective vs. descriptive functions of, 12, 14, 18, 21, 25–26, 53–54, 59, 61, 63, 76, 79–81, 103, 113, 117, 155; public vs. private interests indicated on, 108–9; quantification as central to, 20, 30, 40, 51, 64, 113, 127; relations between, 43, 72–73; relief maps, 21, 22, 136; science linked to, 17, 30, 40, 74, 123; standardization of, 108, 111–12, 120–23; surveys compared to, 43; symbolic conceptions of, 13; as totalizing/synoptic systems, 3, 12, 18, 21, 47, 51, 53, 101, 113, 116, 117, 129, 141–42, 150; urban planning shaped by, 7, 63, 84, 93–94, 96, 103, 109, 116, 117, 153–54. *See also* bird's-eye

views; cadastral maps; cartography; master plans; orthographic plans; pictorial maps; surveys; universality/neutrality/regularity/interchangeability, as goals/outcomes of modern systems of representation and measurement; urbanism and urban planning
Marin, Louis, 127, 160n37
Marrinan, Michael, 21
Martignac, Jean-Baptiste Sylvère Gay, vicomte de, 109
Marville, Charles, 97, 129; *Boulevard Saint-Michel, vers le sud* (Boulevard Saint-Michel, toward the south [from Place Saint-Michel]), 97, *97*, 100; *Carrefour Sainte-Opportune (de la rue des Halles)*, 101, *101*; *Les piliers des Halles, rue de la Tonnellerie*, 98–99, *99*; *Place Edmond-Rostand du boulevard Saint-Michel*, 97, *98*; *Rue de la Lingerie*, 100, *100*; *Rue de la Tonnellerie (de la rue de la Poterie)*, 98, *98*; *Rue de Rivoli, 62* (detail), 99–100, *99*
Marx, Karl, 115
master plans, 7, 13–14, 27, 154
Mazzarella, William, 117
McQueen, Alison, 25
Mead, Christopher, 90
measurement systems, 66–68, 72, 107, 161n9, 162n17, 162n26, 163n19. *See also* space, rationalization and quantification of; universality/neutrality/regularity/interchangeability, as goals/outcomes of modern systems of representation and measurement
Méchain, Pierre-François-André, 67, 143
mediation, 3, 12–13, 76, 117, 122
Menou, Jacques-François, 48
Mérian, Mathieu, *Plan de Paris sous Louis XIII, xiv* (detail), *9* (detail), 18, *18*
meridians, 31–32, 67. *See also* Paris meridian
Merruau, Charles, 1, 12, 14; plan of Paris indicating traces of the new streets initiated by Emperor Napoléon III, 12, *13*
metric system, 63–64, 66–68, 72–73, 77, 107, 108, 111, 161n9, 162n17, 162n26
Meynadier, Hippolyte, 101–2
Michaux, L., 129
Michelet, Jules, 129
military mapping, 21–22, 153
Ministère de l'intérieur, 66, 111, 117
Ministère des travaux publics, 109, 148, 166n38
Mirabeau, Victor Riqueti, marquis de, 114
Mitchell, Timothy, 50
modernity: Benjamin on, 7; bureaucracy linked to, 5, 24, 106; estrangement as feature of, 103; grids associated with, 64–65, 84, 103; objectivity as value of, 65; Paris associated with, 1–3, 65, 103, 155; visuality in, 5–6, 12–13
Monde illustré (newspaper), 91, *92*
Monge, Gaspard, 74, 160n74, 161n9; *Géométrie*, 74
Moniteur universel (newspaper), 92, *92*
Monographie de l'Exposition universelle de 1889, 150
monuments: circulation in relation to, 97; depictions of, 18, 19, 30, 39; urban planning

centered on, 79–80, 84, 87, 90, 94. *See also* embellissements
Morachiello, Paolo, 84
Morizet, André, 12
Münster, Sebastian, 29

Napoléon I (Napoléon Bonaparte): and Egypt, 7, 47–51, 67, 115, 144–45; expositions organized by, 150; and Les Halles Centrales, 85; and measurement standards, 162n17, 162n26; and military maps, 21–22; portrait of, 16–17, *17*; and taxation, 112
Napoléon III: and Franco-Prussian War, 127; and housing in Paris, 22; marriage to Eugénie, 25; officials under, 1; original Paris plan of, ix, 1–2, 11–12, *13*, 27, 151, 155; paintings of, 16–17, *16*, 23–25, *24, 25*, 158n20; role of, in Paris's modernization, 1–2, 7, 8, 11–12, 27, 102–3, 116, 148, 150; and Second Republic, 84
Napoléonville, 84, 162n54
Neri, Pompeo, 110
Nerval, Gérard de, 26
Neufchâteau, François de, 150
neutrality. *See* universality/neutrality/regularity/interchangeability
Nicolay, Benedit Vassallieu dit, 30
Niépce, Nicéphore, 167n88
Noé, Amédée de (pseudonym: "Cham"), *Croquis contemporains* (Contemporary sketches), 55, *56–57*, 57
Nolli, Giambattista, 113
Notre-Dame de Paris, 32

objectivity: ascribed to maps and plans, 1, 3, 5–6, 27, 30, 35–36, 40, 61, 116, 123, 155; bureaucracy and, 119–20; grids linked to, 63; individual and sociopolitical factors obscured by, 2; mechanical reproduction associated with, 142; modernity and, 65; orthographic plans and, 6, 113–14, 154; science linked to, 5, 142; surveys associated with, 35–36; value and authority associated with, 2, 4, 5
Observatoire de Paris, 30–33, 36, 41, 49, 67, 72, 79–81, 142–43
O'Connell, Lauren, 117
Odeleben, Ernst Otto Innocenz von, 22
Olson, Kory, 147, 166n44
Opéra, 26, 84, 90–96, 162n71, 162n74
Orientalism, 49–51
Ortelius, Abraham, 166n46
orthographic plans: advantages of, 4, 21; as basis of Paris, 103; bird's-eye views vs., 113; blank spaces within, 20–21; emergence of, 3, 74–76, 112–13; explained, 20; geometric basis of, 21, 74, 113; governance role of, 112–14; and instrumental value of maps, 20; limitations of, 113–14; objectivity and universality of, 6, 113–14, 154; perspectival vs., 3, 4, 18, 97, 116; urban, 112–13. *See also* straight lines
Ozanam, Jacques, *Méthode de lever les plans et les*

cartes de terre et de mer (Method of mapping plans and maps of the land and the sea), 45–46, *47, 48,* 160n51

Palais-Royale, 26
Palsky, Gilles, 113
Papayanis, Nicholas, 84
paper, 5, 23, 106–7, 123. *See also* documents
Paris: annexation of suburbs by, 23, 53, 59, 84, 117; census of, 115–16; as center of quantification efforts, 3, 81; comprehensive plan of, 18, 22, 31, 77, 97, 102–3; constructing an image of, 5, 12, 40, 59, 61, 65, 101–3, 116, 130–36, 140–41, 145, 150–51, 155; Exposition universelle (1889) centered on, 135–44; geographic center of, 85; and the grid plan, 84, 103; guidebooks for, 42; history of, 129–45; housing in, 22; maps' and plans' role in shaping, 61, 103, 107–8, 110, 117, 129, 155; Marville's photographs of, 97–101; modernity associated with, 1–3, 65, 103, 155; modernization of, 1–3, 7, 8, 11, 53, 65, 107, 116, 122, 148, 150; perspectival images of, 18–20, *18, 20;* hybrid perspectival and orthogonal plan, 19, *19;* Plan of Paris (from *Atlas des travaux de Paris*), 138; pre-Haussmann plans for, 22; scholarship on planning of, 12; standardization and regularization of, 6; straight lines forming the plan of, 2; taxation in, 110, 113; unity as underlying theme of planning of, 77, 128–29. *See also* surveys
Paris Commune, 12, 123, 127–28, 144
Paris meridian, 31–32, *31,* 36–37, 49, 80, 81
Pascal, Blaise, 164n73
Pasquiet, Jean-Baptiste, 77
Patte, Pierre, 75; *Monumens érigés en France à la gloire de Louis XV* (Monuments erected in France to the glory of Louis XV), 75–76, *76; Partie du plan générale de Paris où l'on a tracé les différents emplacemens qui ont été choisis pour placer la statue équestre du roi* (Part of the general plan of Paris where different locations for the equestrian statue of the king are drawn), 75–76, *75*
Pavillon de Flore, 25
Pavillon du roi, 30
Pearce, Margaret Wickens, 129, 166n48
percements (openings), 23, 84, 87, 90, 94, 97, 99, 103, 108, 154
Pérez-Gómez, Alberto, 74
Perrault, Claude, 79, 87, 100
Persigny, Jean-Gilbert-Victor Fialin, duc de, 2, 22, 102, 157n3
perspectival images: Marville's photographs, 97–101; Merian's *Plan de Paris sous Louis XIII,* 18, *18;* orthographic vs., 3, 4, 18, 97, 116; of Paris, 18, *18, 20,* 97; hybrid perspectival and orthogonal plan, 19, *19*
Petit, Pierre, 37
Petit-Radel, Louis-François, 77
photoengraving, 151, 167n88

photography, 97–101
physiocrats, 113, 114
Picard, Alfred, 150
Picard, Jean-Félix, 31
Pichot, Jean-Baptiste, 84
Picon, Antoine, 67
pictorial maps, 3, 4, 19, 20, 37, 61, 68
Pinkney, David, 12
Place de la Bastille, 79
Place de l'Opéra, 26, 84, 90–96
Place du Carrousel, 24
Place Louis XV, 32
Plan de l'Abbé Jean Delagrive, 133
Plan de la cité gauloise (Plan of the Gallic city), *126* (detail), 130, *131,* 132
Plan de la Commission des artistes, 77, 79, *79, 138*
Plan de Roussel, 133
Plan des artistes, 113, 154
Plan des opérations trigonométriques de la Ville de Paris (Plan of the trigonometric operations of the city of Paris), 32, *32,* 36, 40
plan d'intendance (stewardship plan), 113
planimetric views, 3, 19–20, 109, 116, 164n70
plans. *See* maps and plans
politics. *See* bureaucracy; governance; governmentality
Pompe de l'Arsenal, 32
Porter, Theodore, 113, 123
portolan charts, 21
portraits: of kings and maps, 14–17; maps as, 18–19
positivism, 5, 6, 13, 128, 130
precision: in cartographic representation, 17, 33, 63, 154; as a value, 63, 68. *See also* accuracy
Préfecture du Département de la Seine, 123, 144; *Commission des Halles, Document à étudier, no. 2. Plan de la Nouvelle Halle,* 85, *85; Rapport sur l'emplacement des Halles Centrales de Paris, Juin 1851. Plan des Halles Centrales et de leurs abords* (Report on the enlargement of the central markets of Paris, June 1851. Plan of the central markets and their surroundings), 90, *91*
Prony, Gaspard François de, 111
property, 23, 110–14, 117, 163n25, 165n110
property owners, 43, 66, 108, 110–12, 120, 165n110
public utility, 91, 102
Ptolemy, 43

quantification: administrative/bureaucratic value of, 3–4, 107, 113, 115–16, 123, 127; egalitarianism furthered through, 137; in mapmaking, 20, 30, 40, 51, 64, 113, 127; Paris as center of efforts in, 3, 81; of space, 3, 23, 67; value and authority conferred on, 4, 30, 40–41, 45, 51, 117. *See also* geometry; science
Quesnay, François, *Tableau économique,* 114–15, 164n85

Rambuteau, Claude-Philibert Barthelot, comte de, 22, 102

real estate market, 23
Recherches statistiques (census of Paris), 115–16
Recueil des plans des places fortes (Collection of fortification plans), 21
Redont, Édouard, *Comment reconstruire nos cités détruites* (How to rebuild our destroyed cities), 153–54
regularity. *See* universality/neutrality/regularity/ interchangeability
reproduction/reproducibility, 40, 142, 151
République française (journal), 129
réseaux (networks), 116
Revue générale de l'architecture et des travaux publics, 101
Revue municipale (journal), 22
Reyes, Hector, 164n47
Richeprey, Jean-François Henry de, 113
road commission, 33–34, 159n16
Rohault de Fleury, Charles, 91–93, 162n71
Rousseau, Jean-Jacques, 129

Salon, 76
Santana-Acuña, Alvaro, 40
Saqui, Madame (tightrope walker), 57
Sauvigny, Louis-Bénigne-François Bertier de, 113
science: art vs., 4; institutional/discursive underpinning of, 4; maps and plans associated with, 17, 30, 40, 74, 123; and narrative, 134–35; objectivity and neutrality ascribed to, 5, 142. *See also* geometry; quantification
Second Empire, 2, 12, 25, 26, 53, 84, 92, 94, 96, 97, 128, 145–46, 154
Second Republic, 84
Service de l'architecture, Direction du (Direction of architectural services), 3, 117.
Service de la statistique, 137
Service de la voie publique (Service for public roads), 123, 147, 165n107
Service de la voirie (Roads department), 94
Service des eaux et égouts (Water and sewage works), 4, 147
Service des bâtiments (Building works), 117
Service des promenades et plantations (Service for promenades and green spaces), 4, 147, 148, 165n107
Service des travaux (Public works service), 54
Service des travaux historiques (Service for historical works), 129, 135
Service du plan, 4, 54, 59, 116–18, 123, 139, 147
Seurat, Georges, 65
Sewell, William, Jr., 160n69
Siegfried, Susan Locke, 22
Siméon, Henri, 2, 22, 84, 102–3
Simon, Jules, 158n14
Sitte, Camillo, 23
Snellius, Willebrord, 31, 159n9
Société centrale des architectes, 3
Söderström, Ola, 118
space, rationalization and quantification of, 3, 6, 23, 30, 34–35, 61, 67, 73, 74, 111.

space, rationalization and quantification of (*cont.*)
See also space, universality/neutrality/regularity/interchangeability, as goals/outcomes of modern systems of representation and measurement

spatial turn, 161n105

Stafford, Barbara Maria, 164n47

standardization: in building process, 117; bureaucracy and, 107, 111–12, 118–20, *118*, *119*, 123; of the city through maps and plans, 6; as feature of modernity, 8; of maps and plans, 108, 111–12, 120–23; of measurement, 63–64, 66–67, 161n9, 163n19; of surveying practices, 40–41, 43–44

statistics, 136–37, 154

Storcks, Marin Kreenfelt de, 34

straight lines: agency of, 4; authority ascribed to, 2; cautions against, 2, 22–23, 103; function of, in maps and plans, 72–73; in plan of Paris, 2, 23. *See also* orthographic plans

streets: alignment of, 7, 22, 77, 79, 87, 102–3, 107; plans and maps of, 19, 23, 34, 37, 39, 53, 79–80, 87, 96, 101, 103, 109; width of, 2, 22, 79, 103. *See also* circulation

subjectivity: devaluation, 5; of maps, 151; reader's, 119

surveillance, 6, 14, 21, 34, 85

surveying towers, 54–57, *54–57*

surveys: for cadastral maps, 111–12; Delambre and Méchain's, 67, 143; Deschamps's, 7, 23, 53–61, 84, 117; in Egypt, 47–51, 160n74; fidelity established through, 30; Haussmann and, 53–54, 59; history of, 30–34; labor involved in, 35, 40–46; leveling measures for, 59; manuals on, 41–47; maps compared to, 43; under Napoléon I, 21–22; objectivity associated with, 35–36; popular suspicions of, 34, 43; practitioners of, 160n61; process of, 36–37, 59; profession dedicated to, 120; social status associated with, 42, 43; sociopolitical context of, 30; standardization of, 40–41, 43–44; triangulation method of, 30–32; Verniquet's, 7, 32–41, 46, 53, 57–59, 65, 68, 81, 96, 107, 113. *See also* ingénieurs-géographes (engineer-surveyors); maps and plans

Sutcliffe, Anthony, 166n38, 166n79

synoptique (synoptic, "at a glance," views), 113–17, 129, 140–41, 150

Tableau des distances des principaux monuments (Table of distances from principal monuments), *36*, 37

Talleyrand, Charles-Maurice de, 67, 163n19

taxation, in France, 110–14, 117. *See also* cadastral maps

Taxil, Léo, 79, 162n50

technology: curriculum, 74; graphic, 137; history of, 7; photoengraving and reproduction, 151

terrain, maps' representation of, 3, 5, 14–16, 18–22, 29–31, 49, 59, 61, 64, 68, 109–10. *See also* topography

Testelin, Henri, *The Establishment of the Academy of Sciences and the Foundation of the Observatory, 1667, Based on a Drawing by Charles Le Brun*, 15, *15*

texts. *See* image-text relationship

Teyssèdre, A., *Nouveau manuel de l'arpenteur* (New manual for the surveyor), 41–42, *41*, 45

Teyssot, Georges, 84

Thiers, Adolphe, 127

Third Republic, 4, 77, 79, 97, 109, 116, 127–46, 148, 151, 153, 154

Thompson, Victoria, 42

time, maps and, 65, 142

Tisserand, Lazare Maurice, 129, 130

Tissier, Jean-Baptiste-Ange, *Napoléon III and Eugénie Approving the Plans Presented by Visconti*, 24–26, *25*

Tocqué, Louis, *Louis, Dauphin de France*, 14–15, *14*

toise du Châtelet, 66–68, *72*

topography: absence of, in maps, 46, 61; military mapping and, 21–22; surveys' representation of, 32–33, 59. *See also* terrain

Travaux publics. *See* Ministère des travaux publics

triangulation, 30–32, 159n9. *See also* surveys

trigonometry, 31

Tuileries, 24, 26

Turgot, Anne-Robert-Jacques, 110

Turgot, Michel-Étienne: *Plan de Paris*, 20, *20*; *Plan de Turgot* or *Paris au XVIIIe siècle*, 4, 132, *133*, 164n70

universality/neutrality/regularity/interchangeability, as goals/outcomes of modern systems of representation and measurement, 2–3, 6, 13, 30, 40, 61, 66–68, 73, 74, 90, 107, 111–14, 116, 118, 150. *See also* space, rationalization and quantification of

urbanism and urban planning: architecture in relation to, 7; bureaucracy and, 107–23; Chabrol de Volvic and, 84; comprehensive plans and projects, 18, 22, 31, 77, 97, 102–3; geometric basis of, 23; grids and, 84, 97; historical conceptions of, 12; institutional/discursive underpinning of, 118; maps' and plans' role in shaping, 7, 63, 84, 93–94, 96, 103, 109, 116, 117, 153–54; modern, 12, 23; origin of the term "urbanism," 153–54; orthographic plans and, 112–13; photography and, 97–101; and socio-spatial relations, 6; and war, 153–54. *See also* maps and plans

Vacquer, Theodore, 130

Van Zanten, David, 96, 117

Vardi, Liana, 114

Vasserot, Charles, *Plan of the Louvre and Its Surrounding around 1830*, 26, *26*

Vasserot, Philibert, 66, 94, 161n7

Vauban, Sébastien Le Prestre de, 21, 66, 160n54

Vaugondy, Robert de, *Plan de la ville et des faubourgs de Paris divisé en ses vingt quartiers* (Plan of the city and the surroundings of Paris divided in its twenty neighborhoods), 133, *134*

Vérin, Hélène, 113

Verniquet, Edme, 7, 32–37, 40–41, 46, 53, 65, 68, 76, 96, 102, 103, 107, 110, 113, 154; *Atlas du plan général de la Ville de Paris*, 32, *32*, 35, 36, *36*, 40, 57, *58*, 59, 68, *68*, 69, 76–77; *Plan de Verniquet*, 133–35, 139, 144, 151, 167n86

Vidal de la Blache, Paul, 166n64

Viguet, Louis, 162n71

Villot, Frédéric, 116

Viollet-le-Duc, Eugène, 97, 105, 129

Visconti, Ludovico, 24–26, 96; *Plan for the Nouveau Louvre by L. Visconti*, 26, *27*

vision/visuality: cadastral maps and, 112; history of Paris presented by means of, 129–30, 140–42; maps and, 14, 27, 61, 116–17, 150; as means of knowledge, 14–20, 113–14, 142; modernity and, 5–6, 12–13; in nineteenth-century French culture, 12–13, 65; as tool of governance, 112, 114, 117, 127, 136. *See also* image-text relationship

Vismann, Cornelia, 118, 165n115

Vitruvius, 73

Voltaire, 75, 129

Wailly, Charles de, 77; *Plan général du projet des embellissements de Paris* (General plan of the improvement projects of Paris), *76*, 77

Weber, Max, 106, 107, 116, 119

Winterhalter, Franz Xaver, 158n20

Wise, Norton, 157n21

Wood, Denis, 5, 165n30

World War I, 153–54

Wyngaard, Amy, 159n25

Yates, Alexia, 23

Yvon, Adolphe, *Napoléon III Handing to Haussmann the Decree for the Annexation Plan of the Peripheral Districts (16 February 1859)*, *10* (detail), 23–24, *24*, 159n45